Biology of Halophytes

PHYSIOLOGICAL ECOLOGY

A Series of Monographs, Texts, and Treatises

EDITED BY

T. T. KOZLOWSKI

University of Wisconsin
Madison, Wisconsin

Biology of Halophytes

YOAV WAISEL

Department of Botany
Tel-Aviv University
Tel-Aviv, Israel

 1972

ACADEMIC PRESS　*New York and London*

ACADEMIC PRESS, INC.
111 Fifth Avenue, New York, New York 10003

United Kingdom Edition published by
ACADEMIC PRESS, INC. (LONDON) LTD.
24/28 Oval Road, London NW1

LIBRARY OF CONGRESS CATALOG CARD NUMBER: 73-182663

PRINTED IN THE UNITED STATES OF AMERICA

Contents

Chapter 14. **Marine Halophytes**

Chapter 15. **Ecological Notes on Some Terrestrial Halophytes**

Chapter 16. **Ecotypic Differentiation**

Preface

An investigation of the biology of plants, particularly of halophytes, naturally has wide scope. Such a study must cover several divergent scientific fields, among which climatology, geomorphology, soil science, plant anatomy, ecology, sociology, and physiology are only the most obvious ones.

Thus, in our modern era with its narrow scientific specialization, undertaking the task of writing a monograph on the biology of halophytes seemed a somewhat frightening challenge. Nevertheless, having had experience in teaching ecology and plant nutrition, I recognized the need of many students for a text and reference book in such an interdisciplinary topic.

This monograph is intended primarily for advanced students and for teachers of plant and environmental sciences. In institutions in which salinity is taught as a separate topic, the book can be used as a text. It is designed to give the reader a basic knowledge of halophytes in relation to their environment. Sufficient necessary information and ideas in each aspect of the related scientific fields were introduced so as to give the reader a good foundation for the further study of the behavior of plants under saline conditions.

Sources of salinity, as well as the development and nature of salines and salt-affected soils, are discussed first. Classification of halophytes, their mutual relationships, distribution, and sociology are then treated briefly. Much of the book is concerned with physioecological characteristics such as water relations, mineral nutrition, salt transport, salt secretion, reproduction, growth, metabolism, and salt resistance. Ecological information on higher marine plants was divided into three sections which deal, respectively, with submerged angiosperms, mangroves, and high coast plants. Autecological information on some terrestrial halophytes is summarized in a separate chapter. This monograph also includes a short discussion on the formative effects of salinity and ends with a short review on ecotypic differentiation in halophytes.

In all cases, an attempt was made to present a broad view of each topic. However, as one naturally remains close to his own interests, most of the data and illustrations were taken from my own investigations. Some of this information is new and has not been published previously.

So far as possible, attempts were made to retain the terminology used in each of the cited publications. It was inevitable in some cases that different terms were used for similar phenomena and that the same plants were referred to by synonymous names.

My heartiest thanks are due to the indefatigable Professor T. T. Kozlowski, who invited me to write this monograph and who has given me so much advice and encouragement. He read the manuscript and contributed a great deal to the improvement of its style and language. My gratitude is also due to my colleagues Dr. Irene Gruenberg-Fertig and Dr. Nili Liphschitz who helped me transform the data into a manuscript. A great deal of the reported information was gathered in this laboratory and is a result of a group effort. Sincere acknowledgment is made to Mrs. Y. Cohen-Eshel, Mr. A. Eshel, Miss A. Keren, Mr. Z. Kuller, Dr. R. Neumann-Ganmore, Mr. S. Ovadia, Mr. G. Pollak, and to all other members of the laboratory who contributed, directly or indirectly, to the knowledge of the biology of halophytes.

Thanks are due to Mr. S. Shaefer and Mr. Y. Cohen for preparing the illustrations and to Professor D. Yaalon and Dr. U. Kafkafi for reading some of the chapters. I am also grateful to the various copyright holders for their kind permission to reproduce some of their material.

Yoav Waisel

Introduction

Over four-fifths of the surface of our planet is covered with a salt solution containing, among many other constituents, approximately 0.5 M NaCl. Still, only very few groups of higher plants can withstand such conditions. Most terrestrial species are unable to tolerate even one-tenth of the salt concentration of ocean water without a serious setback in their water and nutrient balance or in their metabolism.

The fact that only a small group of higher plants can grow under saline conditions was recognized many years ago. In 1563, Dodoens described *Plantago maritima* as belonging to such a specific group. In 1576, Lobelius mentioned *Salicornia* as a typical plant of sea coasts. In 1695, Sloane emphasized the fact that *Avicennia* is a species of saline habitats and is exceptional in this respect. Nevertheless, only early in the nineteenth century was the name "halophyte" given to that group of plants by Pallas (see Schrader, 1809).

Great variations are known in the plant world regarding responses to salinity. Bacteria not only contain certain groups which are salt sensitive, but also groups with the highest salt tolerance known. Some of them (e.g., *Halobacterium*) survive even in the concentrated salt solution which comprises the waters of the Dead Sea. On the other hand, fungi, mosses, ferns, and gymnosperms are usually, as groups, sensitive to salinity.

Variations in salt tolerance are also found within the angiosperms. Some families have species with a high salt tolerance, e.g., Potamogetonaceae, Hydrocharitaceae, Chenopodiaceae, Zygophyllaceae, Tamaricaceae, Frankeniaceae, Rhizophoraceae, Plumbaginaceae, and the Verbenaceae. Plants of various other families of the angiosperms, e.g., Orchidaceae, Araceae, Rosaceae, and Ericaceae, are known to be salt sensitive.

Usually salines are characterized by high NaCl content and high osmotic potentials. Many salines are also affected by high concentrations of sulfates, carbonates, and bicarbonates, and by high pH of the soil, as well as by poor soil structure and disturbed aeration. These factors have an apparent adverse effect on the nutrition, growth, and reproduction of most plants.

Various adaptive mechanisms have been selected in halophytes in the course of evolution. Some mechanisms are responsible for the restriction of various phases of growth and development to such a combination of factors (climatic seasonal or edaphic) that ensure the germinated seedling with the best chances for a successful completion of its life cycle. Other mechanisms enable plants to resist salinity or tolerate it directly.

Despite the fact that halophytes have attracted scientists for so many years, our knowledge of their biology is extremely limited. In most cases, we are ignorant of the metabolic adaptations and direct physiological processes which enable plants to survive under saline conditions. It is well agreed that sodium chloride is the dominant factor in "halophytism," yet the mechanisms of uptake of such ions, as well of germination, growth, and flowering of plants under saline conditions, are relatively unknown.

Salinity also plays an important, indirect role in the existence, behavior, and distribution of plants. Salinity reduces competition between plants, prevents infection of halophytes, and increases their tolerance to freezing, high temperatures, high water stresses, etc. Even these aspects have not yet been explored.

Controversial data exist regarding the question of whether halophytes require saline conditions for their existence and vigorous growth or merely tolerate them. Most of the halophytic species investigated so far seem to prefer saline conditions. Only a few species seem to require salts obligatorily. Nevertheless, even though most halophytic species are also capable of growing in salt-free media, those plants are usually limited in their distribution to salines. The various modes of adaptation of such plants to saline environments are the scope of the following chapters of this book.

The increasing use of water of poor quality, the continuous addition of waste salts to our environment, as well as an increasing contamination of underground water sources lead to gradual soil salinization. Soberly analyzing the consequences of such a process, one must reach the conclusion that the future of plants lies with some group of halophytes. The question of how long it would take until such a state arises still requires an answer.

Not very much is known of halophytes. The material gathered here is perhaps only part of what is already available on some of the topics, but is extremely deficient in others. This book represents an attempt to expose such gaps in knowledge and is a call for future research, rather than a summation of it.

Understanding the basic processes of adaptation to salinity and the "know-how" of applying such principles to crop plants would be, in the near future, a matter of "life or death" to the growing population of a world 95% of whose water sources are saline. The earlier we comprehend such a need, the wiser we shall be.

---------------------------------- |

Sources of Salinity

Salinity is a common phenomenon and one of the basic features of arid and semiarid regions. Even though information regarding the natural processes leading toward salinization is essential for understanding the salt status in any habitat, the quest for the origin of such salts and their circulation in nature has been practically neglected. Results of only a few investigations dealing with such problems are available (Menchikovsky, 1924; Eriksson, 1958; Junge, 1958; Yaalon, 1963).

Although some of the salts reaching the crust of the earth are derived from such sources as cosmic dust and volcanic activity, soluble salts can presently be added from three major sources: marine, lithogenic, and anthropogenic (Fig. 1.1). Each of these sources contributes its share, whether small or large, to salinization of soils and underground water. Consequently, in places where salts are deposited and accumulated, formation of saline habitats is encountered. Contribution of each source to the general process of land salinization will be reviewed briefly in this chapter.

1

Fig. 1.1. Salt sources—a general scheme.

Marine Sources

Volcanic eruptions release large quantities of soluble salts into the atmosphere. It is generally accepted that vast amounts of chlorine and sulfur from this source have accumulated during the ages in the oceans, either directly by underwater volcanic action, precipitation of volcanic ash, dissolution of gases, or indirectly via streams and rivers which wash soluble salts off the continents. However, salts were not left evenly distributed. After continuous cycling for long geological eras, distribution of various elements and compounds between different localities had distinctively occurred; while the more soluble salts of chlorine were leached and accumulated mostly in the oceans, over half of the compounds containing sulfur, presently found in the geochemical cycle, precipitated and concentrated in rocks sedimented under seawater. A small quantity of sulfur can also be found in igneous rocks.

Salts of marine origin are transported to the continents and deposited there in three major ways:

(a) *Cyclic salts*—salts brought inland from the sea in the form of windborne sea spray, and precipitated by rainwater. Such salts are later redistributed or leached back into the ocean by the common drainage systems (Fig. 1.2).

Fig. 1.2. A general scheme of the marine salt cycle.

(b) *Infiltrating salts*—salts brought into coastal habitats by an underground infiltration of seawater. This is a source of local importance only.

(c) *Fossil salts*—marine salts which were precipitated sometime in the past in certain localities and which are now being dissolved. Such salts are brought above ground level or into the plant root zone by flowing masses of groundwater, by springs, by capillary movement of water, or by surface runoff water.

CYCLIC SALTS

It has been well known for a long time that atmospheric precipitations contain varying amounts of soluble salts. Such precipitations were, at one time, even considered as a source of nutrients because of their high mineral content. However, the high NaCl content of windborne particles diminishes any possible beneficial effects they should have had on plants. High salt concentrations of windborne spray droplets strongly affect both living and nonliving matter along seacoasts. Fast deterioration of unprotected buildings and corrosion of metals are well-known phenomena in open seaward-facing places. Also, plants are affected adversely under such conditions. Leaf scorches and tip burns resulting from absorption of large quantities of NaCl occur commonly in trees. Leaf burns which are encountered even at distances of a few kilometers from the coasts affect tall and exposed plants, and rarely appear in low and protected plants (Karschon, 1958). Such injuries are commonly observed in the Mediterranean region during spring when storms carry in salts, and rains, which wash them off, become fewer. Analyses of the relative ion content of the spray droplets affecting leaves revealed that, in many cases, proportions between ions are similar to the ion content found in seawater (Eriksson, 1958; Yaalon and Lomas, 1970).

Tall objects and large plants, such as trees, cause fast precipitation of the salt particles impinging upon them. Ingham (1950) showed that rainwater collected in South Africa from beneath trees had a higher mineral content than rainwater from elsewhere. Although such an increase may be ascribed to leaching of minerals from the free space of leaves, most of it was shown to be precipitated spray salts.

After being intercepted and precipitated, salts percolate through the soil and are washed into drainage systems and groundwater. Thus, salts of maritime origin eventually form the primary and major source of salts of rivers and wadies (Conway, 1942; Anderson, 1945) or of groundwater (Loewengart, 1958).

Tropical cyclones also pour large masses of seawater over the continent, thus increasing salinization. However, since cyclones usually occur in humid regions and are accompanied by heavy rains, salts are also rapidly leached away.

Precipitated marine salts are quantitatively estimated either on the basis of their concentration in rainwater, or on their concentration in runoff wadi water. Distribution of airborne salts exhibits a typical pattern; their concentrations in rain or wadi water are highest near the coasts and diminish as distances from the coast increase.

Teakle (1937) found that rainwater near the coasts of Australia had 15–50 ppm NaCl, whereas rainwater sampled further inland had only 4–20 ppm NaCl. Salts are carried inland as far as 1000 km or more. At distances beyond about 20 km inland, salt concentrations in wadi and rainwater reach a steady level of 8–10 ppm (Eriksson, 1958; Yaalon, 1963; Yaalon and Lomas, 1970). At this level, salt concentrations in rainwater are retained even over great distances (Fig. 1.3).

Wind intensity and roughness of the sea appear to control the amount of airborne droplets, and consequently the amount of salt deposition in terrestrial habitats. Large airborne droplets are usually precipitated within a distance of less than 1500 m from the coast (Yaalon and Lomas, 1970). Within this range, salt composition is very similar to that of seawater. Beyond 1000 m from the water line the amount of salts precipitated per unit area is steady at about 2 mg/100 cm^2/day (Fig. 1.4).

Salt spray collections made in Barbados by Randall (1970a) also reveal a marked decrease in salinity with increasing distance from the sea. The amounts precipitated were very much influenced by wind velocity and therefore dependent on land topography (Fig. 1.5). Absolute amounts were far above those reported for other places around the world.

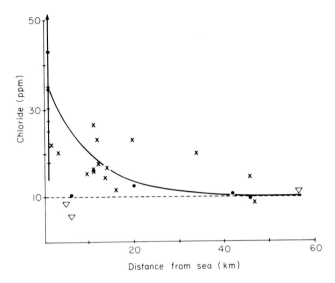

Fig. 1.3. Ion content in rainwater as a function of distance from the coast. x, Mean values of wadi water; ▽, minimum values during storm floods; ●, measurements in precipitation by Menchikovsky (1925). (After Yaalon, 1963).

Chloride salts are mainly transported as droplets, but sometimes they may be carried in a solid form of minute crystals. This is more typical of arid regions, where most of the airborne chlorides are precipitated as solid particles. Such a phenomenon occurs mostly because of the rapid evaporation of the salt spray droplets (Eriksson, 1958).

The small size of salt particles $(0.1-10\,\mu)$ led Cassidy (1971) to believe that

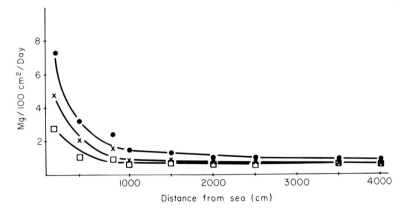

Fig. 1.4. Precipitation of salts as a function of distance from the coast. ●, Na; x, Cl; ◻, SO_4. (Computed after Yaalon and Lomas, 1970.)

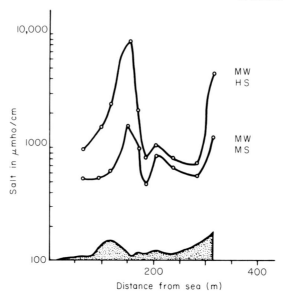

Fig. 1.5. The relationship between salt deposition and distance from the sea at different wave amplitudes with irregular topography. (After Randall, 1970a.) MW, HS, Medium wind, high sea; MW, MS, medium wind, medium sea. Stations 1–10.

such particles are capable of entering leaf stomates. He also suggested that entry of such aerosols into leaves may be the cause of impaired plant growth on coastal habitats. Although entry of salt particles via stomates is feasible, still their exposed surface area is extremely small. The quantities of salt which penetrate leaves through such inlets must thus be of minor importance. Most of the salts penetrate leaves via the cuticle.

The ratio of Na:Cl in rain and runoff water remains usually more or less constant even in different localities. At the same time, the ratio of S:Cl is not as constant and in terrestrial habitats deviations are found in this ratio from that of seawater. Relatively more sulfur is found over terrestrial habitats than near the coasts, because of the high solubility of gaseous SO_2 in the spray droplets and additional contamination with other sources of sulfur in the former. Unlike chlorine, the content of sulfur is not reduced as much with increasing distance from the seashore inland, because sulfur is frequently released into the atmosphere in gaseous form and is thus carried by the wind to greater distances from the sea. On the shore, the ratio of SO_4^{2-} to Cl^- is more than twice its ratio in seawater, and increases gradually with increasing distances from the coast. Besides the relatively higher content of sulfur in airborne particles, more sulfates are found in groundwater or runoff water in arid regions, because of the contribution from sulfate-containing rocks. Solubility of gypsum even increases under saline conditions.

Ratios of Na:Ca are much lower in rainwater than in seawater. Such deviations can be caused by contributions of calcium salts from cosmic or terrestrial sources. As cosmic sources contribute only negligible amounts of calcium, calcareous dust particles from terrestrial origin remain the only explanation for the higher content of Ca in rainwater over the land (Hutton, 1958; Stephens, 1960).

Rainwater also contains various nitrogenous compounds. The concentration of ammonia shows seasonal fluctuations. According to Yaalon (1964a), the content of ammonia in rainwater in Israel varies between 0.17 epm (equivalents per million) in the cold winter months and 0.68 epm in the warmer spring. It is suggested that most of the ammonia evolves from the soil and that warmer temperatures accelerate the ammonification processes. Nitrate concentration is less variable than ammonia and its annual average is 0.04 epm.

Ratios between various cations infiltrating through the soil and reaching groundwater are not kept as constant as those of the anions, and usually differ greatly from those found for rain- or seawater. Such differences are evidently caused by the cation-exchange capacity of the soils through which rainwater infiltrates, and which varies for different soils.

Besides receiving droplets and solid particles of salt, soils also absorb atmospheric gases. Some of these gases may be directly adsorbed on the surface of the soil clay minerals, whereas, others are first absorbed by plants and only later are released into the soil. Very little is known about rates of mineral uptake by plants or their accumulation in soils from the gaseous phase. However, according to Eriksson (1958), it should be of the order of kg/ha/year. Elements accumulated from gaseous sources originate mostly from sulfur dioxide and hydrochloric acid. Sulfur dioxide results from volcanic activity as well as from recent combustion of fossil plant material. Hydrochloric acid is formed in the atmosphere from molecular chlorine under the relatively high pressure of the hydrogen prevailing there (Jung, 1956, cited by Eriksson, 1958). Large quantities of hydrogen sulfide are released from lakes and closed seas where a conspicuous stratification of water occurs. The warm upper layers of water that develop during the summer deprive the lower water layers of oxygen, thus inducing conditions under which hydrogen sulfide is produced. As hydrogen sulfide escapes into the air, the sulfur content of such bodies of water, e.g., the Black Sea, decreases (Anderson, 1945; Beauchamp, 1953; Eriksson, 1958).

After being released from the water into the atmosphere, hydrogen sulfide is oxidized to form sulfur dioxide. In saline water, sulfur dioxide reacts with sodium chloride to release HCl into the air. According to Burkser (1951), the chlorine content of the air just north of the Black Sea is about three times higher $(531.8 \, \mu g/m^3)$ than its content at distances of 100 km and greater $(146.8 \, \mu g/m^3)$. This discrepancy may result from the large quantities of H_2S continuously released from the Black Sea.

INFILTRATING SALTS

Underground infiltration of seawater or other sources of saline water into the capillary fringe of neighboring habitats constitute a continuous supply of salts to such habitats. Surface evaporation causes an increase in concentration of the incoming brine and raises the concentration of salts on the soil surface. The amount of salt contributed by this source to salinization of terrestrial habitats is high in arid regions and negligible in humid ones. Also the overuse of wells near seacoasts increases infiltration rates of seawater into the underground water reservoirs and causes their salinization.

FOSSIL SALTS

Fossil salts comprise an important source of salts on certain sites, but usually salts of this source are limited to a few habitats. Such salts have accumulated in the past on coasts of ancient seas, on sites of salty lakes, or during past penetrations and local accumulation of seawater. Saline shales of the Colorado Plateau of the United States can serve as an example of the effects of sea inundations in the past. Underground waters which presently flow through such deposits of sodium chloride dissolve the salt and carry the brine up to the soil surface and into the plant root zone. Similar phenomena involving various carbonate and sulfate salts are also known.

Saline groundwaters from such sources occur extensively in arid regions. Salinization of groundwater has even increased since humans have interfered with natural salt and water balance, i.e., destruction of native vegetation or large-scale irrigation (Eriksson, 1958).

Lithogenic Sources

Great differences exist in nature between distribution of chlorine and sulfur. Because of its high solubility, chlorine occurs in all rocks in very low quantities. In contrast, a large portion of the total amount of sulfur released, in the past, into the oceans was precipitated into sedimentary rocks.

Chlorine is primarily found in rocks as chlorides of the alkali metals. Presumably, chloride is not incorporated into rock crystals, but is precipitated as intercrystal penetrations. In such a location and form, chlorides can migrate readily within homogenous rocks as well as between various rock layers (Goldberg, 1958).

Low amounts of chloride and relatively high amounts of sulfur are found in rocks (Table 1.1). Among sedimentary rocks, sandstones show the lowest amounts of chloride. Limestones usually contain more chloride and more sulfur than shale. Sulfur is primarily found in rocks in two chemical forms; in hard

Table 1.1

VARIATIONS IN CHLORINE AND SULFUR CONTENT IN VARIOUS
ROCK SOURCES[a]

Type of rock	Chlorine (%)	Sulfur (%)
Primary rocks	0.023	0.090
Sandstones	0.002	0.220
Soft limestones (Campanian)	0.070	0.270
Hard limestones (Cenomanian)	0.001	0.002
Hard limestones (Middle Eocene)	0.240	0.074
Dolomite (Cenomanian)	0.003	0.003
Dolomite (Palaeozoic)	0.045	0.009
Soils[b]	0.010	0.085
Seawater	1.900	0.088

[a] Adapted from Lamar and Shrode, 1953; Goldberg, 1958; Green 1959; and Yaalon, 1963.

[b] Average of well-leached soils.

limestones it occurs as pyrite, and in soft limestones and in marls as sulfates, mostly gypsum.

The extent of salinization of groundwater and soils by chloride or sulfate from rock sources depends very much on the rate of rock weathering, which varies markedly in different climates and regions. The rates of rock weathering in the semiarid climate of the Mediterranean region vary between 10 mm per millennium on the flatlands and 50 mm on steep slopes. If the chloride content of the average dominant rocks is taken as 0.04%, the addition of salt from this source is limited to 16–80 gm/1000 m^2/year. Sulfur is released from rocks in relatively greater quantities than are chlorides. Weathering of rocks also releases large amounts of carbonates of calcium and magnesium. Both divalent cations play an important role in control of salinization of soils. However, as rocks generally are very low contributors of salts, such sources can hardly account for salinization.

Under specific conditions, weathering of saline rocks may also contribute large amounts of soluble salts. According to Bocquier (1964), such sources are primarily important for local formation of salines in inland habitats.

The oceans do not supply all of the salts found in rainwater. Salt-containing dust from terrestrial sources also influences the chemical composition of rainwater (Yaalon and Katz, 1962). Evidence for such contributions is based on the following:

1. Measurable quantities of airborne calcium carbonate are precipitated in rainwater, in addition to sodium chloride.

2. Early rains, or rains following long periods of drought, contain more salts than the average rains.

3. Rainwater of arid regions contains much more carbonates and silicates than rainwater of humid areas.

Soil particles carried from one site to another in the form of atmospheric dust vary in size from 10^{-4} to 10^{-5} cm. The smaller particles can be carried long distances by air currents from the sites of their origin. For example, dust particles presumably originating in the Sahara Desert were reported to be precipitated in the northern Atlantic (Correns, 1937) or in western Europe (Glawion, 1939).

Airborne solid particles consist of salt condensation nuclei or dust particles. The large salt nuclei contain quantities of sodium chloride and sulfate in proportions similar to that found in seawater. In the smaller nuclei, sulfate is present in much higher proportions.

Composition of atmospheric dust represents essentially the composition of particles from the upper layers of soils in arid regions. The insoluble fraction is comprised mostly of quartz, feldspar, and mica. Large quantities of calcium carbonate and calcium sulfate are also found. Particles comprised of sodium chloride and sodium sulfate may also be swept into the air from the upper salty surfaces of salines. Crozier and Seely (1952) reported such phenomena for the Mojave Desert in the United States.

A certain proportion of the salts precipitated by rains is returned to the sea in runoff water. The remaining salts percolate through the soil and into groundwater. Thus, salt concentration in ground and in precipitation water can be correlated using the following equation:

$$C_w = C_p \frac{P - R_{st}}{P - R_{st} - E}$$

where C_w = salt concentration in a given aquifer; P = precipitation (mm); C_p = salt concentration in precipitation water; E = evaporation (mm); R_{st} = storm runoff water.

Thus, when the salt content in groundwater is known, an estimate of the salts precipitated over the drainage basin can be obtained. However, the above equation holds true only for well-defined drainage systems where the groundwater table is deep and amounts of precipitation are greater than potential evapotranspiration.

The distribution and precipitation of salts over the continents depend very much also on the wind and rain regimes. However, despite the similar averages of salt concentrations in rainwater, a correlation between salt content and precipitation does not always hold. Above a certain limit of rainfall, the air is cleaned to such a degree that practically no additional precipitation of salts occurs.

In regions receiving 500 mm mean annual rainfall, at an average concentration of 10 ppm chloride, an annual addition of approximately 8 kg NaCl/1000 m^2 may be expected. Even higher quantities are expected in countries located on coasts of open oceans, and exposed to rough seas and heavy storms.

It was generally accepted that high salinity is a natural phenomenon of arid regions, where salts are continuously added, but precipitation is too low to wash them off. Although in most cases this is so, this assumption is not essentially true everywhere. Soils of certain arid regions which are located at some distance from the ocean and are protected from windborne salts remain with a low salinity, examples of which can be found in land areas of Central Australia (Jackson, 1957).

Dew deposits also contain high quantities of salts. For example, Yaalon and Ganor (1968) showed that dew droplets had a high concentration of calcium, bicarbonate, and sulfate ions. Nevertheless, because of the composition of the salts, it seems that salt precipitation in dew is of terrestrial, rather than of marine, origin.

Besides dew, fogs also carry inland large quantities of salts. Considerable amounts of salts are carried by this means in South America (Eriksson, 1958).

Anthropogenic Sources

Since ancient times, human activity has been adding large quantities of soluble salts to agricultural lands. In some areas, irrigation with unsuitable water under the high evaporative conditions of arid or semiarid climates and the practice of improper agrotechnical procedures have resulted in deposition of large quantities of salts in upper soil layers. For example, irrigation with water from the Jordan River (approx. 500 mg/liter soluble salts) at an annual rate of 1000 m^3 water per 1000 m^2 land, adds at least a net 500 gm of salts per square meter to the soil. Even subsequent leaching of such soils with large volumes of water will only temporarily prevent their salinization. Such leaching merely moves the salts into deeper soil layers or into groundwater, from which they subsequently might return to the root zone. Excess irrigation also causes leaching of salts into low sites and eventually results in formation of salt marshes.

With development of industry and extensive use of fuel, i.e., since the second half of the nineteenth century, another source of soluble salts on land surface has been added. For example, a fair sized crude oil refinery with a capacity of a million tons per annum, releases about 45,000 tons of sulfuric acid into the atmosphere, if the average content of sulfur in the fuel is only 1.5%. According to Loewengart (1958), even if only 10% of such quantity enters into the annual cycle of renewing groundwater, it means an addition of approximately 4 mg sulfate per liter of rainfall each year for a country of the size of Israel. Electric power plants, metal smelters, and automobile fumes may also add considerable amounts of pollutants, mostly of SO_2 to limited areas. Cement factories also spread annually thousands of tons of salts in their surroundings. Although the major contribution of such factories is in the form of calcium carbonate and calcium hydroxide, the added salts are still of local importance.

Urban use of water also results in increase of salinity. Measurements made in Ohio (Bunch and Ettinger, 1964) and in Israel (Rebhun, 1965) indicate that the average increase in water salinity in one single cycle of urban use is approximately 325 mg/liter. Reuse of city sewage evidently causes considerable salinization of groundwater (Bagley, 1967).

Eriksson (1958) summarized the sources and gross composition of salts deposited on continents (Table 1.2). Evidently, compared to the low salt contribution from lithogenic sources and a similar amount from other sources, the oceans predominate as sources of salts spread over the continents.

Table 1.2
ESTIMATION OF THE ANNUAL ADDITION OF SALTS FROM VARIOUS SOURCES[a]

	Element	
Source	Chloride	Sulfur
Human activities		
Production	2.3	1.0
Combustion	0.01	0.6
Weathering of rocks	0.04	0.44
Volcanic (juvenile)	0.2	0.02
Total known sources	2.6	2.1
Cyclic salts	12.9	9.0
Total salt additions per annum (kg/ha)	15.5	11.1

[a] After Eriksson, 1958, from *Arid Zone Research,* Vol. X, p. 156. Reproduced with the permission of UNESCO.

Only approximately two-thirds of the continents are drained to the sea. The remaining areas, e.g., inner deserts, which lack runoff or groundwater flow, are drained into nonmaritime terminal drainage basins. Such systems constitute the source for formation of most terrestrial salines.

—————————————— 2

Formation of Salines

Salt Cycles in Nature

In nature, salts undergo numerous cycles of dissolution, transport, and precipitation before accumulation occurs at certain sites. Five major salt cycles, each having separate and different patterns, were recognized by Kovda (1961):

1. *Continental salt cycles:* Movement and accumulation of chloride, sulfate, and carbonate salts in arid regions, where only small amounts of runoff or percolating water exist. Two subcycles are distinguished:

(a) *Primary cycles:* Movement of salts originating by weathering or from salt-containing primary rocks.

(b) *Secondary cycles:* Movement of salts which originated by weathering of sedimentary rocks.

Only a small portion of such salts reaches the ocean.

2. *Marine cycles:* Movement and accumulation of salts near sea coasts. Such salts return eventually into the ocean.

3. *Delta cycles:* Precipitation and accumulation of salts in deltas of large rivers.

4. *Artesian cycles:* Accumulation of salts from underground saline water

sources on the land surface. Such cycles are long-term ones, and the salts which take part in such cycles are usually fossil.

5. *Anthropogenic cycles:* Movement and accumulation of salts produced and released by human activity.

Water Regimes and Salt Accumulation

Dissolution and movement of salts, which have been once precipitated are greatly dependent upon local climate, topography, and hydrological status of each drainage basin involved (Fig. 2.1). Salts are precipitated and dissolved several times before being deposited in certain spots for long periods. Rates and distances of salt movement are markedly affected by the water relations of each specific habitat, especially by the groundwater level in relation to soil profile (Yaalon, 1963).

The so-called "normal" percolative regime occurs in regions where precipitation is higher than actual evapotranspiration. Habitats with such a regime are well drained and the levels of groundwater are deep. Sources of salinity for such habitats consist of both cyclic and rock salts. However, due to the low retaining capacity of the soil to negatively charged particles, soluble anions, such as chloride, are leached out of the soil. Sodium and other cations are as mobile as the anions, but are adsorbed on the negatively charged soil clay

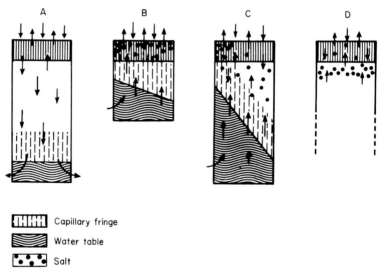

Fig. 2.1. The four dominating water regimes and their correlation with salt accumulation. (After Yaalon, 1963.) A, Percolative; B, epipercolative; C, amphipercolative; D, subpercolative.

minerals and thus are retained to a certain degree in the soil profile. Owing to its high leaching, a soil with a percolative water regime cannot be saline, although some sodic soils may have developed under such a regime.

Epipercolative regimes were described for low habitats in soils with an underlying impermeable rock or hardpan layer. Groundwater levels of such habitats are high and must be at least within the capillary fringe. Usually such a regime is found in climates where evaporation exceeds precipitation. Soluble salts such as NaCl, Na_2SO_4, and $CaSO_4$ are brought in by the inflowing or by rising groundwater. The horizon for highest accumulation of salts occurs on the soil surface, where the underground water flows in owing to the high evaporation. Certain layers of the subsoil may contain high percentages of adsorbed sodium. Frequently, the epipercolative regime can also be found in coasts of arid regions where seawater infiltration continuously adds salts to the upper layers of the soil.

An amphipercolative regime is usually found in habitats with a variable groundwater level, with seasonal fluctuations in water level, different salt sources, as well as a variable capacity for water conductivity. Some sites in such habitats have an impaired water flow and accumulate salts, whereas others are leached readily and become practically salt free.

Subpercolative regimes are typical of arid regions where evaporation greatly exceeds precipitation. The groundwater level is either very deep or does not exist. Salt sources in such cases are mostly of marine origin, either fossil or recent. Because of the low precipitation, salts that reached such areas remained there in larger quantities than can be naturally removed. Under such a water regime, salts tend to accumulate and frequently crystallize in a layer at the depth of wetting (Fig. 2.2).

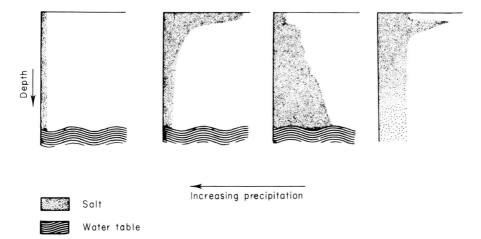

Depth ↓

░░ Salt

≈≈ Water table

→ Increasing precipitation

Fig. 2.2. Accumulation of salts in various soil profiles under different rainfall conditions.

Movement of salt solutions across a soil profile involves reactions of ion exchange between the negatively charged clay minerals of the soil and the saline solution. In fact, even in cases where salts are present in relatively small quantities, their mere movement across the profile may affect plants by changing the ion composition in the adsorption complex of the soil. Sodium chloride solutions have a marked effect on the sodium exchange percentage of soils low in divalent cations, but they have only minor effects on the cation-exchange capacity of soils rich in calcium. Under arid conditions, when concentration of the soil solution increases because of intensive evaporation, calcium is gradually precipitated and the relative adsorption and effect of sodium increase. When sodium adsorption percentages increase beyond 15% of the cation-exchange capacity of the soil, various physical properties of the soil are adversely affected. Physical changes due to high adsorption of sodium, including reduced permeability to water, are frequently encountered in heavy clay soils. High adsorption of sodium has only minor effects in sandy soils.

Accumulation of saline water, as well as high adsorption percentage of sodium in the soil, comprise the initial stages in the formation of salines.

Salines are usually classified into a few types on the basis of their specific nature and location.

Marine Salines

Habitats of hydrohalophytes are segregated also, according to the salinity of the water, to brackish and saline water.

According to Walter (1968), saline water habitats can be divided into four classes: (1) oligohaline, 500–5000 mg/liter; (2) mesohaline, 5000–18,000 mg/liter; (3) polyhaline, 18,000–30,000 mg/liter; and (4) euhaline, 30,000 mg/liter and above.

The first three groups are commonly designated as brackish water. The Baltic Sea water, off the southern coast of Sweden, has a salt content of about 18,000 mg/liter, while off the Finnish coast, salinity is even less than 10,000 mg/liter. Accordingly, water of the Baltic Sea is considered brackish, and various semihalophytic or even glycophytic hydrohalophytes grow there. The salt content of such seas is not constant and changes from a saline sea (8800 B.C.) to a freshwater sea (5000 B.C.) and back to a brackish water sea (since 2000 B.C.) were reported for the various parts of the Baltic Sea (Walter, 1968). Most of the phanerogamous flora of such a sea is composed of semihalophytes or facultative halophytes.

Also the Black Sea has brackish water in its upper layers (18,000 mg/liter), but is much more saline deeper down.

The Caspian Sea, as well as the Aral Sea are remnants of what once was the Tethys. The Caspian Sea is rich in sulfate due to the high input of sulfate salts by the Volga (Walter, 1968). Its submerged angiospermous flora mostly consists of *Zostera nana.*

Coastal Salines

Salines which are formed on coasts and which are directly affected by the proximity of the sea are included in this group. Salt composition of such salines closely resembles that of seawater, although their salt content varies greatly with specific local conditions. The formation of such salines and their special traits depends on the following factors: (a) height, frequency, and duration of tides; (b) soil types and stability of soil level; (c) rainfall; (d) groundwater level; (e) presence of river fans and underground water flows; (f) intensity of winds and spray; (g) vegetation types and cover.

Since composition of vegetation is dependent, among other things, on temperature (air and soil, mean and extremes) and on rainfall (total amount and distribution) those factors contribute also indirectly to the formation and characterization of salines (Chapman, 1954).

Coastal salines are formed all over the world under semiarid and arid regions, as well as under maritime temperate climates and can be classified into the following formations.

SALT WATER SWAMPS

A swamp in the temperate or subtropical regions constitutes a woody plant community occurring in an area where the soil is saturated or covered with surface water during a good part of the growth season (Penfound, 1952). In the tropics, the vegetation of inundated swamps may be comprised of grasses or *Papyrus.*

Swamps with arboreal plant communities in brackish or saline habitats are found only in tropical or subtropical regions and are usually covered with surface water. The most typical ones are mangrove swamps with a plant cover of *Rhizophora* sp., *Avicennia* sp., or *Sonneratia* communities. Nevertheless, true deep-water salt swamps are formed only by *Rhizophora.*

Swamps with shrub communities are typical of the drier sites of the coastal salines. Such sites are covered with surface water, but only for short periods. Shrub swamps are characterized by *Baccharis, Iva,* or *Conocarpus* communities.

SALT MARSHES

A marsh is a grass community occurring in a habitat where the soil is saturated or covered with surface water during part of the growth season. Such marshes are formed in brackish water or in saline habitats with wet soil surface or with a cover of surface water. Marshes are typical of temperate regions and are mostly covered with grass communities such as *Spartina alterniflora* or *Spartina–Distichlis–Juncus* communities.

Where very little or no surface water exists, the marsh is typically covered by *Distichlis* or a similar grass.

SALT BOGS

Peat forming swamps are often called "bogs" (or peaty swamps); these are usually salines with an abundant shrub vegetation that grows in habitats that receive only limited amounts of mineral sediments.

When seawater rises, coastal marshes extend inland, and whenever land surface rises the marshes retreat seaward. On any site, where a few patches of muddy soil are exposed, a salt marsh develops. The only requirement is that the site be well protected from full-strength sea gales, but still exposed to frequent tides. Usually such salt marshes are limited in width. However, whenever an inlet from the sea is in a form of a funnel, tides are much higher and may cover vast areas. In the Bay of Fundy region, incoming tides periodically cover some 70 square miles. On such sites, salines are wide and deep (Teal and Teal, 1969).

Salines reach their highest and most extensive development in regions where evaporation exceeds precipitation $(E > P)$. In humid regions, most salines are coastal where sea levels reach the capillary fringe of the soils. Such salines also are periodically inundated by seawater, or receive saline groundwater. In humid regions, salts are washed off by rains but are repeatedly replenished by the incoming tides. In such regions, the habitats remain saline only when large quantities of salts enter the soil and replenish those that were lost by leaching.

Salt marshes are further divided into upper or high, and lower or low ones. In European terminology, those marshes are designated inner and outer marshes, respectively. Classification is based on the periods of inundation and exposure. Low marshes are inundated daily whereas the upper ones remain exposed for long periods. Upper marshes usually are well leached by rainwater. Such marshes are subjected to diurnal fluctuations in water table caused by the daily tides, and to cyclic fluctuations formed by spring tidal cycles. One of the characteristics of the upper marshes is their long nontidal exposure in summer. As desiccation during this period is severe, plants have to be adapted to high water stresses at that time.

Plants play an important role in the buildup of various types of hydromorphic salt marshes, thereby changing low marshes into high ones. Along the Atlantic coasts, salt marshes occur from well within the Arctic Circle down to the Tropics, where they change into mangrove swamps. For example, it is well known that *Spartina alterniflora* contributes to establishment of the coast line along the Atlantic Ocean, until the marsh level rises to above that of the average high tide. When accumulations of peat and mineral soil raise the bottom of the salt marsh to such a point, optimal ecological conditions for existence of *Spartina alterniflora* are shifted and this species is eventually substituted by *S. patens* (Teal and Teal, 1969).

Development of a coastal marsh usually involves an initial deposition of a sand ridge near the coast line. Beyond such a ridge, a stand of *Spartina* plants begins to develop, helping to slow down the flow of tide water and to precipitate clay minerals brought in by the tides. With development of a richer substrate, the stands of plants extend rapidly, forming a fully covered marsh. Such a process can be repeated several times, and each period is clearly marked by a buried sand ridge. The substrates of such salt marshes change from sandy ridges to muddy soil, to solid peat or clay and are covered by grasses. Sedimentation of fine soil particles only accelerates development of new plant stands. Once a dense stand of plants is established, dead plant fragments accumulate, and a layer of peat is formed underneath. The peat in salt marshes of warm regions is soft and usually referred to as *muck*. Because of its softness, muck areas in humid regions are heavily eroded. In such cases, the area becomes characterized by isolated creeks which eventually are filled with mud. *Spartina* plants appear in such habitats only in patches.

Salt marshes along the Atlantic coasts were formed a few thousand years ago. They expanded northward along the Atlantic coast immediately after the region was exposed by the retreating glacier (Teal and Teal, 1969). Plants were disseminated into the newly formed habitats mostly by hydrochory or by zoochory.

Inland Salines

Inland salines are typical and an integral part of arid and semiarid regions. Only in limited areas can such salines be found in humid regions, mostly on fossil salt depositions, or around salty springs. In different localities, salinity can mean accumulation of different salts and under the heading "saline," we include sites rich in sodium sulfate, sodium carbonate, borax, magnesium sulfate, and alum. Although sodium chloride usually is the dominant component of such salines, salt composition and concentration may vary greatly. In Chile, large coastal areas are affected by sodium nitrate because of a unique combination of physical and biotic factors. In the Death Valley of California (United States), borax was deposited and borates dominate the salts present; around the Dead Sea in Israel, high concentrations of potassium and magnesium chlorides prevail, in addition to NaCl.

Accumulation of salts in inland salines depends on several factors including: (a) amounts of incoming cyclic salts; (b) existence and depth of fossil salt layers; (c) existence of salty springs; (d) quality and quantity of incoming water. Such waters usually include both aboveground runoff water, as well as underground water flows; (e) evaporation–precipitation ratio; the higher the ratio, the more favorable are conditions for saline formation; (f) efficiency of soil drainage.

Inland salt marshes and salt flats have developed in the seepage basins of deserts or of saline water sources. In arid regions, sites with a basin-shaped topography form typical concentric salines because of the inflow of saline water from the surroundings, or as a result of a rise of saline underground water. Nevertheless, inland salines are inundated mostly during part of the year only. During the rest of the season, such salines remain with a dry salty surface. Soil thickness, as well as plant cover of such salines are usually arranged in concentric rings. The surface of the lowest point of such a terminal drainage depression is also the saltiest. However, this differs somewhat under an epipercolative water regime.

There has been some disagreement on classification of inland salines. Confusion exists in local names for different types of salines in different countries (Walter, 1962, 1968; Zohary, 1962). Nevertheless, more or less general agreement exists on the distinction between various types of salines, based on their salt sources or water relations (Zohary, 1962). Accordingly, inland salines are divided into automorphic or lithogenic and hydromorphic salines.

AUTOMORPHIC SALINES

These salines are located in the general circumglobal belt of the great deserts, where precipitation is too low to wash off the incoming airborne salts. The water regime of such salines is thus of the subpercolative type. In cases where underlying rock layers constitute the source of salts, salines are called lithogenic. Typically, they form flat plains strewn with rocks or with a gravel cover on their surface. The Arabic names for those types of salines are *reg* or *hammada* depending on local dialects. These salines are found not only on plains, but also on badlands and in hilly regions. They are often characterized by a high gypsum content.

In this group of automorphic salines are also included salines of semiarid regions which were formed because of accumulation of a gypseous hammada soil. A few other salines found around the Mediterranean were formed due to accumulation of alluvial sodic clays.

HYDROMORPHIC SALINES

Salines of this origin occur primarily in depressions. They accumulate cyclic salts directly, as well as from groundwater or surface water. In the latter case, salts either are continuously supplied from salty springs, or periodically only from saline runoff water. In both cases, the brine concentrates at the soil surface and salts precipitate as a result of high evaporation. The salt content in the upper soil layers of such salines increases during the summer. As a result of seasonal leaching, it tends to be sodic during the winter months. Whenever hydromorphic salines form, the water regimes are of the epipercolative or the subpercolative types. Hydromorphic salines frequently are subdivided into three types: (1) salines of flowing water; (2) terminal drainage basins which are periodically inundated; (3) high groundwater salines.

Salt pans and other types of such salines received local names, some of which are widely used. *Playa* is a Spanish word for shore. However, in salt marsh classification, this term is usually applied to flat terminal drainage basins with hard and impermeable subsoil layers. A playa is a depression with a high water table, into which salts and silts are brought from surrounding areas by runoff water. Playas usually are inundated during the rainy season. Besides salts, water also brings in considerable amounts of clay minerals, which eventually are sedimented and form a heavy fine-textured paddy layer of flat soil. During the hot and dry summer, the surface dries forming a hard cracking top layer, which is topped by a salt crust. The Spanish designation for a salt encrusted playa is *salada.*

In Central Asia, salt pans are also termed *schor* or *takyr.* Both types of salt pans are highly saline, flat plateaus with a very sparse plant cover; most of the water supply of the *schor* infiltrates upward from subsoil sources in contrast to that of the *takyr* which receives more surface runoff water. The *takyr* differs from the *schor* in being flooded more than once a year (Walter, 1968). The *takyr* is a fine-textured sodic soil with low permeability to gases and water. The high moisture retention by such soils prevents air movement in rates sufficient to supply plant roots. Rainwater turns *takyrs* into paddy bogs, thus making them anaerobic habitats. After desiccation, a heavy clay crust is formed on top of the *takyr* soils, which impedes plant growth.

Undrained flat salt pans which sometimes are flooded after rains and which frequently contain a crust of pure salt on their top in summer are termed *sabkhah* in Arabic, *schor* in Mongolian, or *schott* (Walter, 1962).

Salt pans are formed only in regions where enough rain falls in the general area to wash the salts into the pans. In extremely dry regions, where there is not enough water to form hydromorphous salines, salts remain evenly distributed throughout the profile, or form a distinct layer at the depth of wetting.

Depth of groundwater is an important factor in the determination of salinization patterns. When groundwater is within the capillary range, salts move up and are deposited on the surface. Groundwater, in depths below 3 m, does not contribute to the salinization of the soil surface. This excludes cases where salt-secreting plants (see Chapter 8) absorb salts at such depths, and dispose of them on the soil surface.

Soils of hydromorphous salt pans generally are heavier than those of surrounding areas, because the flowing water brings in fine clay particles, together with the salts. The flow of water leaves the water courses leading into salines with a coarse gravel substrate which, to a great extent, is salt free. Hydromorphous salines are formed whenever a barrier, e.g., sand dune or clay alluvium, is formed thus impeding drainage of runoff or groundwater flow (Fig. 2.3).

Fig. 2.3. A terminal drainage basin saline in North Sinai, Israel. Vegetation is dominated by *Halocnemum strobilaceum, Arthrocnemum glaucum,* and *Zygophyllum album.*

Salt and water status of hydromorphous salines vary throughout the year, and in semiarid regions such salines exhibit seasonal fluctuations in water level, as well as in salinity. During the rainy season, water level of such salines rises and salts are diluted. Summer evaporation causes a drop in the water level, and salts concentrate and accumulate on the soil surface (Fig. 2.4). Highest accumulation occurs in arid regions in such a layer of the soil profile, which is in equilibrium with both bidirectional upward and downward seasonal movements. In temperate regions, the magnitude of variation is low, because of the high precipitation and low rates of evaporation.

Precipitation of salts in a drying body of saline water follows a typical pattern. The sequence starts with precipitation of the least soluble salts in the periphery of such salines, and ends with the highly soluble ones on the surface of the saline center. The salt layers precipitated in the circumference of a saline consist mostly of calcium and magnesium carbonates. Inner layers are made up

Fig. 2.4. Aerial view of an inland saline, North Dakota, United States.

of more soluble salts, such as sulfates and bicarbonates. Chlorides usually precipitate last and consequently are found in the center, on top of all other salt layers (Kovda, 1954; Hunt, 1960).

The gradual precipitation of divalent cations in the saline's circumference causes a decrease in their concentration in the center, and consequently, an increase in the adsorption capacity of the monovalent cations. Sodium adsorption percentages are by far higher in soils of the centers of salines than in the periphery.

Under certain conditions, precipitation of calcium sulfates yields crystals up to 2 mm in diameter, which accumulate in sandlike layers. Such layers were reported for many salines of arid regions in Tunisia (Bureau and Roederer, 1961), in the United States (Eardley, 1962), in Rajasthan (Shastri, 1962), and in Israel. To some extent such deposits give an indication of the past boundaries of the salines.

Two additional classes of salines should be mentioned: the aerogenic and anthropogenic salines.

Aerogenic Salines

These salines form as a result of accumulation of salts brought in from eolian sources. Loess soils, for example, which accumulate on desert fringes, usually are saline as a result of the precipitation of airborne soil particles under conditions of high evaporation over rainfall. In such soil types, salts concentrate in the depth of wetting, i.e., in upper layers of soil profiles under low rainfall conditions, but in the subsoil under 200 mm mean annual rainfall or higher. Sodium chloride and other soluble salts are leached into deeper layers, but $CaSO_4$ tends to accumulate near the soil surface. The exact site of accumulation depends on topography, soil permeability, and texture.

Volcanic eruptions also contribute large quantities of various salts, such as borates, and volcanic dust is carried away over long distances to be finally precipitated in salines.

Anthropogenic Salines

Although such salines belong to various types described previously, man-made salines deserve a special category. Faulty irrigation techniques are, of course, a major cause for accumulation of salts in soils and for saline formation. Such salines are well known in arid regions, and large areas in countries such as Iraq, India, Pakistan, and California were and still are affected.

Anthropogenic coastal salines were formed in places where gravel or peat deposits were removed by man. Under such conditions, infiltration and accumulation of salty groundwater resulted in the formation of the saline.

Inland anthropogenic salines usually are also formed in salt-containing refuse dumps or near sewage outlets. Such salines are characterized by an accumulation of specific salts which differ from site to site.

——————————————— 3

Saline and Sodic Soils

General Characteristics of Saline Soils

Wherever potential evapotranspiration exceeds rainfall or faulty agricultural techniques are used, soluble salts accumulate at or near the soil surface. When excess salts are not removed with drainage waters from soil, saline and saline-sodic soils develop. Such soils, which contain large quantities of soluble salts or exchangeable sodium, inhibit growth of most agricultural and wild plants. Only very few species of specially adapted plants can survive under such conditions.

In the past, salinization and sodification of soils caused the destruction of several civilizations which practiced faulty methods of agricultural irrigation. However, even at present, vast areas of saline and sodic soils produce low crop yields (Richards, 1954).

Soil Salinity and Water Relations

Saline soils have multiple effects on plants. Responses of plants to specific ions are restricted primarily to sodic soils or to soils in which carbonates or borates dominate. However, plants generally seem to be affected by the total activity rather than kind of ions in the soil. Thus, the most prominent effect on plants of high salt concentration in a soil is an osmotic one. Such an osmotic potential of a soil solution forms a considerable component of the total moisture stress in soils where high amounts of soluble salts are present (cf., Slatyer, 1958, 1967). The moisture-retaining capacity of saline soils differs from that of nonsaline ones. The wilting point, i.e., the point of equilibrium in water potentials between plants and soil is of a different magnitude in saline soils, being far above the common −15 bars range. Thus, salinity contributes to the forces which prevent water from entering plant roots. A few specific examples were given for halophytes by Walter (1961) and by Bernstein (1961, 1963) and Slatyer (1961, 1967) for glycophytes. Most investigators concluded that water economy should not be a problem for adaptation of plants to salinity because the difficulties in water uptake usually are overcome under saline conditions by accumulation of salts in the cell sap. In fact, halophytes grown under saline conditions have their wilting point at a lower water potential than glycophytes. *Aeluropus litoralis* plants, under certain conditions, reached the wilting point in nonsaline media at a water potential of −7 bars, but at −16 or −22 bars at 200 mM and 400 mM NaCl solutions, respectively (Waisel and Pollak, 1969a). In spite of such adaptation, growth of most common plants under conditions of salinity or dryness is inhibited (Hayward, 1956). This is discussed in more detail in the chapters on growth and on water relations.

Sodium Adsorption and Profile Formation

The presence of excess soluble salts in the soil may affect plants in other ways beside their water economy. As soil clay particles are negatively charged, they adsorb cations. Adsorption is practically a nonspecific first-order reaction that is affected by physical properties of the adsorbing surface of soil clay minerals, physical properties of the cations (charge, radius, etc.), and ionic strength and composition of the soil solution. The capacity of a soil to bind cations is named cation-exchange capacity, and is expressed in mEq/100 gm.

Cations which are adsorbed at a certain moment may thus be replaced by other cations, depending on their properties and relative concentration. When sodium constitutes the dominant cation in the soil solution, the cation-exchange capacity of the soil becomes loaded with sodium ions. In soil solutions containing a mixture of mono- and divalent cations, the expected relative adsorption of sodium ions can be calculated by the following equation:

$$SAR = \frac{Na}{[(Ca+Mg)/2]^{1/2}}$$

where sodium adsorption ratio (SAR) is the ratio between the concentration of sodium in the soil solution divided by the square root of half of the concentration of calcium and magnesium. Concentration of the three ions is expressed in milliequivalents. When sodium is adsorbed on 15% or more of the soil's total cation-adsorption capacity, it has a definite effect on the soil's chemical and physical properties (Gardner, 1945; Whitney and Peech, 1952). A dispersion of soil particles occurs and soil structure deteriorates under such conditions. Because of the high pH that develops under such conditions, hydrolysis of various aluminosilicates is accelerated. The dissolved substances, together with colloid clay minerals, which are washed in, tend to accumulate in the alluvial horizon of the soil and often form a hardpan. Owing to their high degree of swelling, sodium-saturated soils usually develop low permeability to water and air.

The general equation related to water flow through a soil (Darcy's equation) is

$$q = -K \, \Delta\Phi$$

where q is water flux (volume per unit area, per unit time), $\Delta\Phi$ is the difference in the water potential, and K the hydraulic conductivity constant characteristic of the soil. A decrease in K, due to sodification of a soil and dispersal of the upper layer of its clay minerals, thus affect the flow of water from the bulk of the soil into the root zone, even under conditions where the water potential gradients remain constant. Thus, a gradual increase in exchangeable sodium percentage (ESP) in the root medium reduces the availability of water to plants (Eaton and Horton, 1940). Under such conditions, the hydraulic conductivity of a soil also decreases during its depletion cycle. Thus, the decrease in availability of water to roots must exceed the decrease in soil water potential. Formation of impermeable soil layers also impedes gas exchange across the profile and further affects root growth and water absorption capacity (Chang and Dregne, 1955).

An increase in saturation of the cation-exchange capacity of soils with sodium ions usually occurs under conditions which cause percolation of a salt solution through the soil and result in their alkalinization. Thus, infiltration of saline water markedly affects the chemical and physical properties of soils. Heavy clay soils tend to be affected more than sandy ones because of the higher exchange capacity. Heavy soils, mostly those which consist of montmorillonite clay minerals, swell, disaggregate, and exhibit a lower permeability to water. The carrying capacity for plants of such soils, i.e., their maximal ability to produce plant material, decreases considerably. Under extreme conditions of sodium-saturated heavy soils, plant growth is prevented altogether. The presence of calcium ions tends to improve sodic soils. However, in order to exert its effects, calcium must be present in a soluble form in the soil solution. A high content of $CaCO_3$ in the soil does not always contribute enough calcium ions to the soil solution to prevent soils from being saturated with sodium (Yaalon, 1955).

The increase in bicarbonate or carbonate ion content in the soil only accelerates the sodification process. Under such conditions, the less soluble salts such as calcium and magnesium carbonates are precipitated, thus increasing the percentage of adsorbed sodium. According to Wahhab (1961), prediction of ESP values from sodium adsorption ratio (SAR) measurements of the solution give precise results only when the anion is chloride. In experiments conducted with a gradual increase of bicarbonate, Wahhab showed that the actual adsorption percentage of sodium increased in proportion to the increase in carbonates, although the sodium and calcium content in solution remained constant. Similar conclusions were reached by Lewis and Juve (1956) and Wilcox *et al.* (1954). The increase in sodium adsorption is further accelerated when soils become dehydrated.

In regions where no leaching occurs, soils have an excessive content of both cations and anions. Such soils remain saline but their pH is relatively low. Loess soils, accumulating on desert fringes are of such a type. Salinity of such soils is due to the salt content of the precipitated airborne soil particles and additions of cyclic salts, under conditions of high evaporation over precipitation. Under low precipitation conditions, salts concentrate in the upper horizons of the soil profiles (upper 20–50 cm) (Fig. 3.1). With 200 mm mean annual rainfall or more, salts accumulate in the lower horizons (e.g., in depth of about 100 cm). In such soils, NaCl and other soluble salts are leached into the deeper layers of the

Fig. 3.1. Profile of a saline soil showing salt and gypsum layers.

profile, whereas $CaSO_4$ and carbonates tend to accumulate in a higher horizon or on the soil surface.

The different patterns of ion distribution in soils of highly soluble salts, e.g., chlorides, and of poorly soluble ones, e.g., sulfates, are shown in Fig. 3.2. Sulfate moves downward more slowly than chloride and thus remains in higher layers (Yaalon, 1964b). The exact site of accumulation of each mineral depends on topography, soil permeability, and soil texture.

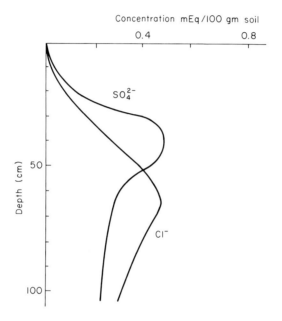

Fig. 3.2. Differences in downward movement of SO_4^{2-} and Cl^- through a soil profile. (After Yaalon, 1964b.)

In humid regions, i.e., in regions with Lang's rain factor* of 50 or higher, and in a percolative water regime, salts are washed out of the soil into groundwater. Saline soils can thus rarely develop under such conditions, excluding sites where the addition is so high as to overshade the leaching rate. Under arid conditions (Lang's rain factor <30) salts remain in the soil and accumulate in a certain layer whose depth varies according to the water regime. Saline soils also develop in areas with a heterogeneous topography or in temporarily closed basins into which salts are washed in from higher spots, or as a result of retreating banks of saline bodies of water.

* Lang's rain factor = precipitation (mm)/mean annual temperature (C°).

Classification of Saline and Sodic Soils

Among the major groups of soils, the saline and sodic soils belong to the intrazonal group, i.e., to that group of soils whose special characteristics are affected by local specific factors rather than by the general climatic conditions.

Terminology regarding salt-affected soils is not uniform and sometimes confusing. While some investigators named all halomorphic soils "alkali" soils (Kelley, 1951), others tended to distinguish between saline (salt-affected) and alkali soils (soils affected by a high percentage of exchangeable sodium) (de Sigmond, 1927; Richards, 1954). As the term "alkali" seems to have many meanings, and alkalinity may be caused by a multiplicity of factors, the use of "sodium" or "sodic soils" instead of alkali soils seems to be preferable (Dregne, 1968).

Soils with excess salts—saline soils—are also called *white alkali* (Hilgard, 1906) or *solonchak* soils (a Russian term: Sol = salt). Kearney and Scofield (1936) and de Sigmond (1938) regard saline soils, as those that contain salts above the 0.1% level. According to Stocker (1933), the critical level of salinity for plants is 0.5% of the dry soil weight. Saline soils are typical of arid regions in which soluble salts are accumulated primarily in the A horizon of the profile.

Salts in saline soils are comprised of a mixture of chloride, sulfate, sodium, magnesium, and calcium ions, with sodium chloride being dominant. Gypsum may also comprise, in local sites, a considerable part of the soil's salt content. In arid regions, gypsum concentrates near the soil surface or throughout the profile. In humid regions, it is usually leached into a certain depth below soil surface. In most cases, the profile of such soils is underdeveloped and vertical movement of salts is considerably greater than that of colloid particles, clay minerals, or sesquioxides. Saline soils usually are low in organic matter because of their low carrying capacity and failure to support a dense stand of plants.

Joffe (1949) distinguished three types of such saline soils:

1. *Primary solonchak:* soils of arid and semiarid regions in which the development of the soil profile is poor, and which are deficient in organic matter.

2. *Secondary solonchak:* soils with a developed profile which were affected by salinity only at a late stage of their development. Secondary solonchaks can be superimposed on any type of regional soil, i.e., one can find laterite solonchak, podsol solonchak, etc.

3. *Hidden solonchak:* soils of arid and semiarid regions in which layers of salt have accumulated in the subsoil forming a typical salt horizon in their profile. The upper horizons of such soils are salt free.

In addition to the morphological subdivision of saline soils—solonchaks, such soils were also classified mostly by Russian scientists on the basis of their chemical composition. Thus, solonchaks are subdivided also according to their dominant anions into chloride solonchak, sulfate solonchak, etc. (cf. de Sigmond, 1927).

According to Dan *et al.* (1962), halomorphic soils of Israel can be divided into four subgroups: organic solonchak; alluvial and eolian solonchak; marly solonchak; sterile solonchak. This is a classification which considers both substrate and source of salinity, as well as plant cover.

Under some conditions, salts are washed out from halomorphic soils. When the water table is lowered, when an increase in rainfall occurs, or when soils are intentionally leached by man, the saline soil solution moves gradually down the soil profile and by so doing causes a replacement of a large part of the adsorbed nutritive cations, potassium, magnesium, and calcium by sodium ions. Dissociation of adsorbed sodium causes an increase in pH of the soil solution; more CO_2 is dissolved, and finally sodium bicarbonate and sodium carbonate are formed. The relative content of carbonate or of bicarbonate in the soil is pH dependent. Below pH 9 bicarbonate ions predominate; above this pH carbonate ions also occur. A soil which has a columnar structure in the B horizon with over 15% of its cation-exchange capacity saturated with sodium, and has a pH higher than 8.5 is called *solonetz, alkali,* or *sodic soil.* In cases where the soil still contains large amounts of soluble salts, and its pH is thus lower, it is defined as a *saline–alkali soil.*

The presence of sodium carbonate is associated with high pH of the soil solution. Under such conditions, the soil's organic matter is decomposed and dissolved. As a result, the soil obtains a typical black color, and in certain cases was named *black alkali.* Under conditions of a relatively high NaOH or Na_2CO_3 content, certain clay minerals of the soil are also dissolved, forming sodium aluminates, sodium silicates, and silica gel.

Disintegration of such clay minerals releases various substances into the soil. Some of them are nutritive (e.g., potassium or magnesium) and have a positive effect on plants, whereas others such as boron or aluminium are toxic.

Boron is mostly released from rock minerals where it comprises part of the mineral *tourmaline.* Boron concentrations in the saturated soil extract of 0.7–1.5 ppm may already be toxic to some plants. Above a concentration of 1.5 ppm boron is toxic to most plants. At high soil pH, boron is adsorbed to a certain extent on the clay particles. Soluble aluminate (AlO_2^-) is also toxic to plants under such pH conditions. However, preliminary experiments comparing halophytes and glycophytes reveal that halophytes such as *Aeluropus litoralis* or *Suaeda monoica* are much more tolerant to aluminate than many glycophytes.

The higher the exchangeable sodium percentage becomes, and the higher the soil alkalinity is, the clay minerals of such soils tend to disperse and the A horizon becomes structureless (Fig. 3.3). Permeability of water in sodic soils is low. Light rains are hardly absorbed and the downward drainage of water is impeded. Nevertheless, in due time, the dissolved aluminates and silicates,

Fig. 3.3. Structureless layer on top of a salt-infected soil.

together with a colloid suspension of clay minerals, are gradually leached into B horizon where they are precipitated and form an impermeable hardpan. Horizon A of sodic soils may vary between 3–30 cm in thickness. It is characterized by scarcity of organic matter and, in many cases, its upper layers are in a laminated form. In some sodic soils which have a structureless and loose A horizon, the upper layer is washed away. In such a case, the exposed compact layers of the B horizon form bare spots, known in North America as slick spots. Horizon B of sodic soils has generally a columnar structure. Some of such soil prisms have flat tops while others are cap-shaped. The B horizon in sodic soils may be up to 1 m deep and consist of several layers of prisms.

Sodic soils are not uniform, and five soil groups were defined by Kovda (1935). Such distinction was based on salt, carbonate, and gypsum content, on content of organic matter, and on location of the lime horizon in the profile.

Continuation of the leaching processes of sodic soils turns the surface layers eventually into a light gray acid horizon and the soil is named *solod*. This is the last stage in the genetic series of halogenic soils out of which the soluble salts and adsorbed sodium were exchanged and washed away. A solod soil profile very much resembles soils of the podsol type, but differs from them in the state in which its silicates are precipitated. While in podsols silicates are crystalline and silica (quartz) is residual in the B horizon, in soils of the solod type silicates are precipitated in an amorphic form. The claypans of the B horizons in the solod types of soil break into prismatic columns with caplike tops. The acid-leached horizon then extends downward often like tongues. Whereas saline and sodic soils are common, soils of the solod type are rare.

Usually saline, sodic, and saline–sodic types of soils appear together in a salty spot, each occupying a niche in the topographically divergent site. Saline soils are also formed in the centers of terminal drainage basins while the various types of sodic soils are located more toward their periphery.

In addition to the morphological–pedological classification described previously, saline soils were also defined and classified by the Staff of the U.S. Salinity Laboratory, Riverside, California, on the basis of their agricultural value (Richards, 1954). Accordingly, a few types of salt-affected soils were distinguished, with their classification based on the conductivity and percentage of exchangeable sodium.

1. *Saline soils:* nonalkali soils in which the soluble salt content is high enough to inhibit growth of most agricultural crops. Electrical conductivity of the saturated extract of such a soil is above 4 mmhos/cm, pH is below 8.5, and exchangeable sodium percentage is below 15%.

2. *Alkali sodic soils:* soils with enough adsorbed sodium to inhibit growth of most agricultural crops. Usually, pH of those soils is above 8.5 and most commonly may reach values of 10 or higher. Two types of alkali soils were determined:

(a) *Nonsaline alkali soils:* soils in which conductivity of the saturated extract is below 4 mmhos/cm, but which have an exchangeable sodium percentage above 15%.

(b) *Saline–alkali soils:* soils with an exchangeable sodium percentage of 15% or more, but which, in addition, have a high content of soluble salts. Electrical conductivity of the saturated soil extract is above 4 mmhos/cm. Soils of this type frequently contain considerable amounts of gypsum.

Very little is known of the specific effects of various types of saline or sodic soils on growth and behavior of halophytes. However, the scattered data available suggest that as a group, halophytes do not respond only to salinity but also are more tolerant to the entire complex of physical, chemical, and biological modifications induced in the soil by salinity. Among such modifications, aeration and microbiological activity seem to be most important.

The Ecology of Waterlogged Halomorphic Soils

Plant growth on hydromorphous salines is affected by high salinity as well as by the low partial pressure of oxygen, high CO_2 concentration, a different oxidation–reduction status of various soil minerals, high methane content, specific soil microflora, etc. Only some plants can survive such conditions. Some plants can tolerate reducing and semianaerobic conditions, as long as these conditions are kept constant. Changing conditions sometimes have negative effects. For example when the tide is coming in and water covers the marsh, air is either trapped between the two water layers, or is slowly bubbling out. In the latter case, a diurnal tidal cycle involves a periodic flush of the soil with fresh air. When the water level in a swamp with a high sulfide content is lowered and the upper parts are exposed to air, oxygen penetrates the mud and oxidizes sulfides to sulfates. Sometimes enough H_2SO_4 is formed to acidify the soil and kill plants.

This is also true with respect to temperature. With changes in water levels of coastal salt marshes of the temperate region, plants are subjected also to periodic changes in temperature. In winter, the incoming tide water usually is much warmer than the frozen soil surface; in summer the warm soil surface becomes inundated below a layer of cold water. In April, soil surface temperatures of certain marshes in New England may reach $30°$ C, while temperature of the incoming tide water is only $5°$ C. Thus plants of such habitats must be adapted to sudden fluctuations in temperature (Teal and Teal, 1969).

Hygrohalophytes also seem to be adapted to long periods of inundation. Thus, *Plantago maritima* and *Aster subulatus* inhabit the waterlogged portion of salines because of their high tolerance to inundation (Chapman, 1960). According to Adams (1963), *Salicornia foliosa* not only tolerates waterlogged habitats, but seems to derive benefit from them, because of an increased capacity to obtain iron under such conditions and avoid chlorosis.

On the other hand, shoot growth in terrestrial halophytes is inhibited by inundation. Following prolonged periods of root submersion, leaf abscission and early death of *Nitraria retusa* plants were observed. Salinity of the water did not change the pattern of such effects, but the physiological status of the plant did, with slow-growing and old plants of *Nitraria* affected less by inundation than young, fast-growing, and metabolically active ones (Keren, 1970).

Tidal immersion of salt marshes alters aeration of the substrate, its temperature, salinity, light intensity, composition, and duration.

Data regarding the ability of various plant species to remain submerged do not always show agreement. According to Oliver (1925), stands of *Spartina townsendii* cannot extend below 3 feet of the high spring tides. On the other hand, it was shown by others (Goodman *et al.,* 1959) that *Spartina* may survive down to 2 m below spring tides with daily immersion periods of over 4 hours. According to Hubbard (1969), the seaward distribution of *Spartina anglica* is governed by immersion periods during neap tides. Its resistance for low oxygen conditions is high and this species can survive up to 23½ hours of immersion in a neap tide cycle. Conditions of immersion differ in the field and in the laboratory. Seedlings of *Spartina anglica* survived 4½ months of continuous submersion in clear seawater. However, under natural conditions, the tidal waters contained enough suspended sediments to inhibit growth of the submerged plants. Periodic inundation in unclear water alters both the light intensity and the photoperiod to which plants are exposed. The conflicting conclusions which have been reported probably result from differences in ecotypes investigated as well as from differences in environmental conditions.

Methane content of the soil also seems to play an important role in survival ability of some plants. Methane appeared to have an adverse effect on plants, such as barley and tomato, but did not affect swamp plants such as rice. No information regarding responses of halophytes to methane is available.

Certain plants are able to evade anaerobiosis by altering the aeration conditions of the soil around their roots. *Spartina* plants, as well as most other hydrophytes and hydrohalophytes have large-sized air ducts in which oxygen moves from the shoot downward to the roots. Oxygen may even leak from the roots outward (cf., van Raalte, 1940). In fact, due to such oxidative conditions, the soil around *Spartina* roots turns reddish in color (iron oxides) instead of the common black color of the marsh soil which is formed by sulfides.

THE BIOTIC FACTORS

Besides the specific chemical and physical characteristics of salt-affected soils, their microflora and microfauna also change considerably. Indirectly, such changes may have an important impact upon growth of higher halophytes.

The soil microflora plays an important role in improving soil structure, in decomposition of its organic matter, and in the nitrogen and sulfur cycles.

Saline soils affect their microflora both directly and indirectly. Direct effects induce a better growth of halophytic bacteria (e.g., sulfur-reducing bacteria) but inhibition of the salt-sensitive one (e.g., *Myxobacteria*). pH also has a very pronounced effect on the microflora. *Azotobacter* for example appears abundantly on saline and sodic soils, where the pH is approximately 8. When the pH is lowered to 6 or below, populations of *Azotobacter* disappear. Change in pH due to the activity of sulfate-reducing bacteria may be detrimental for acidophilic phanerogamous halophytes on sulfate-affected soils (Abd el Malek and Rizk, 1963). The indirect effects of salinity on microorganisms are mostly due to root secretions. Under saline conditions shifts in various metabolic processes occur, secretion of root metabolites differs in quantity and quality, and, consequently, also the composition and size of the soil microflora. Saline soils inhabit a large group of halophytic microorganisms. Halophilous sulfur bacteria and *Chlamydomonas* may survive solutions with up to 25% salt. Some of them can live even in water containing 30% NaCl or more, as in the Dead Sea, Israel, or in the Great Salt Lake, Utah (Zobell *et al.,* 1937; Elazari-Volcani, 1940).

Salt-dependent and salt-resistant microorganisms are widely distributed in salt-affected as well as in salt-free soils (Stuart, 1938; Henis and Eren, 1963).

However, bacteria isolated from saline soils were found to be more salt-tolerant than similar strains of bacteria isolated from nonsaline soils (Keller and Henis, 1970). Halophytic microorganisms of saline soils belong to different groups. Microorganisms isolated from the soil of a hydrohalomorphic saline in Eilat, Israel, contained populations of cellulose decomposers, nitrate reducers, and anaerobic nitrogen fixers. However, it is interesting to note that no ammonia, sulfur oxidizers, or aerobic nitrogen fixers were detected in this specific site. Some of those bacteria grew better on saline media, whereas others grew similarly on saline or on salt-deficient media (Keller and Henis, 1970).

Not all bacteria isolated from saline soils were halophytic. Some of these bacteria were unable to multiply under saline conditions (Henis and Eren, 1963).

Bacteria constitute important competitors for oxygen with other organisms in waterlogged soils or even on the mud surface (Zobell and Feltham, 1942). Aquatic bacteria were found to consume, in seawater at $22°C$, an average of 14.9×10^{-12} cm^3 oxygen per cell per hour. Even higher values of consumption were found when temperatures either became higher or when the substrate contained more oxidizable organic matter. It is believed that bacterial respiration is mainly responsible for the lack of O_2 in the mud surface of most inundated habitats, including salt swamps and salt marshes.

Bacterial populations in bay water vary between several thousand and a few million per cm^3. The mud in the bottom contains a few million bacteria per gram. The bacteria consume enough O_2 even to reduce its tension in the lower layers of seawater in certain bays (Table 3.1). Bacteria may also cause a change in pH of such muds averaging some 0.2–0.6 pH units. A slight acidification of mud below seawater (pH of 7.4–8.0 as compared to pH 8.0–8.3) presumably is caused by intensive bacterial activity.

Soil microorganisms have still other functions. Biological transformations have been shown to be involved in development of carbonates and bicarbonates

Table 3.1

AMOUNT OF OXYGEN CONSUMED BY BACTERIA (ml/100 gm) OF MISSION BAY MUD, SAN DIEGO, CALIFORNIA, AT $22°C$[a]

Period of incubation	No. of sample			
	1	2	3	4
1 day	0.13	0.04	0.09	0.21
2 days	0.29	0.17	0.17	0.36
5 days	0.53	0.31	0.35	0.92
10 days	0.90	0.51	0.64	1.42
Average per day	0.11	0.06	0.08	0.18

[a]After Zobell and Feltham, 1942.

in various saline soils (Whittig and Janitzky, 1963; Janitzky and Whittig, 1964). Such developments occur in waterlogged soils which contain both sulfates and organic matter. The high water table induces anaerobic conditions, while the organic matter is needed for the supply of CO_2 and support of the microflora. Under such conditions, the anaerobic sulfate-reducing bacteria get their oxygen from reduction of sulfate to sulfides and energy from organic matter. Consequent precipitation of the sulfide as FeS lowers the concentration of the anion, which was originally associated with sodium, enables an increase in pH, and accelerates the formation of bicarbonates. During dehydration cycles, bicarbonates are transformed into carbonates and precipitate as the calcium or magnesium salts. This was experimentally shown by Janitzky and Whittig (1964) (Table 3.2).

Data presented indicate that accumulation of carbonates and bicarbonates, as well as an increase in pH, were observed only in soil samples to which Na_2SO_4 as well as organic matter were added. Under anaerobic conditions, a concomitant decrease in exchangeable calcium and increase in the content of sulfide were found. Bacterial activity may thus initiate certain reactions which will eventually transform saline soils into sodic ones.

Another example which illustrates the effects of microorganisms on the behavior of phanerogamous halophytes can be seen in coastal habitats. Fore-shore dunes are known to be deficient in many nutrients among which nitrogen is the most obvious one (Willis *et al.*, 1959). Nevertheless, native coastal psamohalophytes grow vigorously in such habitats and seem to be sufficiently supplied with nitrogen. Experiments conducted by Hassouna and Wareing

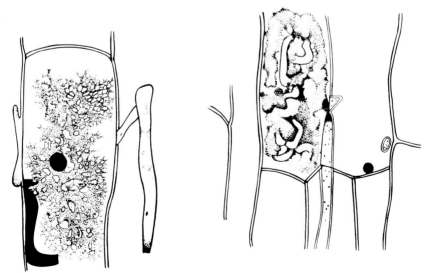

Fig. 3.4. Infected cells in a mycorrhizal root of *Armeria maritima.* (After Mason, 1928.)

Table 3.2

EFFECTS OF SODIUM SULFATE AND ORGANIC MATTER CONCENTRATIONS ON CHEMICAL CHANGES INDUCED DURING ANAEROBIC INCUBATION OF YOLO SOIL SAMPLES[a,b]

Na$_2$SO$_4$ added (mEq/100 gm)	pH after incubation	ΔNa Soluble	ΔNa Exchangeable	Δ(Ca + Mg) Exchangeable	Δ(CO$_3$ + HCO$_3$) Total	SO$_4$ Soluble	SO$_4$ Total	ΔSO$_4$[c]	ΔS^{2-}
Group 2A: no organic matter added									
1	7.6	+0.2	+0.7	+0.6	+0.2	0.1	1.2	+0.2	+0.3
5	7.6	+2.0	+2.9	-1.6	+1.7	1.8	5.0	0	+0.7
25	7.5	+15.2	+9.2	-6.1	-1.5	18.8	24.4	-0.6	+0.4
Group 2B: 0.5% organic matter added									
1	7.2	+0.4	+0.7	+0.2	+0.8	0.2	2.2	+1.2	+0.3
5	8.1	+1.2	+4.0	-2.2	+5.2	0.4	1.9	-3.1	+2.6
25	8.1	+15.3	+9.1	-6.7	+4.5	17.6	18.7	-6.3	+3.1
Group 2C: 3% organic matter added									
1	7.2	+1.0	+1.0	-1.1	+0.1	0.2	2.3	+1.3	+0.4
5	7.8	+2.3	+3.8	-2.6	+3.5	0.4	2.0	-3.0	+2.1
25	9.1	+11.3	+13.9	-7.4	+11.0	7.1	7.8	-17.2	+12.5

[a] Data adapted from Janitzky and Whittig (*J. Soil Sci.* **15**, 145–154, 1964) with permission of the authors and Williams & Wilkins, Baltimore Maryland.

[b] Chemical changes expressed in terms of mEq/100 gm soil. Values represent difference from original soil.

[c] Values represent net loss or gain of SO$_4^{2-}$ added to systems.

(1964) suggest that fixation of atmospheric nitrogen contributed an important part of the nitrogen supply of such plants. Thus, a dense population of microorganisms, mostly *Azotobacter,* in the rhizosphere of *Ammophila* roots seems to be among the factors which contribute to the adaptability of those plants to coastal habitats.

Presence of mycorrhizae may also influence the presence and behavior of plants in saline soils. Several salt marsh plants, including *Plantago coronopus, P. maritima, Aster tripolium, Glaux maritima,* and *Armeria maritima* have mycorrhizal roots (Mason, 1928). The mycorrhizal fungi produce typical "arbusculi" inside the cells of the young and fine roots of those species (Fig. 3.4). No mycorrhizal infections were observed in the old roots of the same species. The mycelium was described as being of the Phycomycete type. However, mycorrhizae seem to be restricted to some halophytic species only and no mycorrhizae were found in roots of *Salicornia europea, Spergularia marginata, Triglochin maritimum, Juncus maritimus,* or *J. gerardii.* A general survey of the occurrence of mycorrhizae in halophytes and thorough investigation into their physiological significance are still needed.

4

Classification

Plants which grow and complete their life cycle in habitats with a high salt content are called salt plants or *halophytes*. Usually that term is reserved only for plants which appear in salty habitats constantly and specifically. Plants occupying only local nonsalty ecological niches in an overall saline environment, or those which appear in such habitats only for short periods, i.e., during the rainy season, are called *pseudohalophytes* or false halophytes.

Despite the wide distribution of halophytes in various ecological or climatic regions, their taxonomic, structural, and behavioral uniformity is striking. In many cases we find that the same genera, and even the same species (e.g., *Atriplex halimus, Salicornia herbacea,* and *Juncus maritimus*), appear in salt marshes of the temperate as well as of the tropical and subtropical regions. Such a distribution under diverse climatic conditions can be understood only if salinity *per se* is the major factor determining the character of the habitats, and if natural selection is affected by salts more than by any other single factor.

The precise determination of the effects of salinity and the sites where salinity may affect plants are not easily assessed. First, since both salt composition and salt concentration differ from one habitat to another, the use of the term "salinity" is usually in its loosest meaning. In certain cases, it is not the absolute amount of a certain ion which may affect plants, but rather the composition and total concentration of salts. Certain species of plants may be found in sites where the sodium chloride concentration is beyond their theoretical tolerance, but where high concentrations of calcium, potassium, or sulfate are found as supplementary ions. Those ions moderate the toxic effects of sodium and chloride, thus enabling the plant's existence.

Second, the contact between the salt and the plant may involve different tissues of the plant body. In terrestrial habitats, contact occurs between the plant roots and the saline soil (*terrestrial halophytes*). In marine habitats, or in salt marshes, salts contact either the plant roots (*emerged halophytes* or *hygrohalophytes*) or the entire plant body (*submerged halophytes* or *hydrohalophytes*). Coastal plants or plants of dust deserts whose aerial organs are affected by airborne salt droplets or salt particles are called *aerohalophytes.*

In determining salt concentrations, both the total amounts of salt as well as the soil water content must be taken into account. Certain halophytes (*hydrohalophytes* and *hygrohalophytes*) can exist only in undrained, constantly wet habitats. On the other hand, other species exist only in salty, but relatively dry habitats (*xerohalophytes*).

Evidently as the ecological conditions in saline habitats are so complex also the definition of halophytes is extremely difficult. Several attempts have been aimed at classification of the divergent halophytic species. These attempts can be divided into three major criteria. Some investigators based their classification upon the soil–salt content of the native habitats. Others tended to emphasize the importance of the sources of salt and classified the plants accordingly. A third approach, preferred lately, emphasizes the importance of plant responses to salinity. None of the classification systems seems to be entirely satisfactory. In most cases, classification is based on quantitative rather than on qualitative traits. Distinctions between groups are thus artificially made and dividing lines are drawn sometimes across a continuous cline of gradually changing characteristics. Nevertheless, each approach has some of its own merits.

One of the first modern classifications of saline habitats was proposed by Stocker (1928) as follows:

1. *Aquatic–haline*
2. *Terrestro–haline*
   Hygrohaline
   Mesohaline
   Xerohaline
3. *Aero–haline*
   Habitats affected by salt spray (maritime).
   Habitats affected by salt dust (salt deserts).

This classification is based on the sources of salts and on the plant organs which they affect. Plants are divided according to their respective habitats and later subdivided according to their water relations. In all cases, salinity is defined as the soil salt content above 0.5% NaCl calculated on the soil dry weight basis. This definition is arbitrary and no mention is made of the ion activity which varies in different soils and under different water conditions.

The classification of Iversen (1936) is also based mainly on salt content of the habitat. Iversen distinguished a number of salinity ranges and claimed that these ranges overlapped the commonly found ranges of the plants' salt tolerances. The habitats were thus classified into the following categories: (1) Oligohaline, habitats containing 0.01–0.1% NaCl; (2) mesohaline, habitats containing 0.1–1.0% NaCl range; (3) polyhaline, habitats containing 1% NaCl and up.

Plants occupying these habitats were named accordingly. Various combinations were suggested for plants occupying habitats with wider salinity ranges (e.g., *oligomesohalophytes*). Plants existing in all three types of habitats were classified as *euryhalophytes* and those specific to the polyhaline habitats were named *euhalophytes*.

Chapman (1942) generally followed the classification of Iversen (1936), but adopted Stocker's limit of salinity (0.5% NaCl). According to Chapman, halophytes were classified as: (1) Miohalophytes, plants growing in habitats of low salinity (below 0.5% NaCl); and (2) Euhalophytes, plants growing in highly saline habitats. The euhalophytes were subdivided into three categories: (a) *mesohalophytes,* plants growing in habitats with a salinity range of 0.5–1%; (b) *mesoeuhalophytes,* plants growing in habitats with a salinity range of 0.5% and higher; and (c) *eueuhalophytes,* plants growing in habitats with a minimal salinity of 1%.

Van Eijk (1939), using a classification based on plant distribution as well as on some plant responses, divided halophytes into the following categories: (1) plants which tolerate salts, but whose optimal development is in nonsaline habitats and (2) plants which have an optimal development in saline habitats.

Each of these two groups were subdivided as (a) plants which appear only in saline habitats and (b) plants which appear also in nonsaline habitats.

One of the first attempts to classify halophytes solely according to their response to salinity was made by Tsopa (1939). His classification included four categories: (1) *obligatory halophytes,* plants requiring salinity throughout their life; (2) *preferential halophytes,* plants exhibiting optimal growth in saline environments, despite their appearance in nonsaline habitats; (3) *supporting halophytes,* nonaggressive plants which are capable of growing in saline habitats; and (4) *accidental halophytes,* plants which appear in saline habitats only occasionally.

One of the difficulties in defining halophytes, stems back to the difficulty in defining salinity and limitations of saline habitats. Thus, in Tsopa's classification salinity ranges for each of those categories were left undefined.

According to Weissenbock (1969), the older terms "obligatory halophyte" and "facultative halophyte" should have been replaced by more precise physiological terms. Weissenbock suggested that the term "facultative" should be applied to plants whose growth is favorably affected by NaCl but where Na can be substituted by K (e.g., *Aster tripolium, Artemisia maritima, Plantago maritima,* and *Suaeda maritima*). Obligatory halophytes are plants which are favorably affected by sodium chloride and by sodium chloride alone (e.g., *Salicornia herbacea* and *Atriplex vesicaria*).

According to Kreeb (1964), halophytes are distinguished from glycophytes mainly by their growth response to salinity. While growth of glycophytes decreases with increasing salinity of the habitat, growth of halophytes seems to be presented by an optimum type of curve. However, such a classification cannot be based on growth alone because growth depends on many other nutritional conditions. For example, according to Harmer *et al.* (1953), glycophytic and semihalophytic agricultural species can be classified into four categories depending upon their response to salinity: (a) plants which show no response to sodium under conditions of potassium deficiency; (b) plants which respond positively to sodium under conditions of potassium deficiency; (c) plants which show a slight positive response to sodium even when potassium is abundant; (d) plants which show a marked positive growth response to sodium when potassium is abundant.

Those last classifications were aimed at determining the effects of salinity on nonhalophytic rather than on halophytic species.

An attempt to classify halophytes according to their response to internal salt content was made by Steiner (1935). According to him, salt marsh plants can be classified into three types.

1. Succulent halophytes: plants which can tolerate high concentrations of chloride in their cell sap due to an increase in succulence (e.g., *Salicornia herbacea*).

2. Nonsucculent halophytes: plants which resist salts by desalinization of their tissues and secrete salt excesses through salt glands (e.g., *Spartina alterniflora*).

3. Accumulating type: plants with no special mechanism for salt removal. Salt concentrations in the tissues of such plants increase until the plant's death (e.g., *Juncus gerardii, Suaeda fruticosa*).

A similar classification of halophytes on the basis of their osmotic regulation was also proposed (cf. Adriani, 1956). Also Henkel and Shakhov (1945) distinguish three categories of halophytes on the basis of their systems of salt regulation:

1. Plants accumulating salts—*euhalophytes.*

2. Plants eliminating salts after salts have been absorbed—*crinohalophytes.*

3. Plants which restrict salt accumulation—*glycohalophytes.*

4. Plants which accumulate salts and localize them in specific tissues. No

Table 4.1
CLASSIFICATION OF PLANTS OF SALINE HABITATS

Euhalophytes				Pseudohalophytes
Salt-requiring halophytes		Salt-resisting halophytes		
Obligatory halophytes	Preferential halophytes	Salt-enduring halophytes (Salt tolerant)	Salt-excluding halophytes	Salt-evading halophytes	Salt-avoiding plants
Plants dependent upon salts for their survival, e.g., *Salicornia* sp. and various bacteria and algae	Plants whose growth and development are improved in the presence of salts, e.g., *Arthrocnemum* sp., *Aster* sp., *Nitraria* sp., *Salicornia* sp., *Suaeda* sp.	Plants enduring a high protoplasmatic salt content, e.g., *Suaeda monoica*	Plants accumulating salts in special hairs, e.g., *Arriplex* sp. Plants secreting salts from their shoot, e.g., *Aeluropus* sp., *Limonium* sp., *Tamarix* sp. Plants retransporting salts from the shoot into the root, e.g., *Salicornia* sp.	Plants evading salt uptake, e.g., *Rhizophora* sp. Plants evading salt transport into the leaves, e.g., *Prosopis farcta*	Ephemers Niche plants

mention was made by Henkel and Shakhov of the various effects of salts on plant growth and behavior.

In the classification presented henceforth (Table 4.1), an attempt was made to unite the plant–salt relationships into one scheme and to put more stress on salt-resistance mechanisms and on internal rather than on external salt relationships.

For the proposed classification, a saline habitat is defined as a habitat where the minimal NaCl content of the saturated soil extract is 100 mEq (conductivity of approximately 10 mmhos/cm at 25° C; osmotic potential of –4 bars) at a pH below 8.5. This definition of salinity is close to the limits suggested by the staff of the U.S. Salinity Laboratory, Riverside, California, for agricultural crops (Richards, 1954). According to our experience, this limit of salinity fits well the distribution limits and behavior of most native halophytes. Similar schemes can be constructed for specific groups of halophytes, i.e., for sulfate halophytes or for alkaline halophytes.

In many cases, plants may possess more than one mechanism for salt resistance and thus belong to more than one group. Furthermore, plants may be grouped into one class at an early stage of development and to another class at later stages. In such cases, classification of the species is determined by the dominance of their characteristics starting with obligatory halophytes and ending with salt-avoiding pseudohalophytes.

The various categories can be further subdivided according to the salt sources (hydrohalophytes, terrestrohalophytes, or aerohalophytes) and/or according to the plant–water relationships (hydrohalophytes, mesohalophytes, xero-halophytes, etc.). Plants may also be classified according to their ability to withstand a narrow range of salinity (stenohaline) or a wide one (euryhaline). Thus, *Atriplex halimus,* for example, which is classified as a salt-excluding plant may still be further defined as a salt-excluding terrestro-xero-euryhalophyte.

—————————————————————5

Distribution and Synecology of Halophytes

Determination and description of vegetation units in salt marshes are intricate and rather difficult. Zonation in most saline habitats is extremely narrow and distribution of each community is thus limited to an area that sometimes is restricted to a belt only a few centimeters wide. Thus, the exact limitation of plant distribution and specific effects of individual environmental factors on zonation of plants in a saline plant community are not easily assessed.

Classification of Halophytic Communities

Because of the difficulties, as well as the lack of standardization of terminology, classification of halophytic plant communities also poses difficulties (Chapman, 1958). At least several systems of classification are in use, but most of these deal only with hydrohalophytes and coastal salt marsh plants. For example, according to Knapp (1958), all phanerogamous halophytes belong to two orders of marsh plants: Salicornietalia and Juncetalia maritimi. Xerohalophytes either are excluded from such a classification or are grouped into one subunit of one order.

The main systems of classification of halophytic communities were incorporated by Chapman (1958) into a single scheme, which included static as well as dynamic information. The prefix *Coeno* was introduced for higher vegetation units in which all the communities had a common dominant genus but different species in different places around the world, e.g., *Juncus* sp. or *Salicornia* sp. The dynamic status of the communities was indicated by addition of the prefix *Eco*. According to this classification, halophytic plant communities were divided into the following nine orders:

1. *Halobenthalia* Chap.

An order of submerged marine plant communities. Usually, communities dominated by *Zostera* sp. are found in colder waters, while those dominated by *Cymodocea* sp. are to be found rather in tropical and subtropical seas. Various communities dominated by the genus *Ruppia* are commonly found both in coastal and inland saline habitats.

This order is subdivided into smaller units each comprised of several communities: *Ruppion maritimae* Br.Bl. and *Posidonion* Br.Bl.

2. *Coeno-Salicornietalia* Chap.

An order of pioneer coastal halophytic communities inhabiting wet or periodically inundated heavy clay substrates. The dominant species are mostly annual or perennial articulated Chenopodiaceae. Various subunits of this order are also typical of wet inland salt marshes. The subunits are classified as follows: *Salicornion* Chap. and *Suaedion* Chap.

Though *Suaedion* is ranked as an alliance in this classification, it is our opinion that the communities of the *Suaedion* should be grouped into a separate order.

3. *Coeno-Puccinellietalia* Chap.

This order includes plant communities similar to those of the previous order, but which inhabit more sandy and aerobic sites. These communities are dominated by grasses rather than by succulent chenopods. The order is subdivided into *Eco-Puccinellion* Chap. and *Distichlion* Chap.

4. *Coeno-Spartinetalia* Chap.

An order of plant communities, mostly grasses, inhabiting the coastal salt marshes of the temperate regions in the northern hemisphere. The order includes one subunit: *Eco-Spartinion* Chap.

5. *Staticetalia* Chap.

An order of mixed herbaceous and grass plant communities inhabiting the midlevel or the upper parts of the coastal salines. The order includes also a few communities of inland salines. It is subdivided into *Staticion galloprovincialis* Br.Bl.; *Staticion gmelinii* Chap.; *Eco-Staticion septemtrionalis* Chap.; *Puccinellio-Staticion gmelinii* Topa; *Crithmo-Staticion* Mollinier; *Staticion orientalis* Chap.; *Eco-Armerion* Chap.; *Petrosimonietum crassifoliae* Chap.; *Eco-Sesuvion* Chap.

6. *Coeno-Festucetalia* Chap.

An order of communities typical of high and well-drained coastal salines. All communities are dominated by species of the genus *Festuca.* The order is subdivided into: *(Eco) Festucion maritimae* Chap.; *Halimionion* Chap.; *Artemision maritimae* Chap.; *Zoysion maritimae* Chap.

7. *Halostachyetalia* Topa

An order of herbs and low shrubs. Usually found on sites with a high water table. It is subdivided into two alliances: *Halostachyon* Topa; *Heterostachyon* Chap.

8. *Coeno-Juncetalia* Chap.

A world-wide order of plant communities of brackish water. Found in high coast marshes as well as on the banks of inland salty springs. In most cases, the communities of this order are long persistent. The order is subdivided as follows: *Juncion maritimi* Chap.; *Trifolion maritimi* Br.Bl.; *Baccharion* Chap.; *Leptocarpion* Chap.

9. *Phragmitetalia* W. Koch

An order of plant communities of fresh or brackish water. It is subdivided into the following: *Eco-Phragmition communis* W. Koch; *Eco-Scirpion maritimi* Dahl et Hadac; *Eco-Magnocaricion paleaceae* Dahl et Hadac; *Halocladion* Chap.

Chapman's system of salt marsh plant classification is extremely valuable and certainly covers most groups of those plants. Nevertheless, this system treated terrestrial hygro- or xero-halophytic communities only slightly and did not give them their full weight. Thus, in order to make Chapman's system circumglobal, an additional order of plant communities and a few other minor changes should be introduced.

The addition of an order, *Suaedetalia* is urged by the present author. It is an order of chamaephytic plant communities of highly saline habitats. The dominant species are annual or perennial articulated Chenopodiaceae as well as species of the Zygophyllaceae, Tamaricaceae, etc. The order is mostly constituted of long-persisting communities in dry desert salines. Dependence on soil water regime of various communities of this order varies under different climates.

The suggested modified scheme, constituted of ten orders is presented in Table 5.1.

The prefix *Eco* denoting the dynamic status of the same communities in Chapman's scheme was omitted. Since by definition no other plant community except the climax is stable, it follows that all halophytic communities are, to a certain degree, dynamic. Some of those communities may change at extremely slow rates, but none are stable. Prefixing the term *Eco* to only some communities might cause misunderstanding.

Typical Halophytic Communities

Relatively little is known of the limitations of plant communities to specific ecological conditions. Usually submerged marine plants, i.e., plants which belong to Halobenthalia communities are salt obligatory. For example, plants such as *Zostera* require a steady and constant level of salinity. If salinity levels are lowered by dilution of seawater with freshwater, yellowing of leaves occurs and is followed by necrosis and death of plants. Zonation of plants in marine habitats is determined mostly by factors other than salinity, in contrast to zonation of halophytes of terrestrial habitats which depends on salt–water relations.

Correlation between soil salinity and distribution of certain plant communities in Holland (Fig. 5.1) was reported by Adriani (1945). Similar

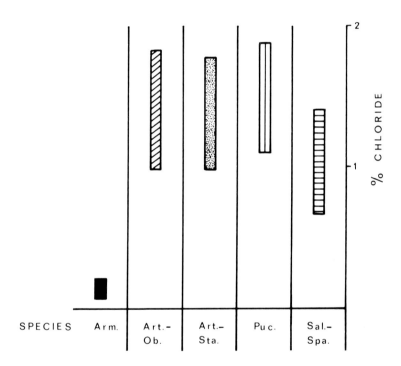

Fig. 5.1. Range of chloride content of the soil solution in various halophytic plant communities in Holland. (After Adriani, 1945.) Sal.–Spa., Salicornieto-Spartinietum; Puc., Puccinellietum maritimae; Art.–Sta., Artemisietum maritimae, *Statice*–facies; Art.–Ob., Artemisietum maritimae, *Obione*–facies; Arm, Amerion maritimae.

results were obtained for distribution of plants in saline sites of Sedom, Israel (Fig. 5.2). Despite wide overlapping in ecological requirements, clear segregation exists between requirements of individual species as well as conditions under which the various communities developed (Burvin, 1963). Most halophytes have a wide amplitude as to soil salinity, but a narrow one regarding soil humidity. Thus, it is interesting to note that some of the species have a common range of

Table 5.1

Principal orders	East coast England	Scandinavia	Central Europe		Low Countries and France
			Coastal	Inland	
1. Halobenthalia	*Zosteretum*	*Zosteretum*	*Zosteretum*	*Parvipotameto–Zannichellietum*	*Zosteretum*
2. Salicornietalia	*Salicornietum Asteretum*	*Salicornietum*	*Salicornietum*		*Salicornietum*
3. Suaedetalia	*Suaedeto–Salicornietum Suaedetum*	*Suaedetum Asteretum Suadeto–Spergularietum*		*Suaedetum maritime Lepidio–Camphorosmetum typicum*	
4. Puccinellietalia	*Puccinellietum Puccinellio–Asteretum Puccinellio–Staticetum*	*Puccinellietum Puccinellio–Asteretum Puccinellio–Suaedetum Puccinellio–Spergularietum*	*Puccinellietum*	*Lepidio–Puccinellietum asteretosum*	*Puccinellietum*
5. Spartinetalia	*Spartinetum townsendii Spartinetum strictae Astereto–Spartinetum*		*Spartinetum townsendii*	*Agrostideto albae–Asteretosum*	*Spartinetum townsendii*
6. Staticetalia	General salt marsh *Plantaginetum Suaedeto–Staticetum Staticetum reticulatae*	General salt marsh *Armerietum maritime Plantaginetum Glaucetum maritime Staticetum*			*General salt marsh*
7. Festucetalia	*Festucetum rubrae Festuceto–Agrostidetum Halimionetum*	*Festucetum rubrae Festuceto–Agrostidetum Festuceto–Poaetum pratensae Festucetum arundi arundinaceum Agrostidetum Artemisietum*	*Festucetum rubrae Festucetum thalassicolae Agrostidetum*		*Agrostidetum Halimionetum Artemisietum*

RELATIONSHIPS BETWEEN VARIOUS REGIONAL SALT MARSH COMMUNITIES AND PRINCIPAL ORDERS OF THE MAJOR SERES[a]

Mediterranean		Australasia		North America		
Coastal	Inland and desert	Australia	New Zealand	New England	Central U.S.	Pacific U.S.
Cymodoceetum Zosteretum	*Najadetum Ruppietum*	*Zosteretum Cymodoceetum*	*Zosteretum*	*Zosteretum Ruppietum*		
Salicornietum Arthrocnemetum Arthrocnemeto– Sphenopetum Kochieto– Suaedetum Salsoleto– Suaedetum		*Salicornietum australis Salicornieto– Hemich– roetum Arthrocne– metum*	*Salicornietum australis*	*Salicornietum*		*Salicornietum ambiguae Salicornieto– Distichli- detum Salicornietum bigelovii Batidetum maritimae*
	Suaedetum fruiticosae Suaedetum monoicae Nitrarietum retusae Suaedetum palaestinae Salsoletum rosmarini Atriplicetum halimi Salsoletum tetrandrae				*Suaedetum depressae Distichlideto- Suaedetum*	
		Sporoboletum Distichlidetum	*Puccinellietum strictae*	*Distichlidetum Spartineto– Puccinel- lietum*		
				Spartinetum glabrae Spartinetum pratensis Spartinetum peltinatae		*Spartinetum leianthae*
Camphorosmeto– Frankenietum Lenzieto– Oenanthetum Staticeto– Limoniastrum Trifolietum maritimii Halocnemetum Staticeto– Artemisietum Halimioneto– Artemisietum Triglochineto– Asteretum			*Triglochideto– Scirpetum Sellierieto– Cotuletum*	*General salt marsh Plantagineto– Spergula- rietum*		*Staticetum General salt marsh*
Halimioneto– Crypsidetum schoenoides Halimionetum						

Table 5.1—*continued*

Principle orders	East coast England	Scandinavia	Central Europe		Low Countries and France
			Coastal	Inland	
8. Halostachyetalia					
9. Juncetalia	*Juncetum gerardii*	*Juncetum gerardii* *Juncetum bufonii* *Festuceto– Juncetum*	*Juncetum gerardii*	*Juncetum gerardii*	*Juncetum gerardii*
10. Phragmitetalia	*Phragitetum Schirpetum*	*Phragmitetum Scirpetum Typhetum Caricetum paleaceae*	*Scirpeto– Phragmi- tetum Scirpetum*		*Scirpetum*

^a Modified after Chapman, 1958.

salinity, but many others are segregated on the basis of their water requirements (cf. Fig. 5.3). In most cases segregation of plant communities depends on simultaneous changes of most of the ecological factors involved (cf. Bodrogközy, 1962). Such multiple effects on zonation of plants are presented in Fig. 5.4.

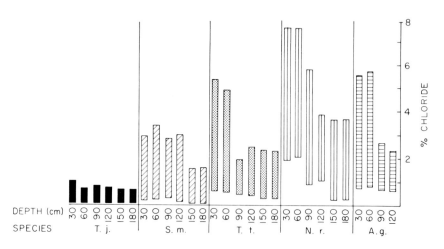

Fig. 5.2. The correlation between soil salinity and plant distribution in the saline of Sedom, Israel. (After Burvin, 1963.) T.j., *Tamarix jordanis;* S.m., *Suaeda monoica;* T.t., *Tamarix tetragyna;* N.r., *Nitraria retusa;* A.g., *Arthrocnemum glaucum.*

Mediterranean		Australasia		North America		
Coastal	Inland and desert	Australia	New Zealand	New England	Central U.S.	Pacific U.S.
Camphorosmetum annuae Plantagineto–Schoenetum nigricantis Camphorosmetum Agropyreto–Lepidieto crassifoliae						
Juncetum acuti Junceto–Triglochinetum Cariceto–Lotetum	*Juncetum*	*Stipeto–Juncetum*	*Stipeto–Juncetum Juncetum maritimii Leptocarpetum simplicae*	*Juncetum gerardii Asteretum subulati Iveto–Bacchari-detum*		
	Phragmitetum	*Typhetum Glycerietum aquaticae Plagianthus–Cladium juncetum*		*Phragmitetum Scirpetum Typhetum*	*Eleocharetum rostellatae Scirpetum americanae*	*Typhetum Scirpetum*

The halophytic coastal and inland floras of northern America were described by several investigators (Schaffner, 1898; Kearney *et al.*, 1914; Weaver, 1918; Ortenburger and Bird, 1931; Flowers, 1934; Coupland, 1950; Love and Love, 1954; Keith, 1958; Baalman, 1965; Ungar, 1965, 1967). Grass-covered salt marshes extend from Canada southward along the eastern coast to northern

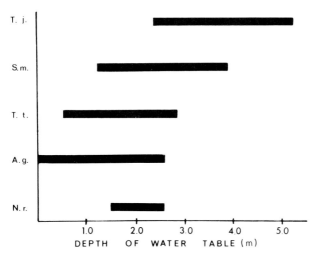

Fig. 5.3. The correlation between water level and the range of plant distribution in the saline of Sedom, Israel. (After Burvin, 1963.) T.j., *Tamarix jordanis;* S.m., *Suaeda monoica;* T.t., *Tamarix tetragyna;* A.g., *Arthrocnemum glaucum;* N.r., *Nitraria retusa.*

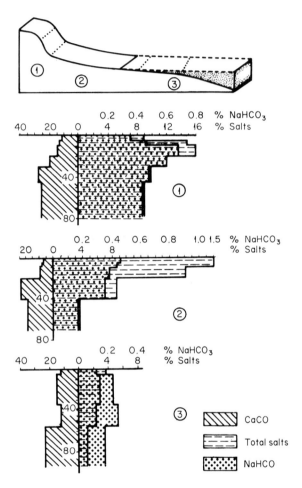

Fig. 5.4. Schematic distribution of certain plant communities in central Hungary and some characteristics of their native habitats. (After Bodrogközy, 1962.) 1. Lepidio Camphorosmetum; 2. Suaedetum maritimae; 3. Bolboschoenetum–Phragmitetosum complex. Depth in cm.

Florida. At that point, mangroves replace the grasses in formation of the dominant coastal plant community.

Among inland halophytic communities of this continent Suaedetum depressae appears to have the highest salt tolerance. This pioneer community invades barren salty flats in large areas over the plains of North America, and is characterized by a scattered appearance of plants. Whereas many seedlings develop annually around the parent plants, only a few survive. In the southern parts of the midwestern United States, Suaedetum includes *Sesuvium verrucosum,* whereas in the northern midwest and Canada, *Salicornia rubra* is the codominant. Other species which may be found invading open salt flats include *Distichlis stricta, Sporobolus texanus, Sporobolus airoides, Tamarix pentandra,* etc.

With decreasing salinity, the Suaedetum turns into a Scirpetum. If such an area remains wet, a sedgemeadow community with *Scirpus americanus, Eleocharis palustris,* as well as other species, prevails (Ungar, 1965, 1967). Where lowering of standing water level occurs, the community turns into a *Distichlis–Hordeum* complex, with *Distichlis* having higher frequency and density in areas of higher salinity. Comparatively well-drained sites are invaded by *Sporobolus* and common prairie species.

A rise in water table around artificial impoundments, in southeastern Alberta, Canada, enabled rapid upward movement of soluble salts to the surface from lower soil layers. Such salinization caused a drastic change in vegetation (Keith, 1958). In early stages of vegetation development, a *Hordeum jubatum* community colonized the site. It was later replaced by a *Distichlis stricta* community. Older impoundments were characterized primarily by *Suaeda depressa* and *Salicornia rubra,* but Keith (1958) was of the opinion that the latter did not substitute for *Distichlis* in the halosere and that other lines of development were involved in their appearance.

In the Tooele Valley region, the presence of certain plant communities directly reflects the degree of soil salinization (Kearney *et al.,* 1914). The *Sarcobatus vermiculatus* and *Atriplex confertifolia* association is found on soils which are saline at depths greater than 30 cm. Sometimes this association also occurs on soils with saline surface layers. The *Sporobolus–Distichlis* association is typical of saline soils with high water content. The *Allenrolfea–Salicornia* association occupies salt flats. The soils of such habitats are wet to the surface throughout most of the year.

On the eastern coast of America, the role played by algae in the composition of salt marshes is insignificant, whereas European and especially English salt marshes are characterized by phanerogamous as well as by a prominent vegetation of algae (Chapman, 1938).

Three types of vegetation are characteristic of coastal salt marshes in Europe (Chapman, 1960):

1. Marshes inhabited by grasses where soils consist of muddy sand (England, The Netherlands, and Sweden).

2. Marshes covered by *Spartina townsendii,* where the soil is a soft muddy clay.

3. Northern European marshes covered mostly by herbaceous species such as *Aster tripolium, Limonium latifolium,* and *Plantago maritima* where the substrates are heavy silts and clays.

Pioneer plants in these sites are *Salicornia herbacea* (= *Salicornia* europaea).

The mud zone may also remain bare until it is stabilized by invading *Puccinellia* sp. plants. Following stabilization under Puccinellietum, other herbaceous halophytes (*Aster tripolium, Plantago maritima,* etc.) enter these newly made habitats. *Spartina* salt marshes in Europe constitute, usually, only a short stage in the halosere. According to Ranwell (1964), a stand of *Spartina* in southern England remained pure for only 22 years. After that period, other plants, mostly herbaceous halophytes, started to invade the habitat and within the next 12 years replaced about 50% of the *Spartina* stand.

Distribution patterns of many halophytic or sodic plant communities in central Europe, depend, to a great extent, on the degree of inundation of their respective habitats. According to Bodrogközy (1958), the total cover of *Puccinellia* in a Lepidio–Puccinellietum limosae stand growing on alkali ("szik") soils in Hungary—increased proportionally to the water depth and time period during which the stand was submerged. Such changes were reversible and occurred even when the content of salts, as well as alkalinity, decreased.

Around the Mediterranean region, the muddy coasts are dominated by highly salt-tolerant, articulated Chenopodiaceae (*Arthrocnemum* sp., *Salicornia* sp., and *Halocnemum strobilaceum*).

Vegetation units in the Mediterranean were divided into two classes (Zohary, 1962) according to their habitats; (1) the Salicornietea fruticosae class of the coastal salt marshes and (2) the Suaedetea deserta of various salty inland and desert habitats. Each of those classes was divided into smaller subunits— Salicornietea fruticosae: Thero-Salicornion; Salicornion fruticosae; Junco Phragmition Mediterraneum. Suaedetea deserta: Junco-Phragmition; Tamaricion tetragynae; Atriplico-Suaedion; Salsolion villosae.

Plants of the Salicornietea are usually found on wet habitats. Here, fluctuations in salinity are caused by interchanging periods of inundation and high surface evaporation, but occur mostly on the top soil layers. Among the inhabiting plants, those associated with *Salicornia herbacea* are located on wetter but more saline spots than those associated with *Salicornia fruticosa.* All species can endure both high and wide fluctuations in salinity. Plants of the Suaedetea occur mostly on drier sites.

Plants of Halimionetum portulacoides are usually exposed to lower salinities than those of Salicornietum, but the former still require wet conditions. Similar plant communities with taxonomically related genera, although with different species, are also typical under a similar climate of South America and Australia. In the eastern United States, such sites are also inhabited by species of *Distichlis* or *Allenrolfea.*

Salt and water relations of the habitat comprise the dominant ecological conditions determining plant distribution in a saline. Such relationships vary from the periphery toward the center, and control the order of appearance and the zonation of the various plant communities. Plants which tolerate high salinity and inundation are usually found in the center of the saline (e.g., *Arthrocnemum glaucum; Salicornia herbacea*). On the contrary, plants sensitive to inundation (e.g., *Nitraria retusa*) occupy the periphery of the saline, although salt concentrations there may even be higher than in the center.

The order of zonation may vary between different types of salines, even when the same species are involved.

On saline or highly gypseous soils of inland desert salines around the Mediterranean, various species of *Suaeda* (e.g., *Suaeda asphaltica*) or *Frankenia* form the major cover. Drier habitats are characterized by *Zygophyllum* sp., *Atriplex* sp., *Reaumuria* sp., and other xerohalophytes (Walter, 1962; Zohary, 1962).

In most deserts, salts are evenly distributed over the entire area, forming dry but saline hammadas. Usually such habitats are uniform and are sparsely covered by populations of xerohalophytes. Nevertheless, in the general areas of regs and hammadas, certain depressions with good drainage (dolina types) can be found. Such pans are very low in salts and their plant cover is dense. A few plant communities (*Althaea ludwigii–Trigonella anguina* ass., *Lotonis dichotoma* ass., and *Ashphodelus pendulinus* ass.) characterize those pans. Because of prevailing low salinity, many arboreal species invade such depressions (Walter, 1962).

Succession in Salt-Affected Habitats

The ecological limitations for distribution of a plant community are chemical, physical, or biotic. Thus, the distribution of a halophytic community generally appears to be limited by salinity and depth of water table, as well as by competitive ability of members of the next community in the halosere (Reed, 1947).

Halophytic plant communities are considered highly dynamic because of their agressiveness and due to the fact that saline habitats are rapidly changing.

Succession in salt marsh vegetation is triggered by modifications induced from outside the habitat as well as by those induced by the plants themselves. Although usually both processes are intercorrelated, reductions in salinity and in level of standing water are regarded mainly as environmental inductions. When only one factor has changed, e.g., salinity, poor aeration still eliminates invasion of normal glycophytic species and enables establishment of a freshwater marsh flora. Changes induced by the plants are comparatively slow, and without external modifications in salinity and water level, the vegetation of salt marshes remains static (Polunin, 1960; Ungar, 1965). Different opinions were expressed by Adriani (1945), who stated that interrelationships between plants affected succession in the Dutch salt marshes even under a constant salinity. Thus, he concluded that succession in a salt marsh was determined not only by salinity changes but also by competition or other interrelationships between plants.

A sere, in which succession occurs because of the activity of the inhabiting plants as well as because of changes induced by plants in the habitat, is called autogenic. Salt marsh seres cannot be regarded as autogenic because of changes in the habitat as a result of a continuous inflow of nutrients and clay minerals, as well as changes induced by plants.

When topographical or hydrological conditions permit, succession lines of halophytes (halosere) end in communities of nonsaline habitats and eventually reach a true climatic climax. However, in most saline habitats, especially in those in which salts are continuously replenished, a long-persisting community, i.e., a sere climax, prevails.

Definition of climax communities in halophytic habitats is a matter of dispute. Oosting (1956) favored the use of the concepts of "polyclimax" or "prevailing climax." Climax in this sense meant a self-perpetuating plant community, which comprised the highest developmental stage of a succession. Such a stage could be reached only under constant environmental conditions. Most salt marshes exhibit dynamic changes in water level, salinity, soil, sedimentation, organic matter deposition, currents, etc. In such habitats, a strict definition of climax cannot be given. However, a community which shows minor changes in structure and composition may be regarded as a "sere climax." Nevertheless, since no evaluation of the time factor involved in the changes is incorporated, such a definition is incomplete.

Chapman (1960) used the term "edaphic climax" for vegetation of inland marshes where clear zonation exists. Such vegetation units change only when the salts are washed away.

Succession of angiospermous halophytic plant communities varies also in each habitat in accordance with other ecological conditions besides salinity. Thus, there is no general trend for development of various halophytic communities around the world, and local variations are encountered in each specific site. Several diversities in structure of plant communities and in their successional stages can be observed under various climates. Chapman (1958) distinguished seven major geographical seres with the following subdivisions:

1. Arctic sere
2. North European sere
   a. Scandinavian sere
   b. North Sea sere
   c. Baltic sere
   d. Channel sere
   e. Irish sere
3. Mediterranean sere
4. Western Atlantic sere
   a. Bay of Fundy sere

        b. New England sere
        c. Coastal plain sere
    5. Pacific American sere
    6. Sino-Japanese sere
    7. Australasian sere
        a. Australian sere
        b. New Zealand sere

The specific communities which are associated with some of these seres are given by Chapman (1958). With slight modifications, this information is incorporated in Table 5.1.

The simplest succession of halophytic communities is that found in the Arctic region. Here, succession proceeds from saline communities dominated by *Puccinellia* (*P. phryganodes* and *P. maritima*) through several communities of *Carex* sp., to a freshwater marsh of *Agrostis* and *Festuca*.

Further southward, succession of coastal salt marshes of eastern North America (New England) proceeds from the submerged community, Zosteretum, to a sterile bare mud and to Spartinetum alterniflorae, Spartino-Distichlidetum, and finally to Juncetum gerardii (Chapman, 1940a). At this stage, Juncetum forms a sere climax, i.e., a permanent community which remains as long as water table and salt level are constant.

From Juncetum, the sere may take various lines of development either into a freshwater marsh vegetation (Typhetum latifoliae or Phragmitetum) or into terrestrial plant communities.

When succession starts in the bare mud stage, Spartinetum may be preceded by Salicornietum herbaceae or by Suaedeto-Puccinellietum (Chapman, 1940b). A similar scheme was presented also by Steiner (1935).

In certain cases, Spartinetum may also form a sere climax. Succession from this point may follow two directions. The plant communities may change into glycophytic communities if salts are washed off, but may retreat to lower successional stages of the sere if a rise in water table or salt level occurs.

A formation of floating-leaved plants is missing in the seres of hydrohalophytes. Thus, unlike succession in freshwater, submerged halophytes (e.g., *Ruppia* sp., *Myriophyllum* sp., or *Najas* sp.) usually are shaded out by the invading swamp or marsh plants and succession proceeds directly from a submerged plant formation into an emerged one. In vegetation of warm water habitats (e.g., mangroves), there are many cases where trees invade deep water stands directly and succeed the submerged marine communities.

In habitats with a rising bottom, in the eastern United States, succession follows more or less, the following major sequence: *Zostera* sp., *Salicornia* sp., *Puccinellia* sp., *Juncus* sp. (Gibb, 1957; Gillham, 1957; Chapman, 1960). The *Juncus* stage is a long-persisting one, but eventually shows a transition to brackish water communities with *Scirpus* sp. as a dominating species. Further changes in succession depend on transition to nonsaline conditions.

An example of such a change in vegetation in Long Island, New York was brought by Conard and Galligar (1929). They showed that halophytes of salt marshes were gradually replaced by freshwater marsh plants (*Aspidium thelypteris, Scirpus americanus, Typha latifolia,* and *Lysimachia terrestris*) which was probably due to an uplift of the marsh bottom. Stands of *Juncus gerardii* or of *Spartina patens* were retreating seaward. Stands of freshwater plants tended to develop into a marsh of *Festuca* sp. which eventually would be transformed into climatic climax communities.

On the east coast of Connecticut, the widely spread pioneer community is Spartinetum. However, on the northern sections of this coast *Spartina alterniflora* prevails, whereas *Spartina patens* is found mainly in its southern sections. Where the geographical distribution of the two species overlaps, *S. alterniflora* inhabits the low marshes, and *S. patens* occupies high marsh habitats (Teal and Teal, 1969). *Spartina* salt marshes are frequently inhabited also by *Distichlis spicata* and *Juncus roemerianus* as associates. In later stages of succession, Spartinetum is followed by communities dominated by various species of *Distichlis* or *Juncus*.

Two species of *Juncus* are typical of the salt marshes of North America; *Juncus gerardii* in the north and *Juncus roemerianus* in the south. Usually, those plants also occupy habitats with freshwater flows.

Succession of plant communities similar to those of the eastern United States was reported by Miyewaki and Ohba (1965) from Japan. The halosere in that country can be described by the following dominants: *Zostera marina* and *Z. nana, Salicornia brachystachya, Puccinellia kurilensis, Glaux maritima, Juncus gracillimus* and *Phragmites communis*.

A reversed sequence of succession was reported by Adams (1963) for the east coast of North America, where the coast is sinking. Along the coast of North Carolina, stumps of *Juniperus virginiana* trees can be found inside *Spartina* marshes (Adams, 1963). Since neither deposition of a new and deep soil nor erosion was observed in this habitat, Adams assumed that a rise in sea level accounted for salinization and for gradual replacement of coastal forests by salt marsh vegetation.

Effects on vegetation by changes in sea level were noted in other places as well. In 1950, several stumps of *Tamarix* trees were exposed in the Dead Sea region (Israel) following lowering of sea level (Fig. 5.5). The trees died during the end of the 19th century (approximately 1890), after the level of the Dead Sea rose. Since then, the sea had retreated again and a renewal of vegetation of the newly exposed habitats was observed. However, because of heavy salt depositions left by the retreating sea, plants follow the retreating coast line but only at a pace of about 50 m. Although the new inhabitants of the exposed coast are highly salt-tolerant shrubs (e.g., *Suaeda fruticosa*) it takes a few years before enough salt is leached to ensure their survival. Trees of various species of

Fig. 5.5. Stumps of trees exposed by the retreating Dead Sea.

Tamarix also become established again in the exposed habitats, but only on well-leached and wet sites.

Succession in coastal salt marshes of the Mediterranean region is very slow. Plant associations change from pioneer communities dominated by *Salicornia* sp. (*Salicornia herbacea, Salicornia fruticosa,* or *Salicornia radicans*) and proceed through communities dominated by *Halimione portulacoides, Inula crithmoides,* and *Limonium latifolium* to Juncetum maritimi. On elevated coasts and under relatively high rainfall, the halosere ends with Phragmitetea communities. In certain habitats, mainly warmer ones, *Halocnemum strobilaceum* rather than *Salicornia* sp. forms the pioneer community. In such a case, *Zygophyllum album* plants constitute part of the community and inhabit the small sandy patches scattered on top of the main salty plateau.

Coastal salt marshes of tropical regions are usually characterized by mangrove communities rather than by grass associations. Such swamp communities (e.g., swamps with *Rhizophora* sp. or *Avicennia* sp.) constitute typical woody plant communities of saline or brackish water. Those communities are always located inside the tidal zone. In fact, *Rhizophora* trees comprise the only true plant formation of deep salt water swamps. On drier habitats, i.e., habitats without surface water, shrub communities are dominated by *Baccharis, Iva,* or *Conocarpus.* Whenever those plants are occasionally destroyed by a disaster, secondary communities may take their place. In the Indian peninsula and in east Africa, such secondary communities are comprised mostly of the articulated

Chenopodiaceae. In Brazil, *Spartina brasiliensis* intermingled with *Scirpus domingensis* enters such sites (Dansereau, 1947).

Human activities also cause divergence from the normal succession in salt marsh vegetation. *Spartina patens* used to be economically important and was cut for hay. Such "salt hay" was a major source for feeding cattle for hundreds of years. Under severe and repeated mowing in Nova Scotia, *Spartina* was replaced by *Puccinellia*. In other places Spartinetum remained unaffected. Following human disturbance in Sweden, *Phragmites* and *Scirpus* were replaced by various species of *Salicornia, Spergularia,* and *Suaeda.*

Soils with sulfate salinity (Na_2SO_4; $MgSO_4$; $CaSO_4$) are very rare in coastal marshes, but common in inland salines. According to Walter (1968) gypsum deserts are mostly found in arid regions. Vegetation cover of such sites is sparse and consists of many annuals. Plants have a xeromorphic structure. Species composition and community structure of such soils differed from that of NaCl-infected soil (Kearney *et al.,* 1914; Keller, 1925). In Central Asia, sulfate-affected habitats are dominated by various species of *Salicornia, Suaeda, Halocnemum* etc. In northern America, *Atriplex polycarpa, Allenrolfea occidentalis,* and *Pluchea* sp., etc., prevail on soils with a high sulfate content. Among the dominant species of gypseous deserts in central Asia is *Ceratocarpus arenarius* of the Chenopodiaceae. It is an annual species which remains green throughout most of the summer. *Ceratocarpus arenarius* is one of the important forage grasses for sheep and natural production may amount to 500 kg/ha dry matter (Walter, 1968). In Somaliland, large areas of pure gypsum carry a specific type of vegetation. The plant cover in such regions consists of *Acacia sucotrana* and various species of *Commiphora, Euphorbia,* etc. (Gilliland, 1952).

Zonation of plants is not always caused by changes in salinity, but may also be caused by plants themselves. Secretion of salts by various halophytes causes a specific halophytic flora to grow nearby. Litwak (1957) showed that *Tamarix* trees influenced the soil salinity underneath (Fig. 5.6). The total content of soluble salts under the canopy of such trees is much greater than the salt content of a neighboring bare sand dune. Chlorides make up an important part of the added soluble salts. This is to be expected as litter and secreted drops of brine falling from trees constitute the sources of salts. Similar effects of *Sarcobatus vermiculatus* plants on other plants in Escalante Desert, Utah, were reported by Fireman and Hayward (1952).

Increase in total soluble salts under the canopies of *Tamarix* trees is related to their age. Below trees approximately 150 years old, the salt content of the upper soil layer reached 2.5% and was presumably still increasing. Salinization of the soil under old trees had a marked influence on ground vegetation, with no plants of any kind under the canopy. At the periphery of the tree, still within partial influence of the falling brine drops, exclusive halophytic and ruderal plant communities appeared. Such communities were not found elsewhere in the

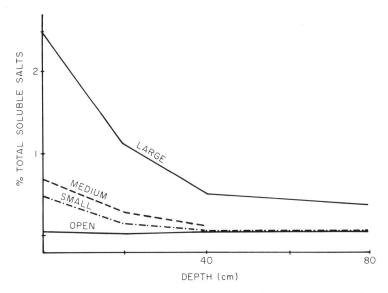

Fig. 5.6. The correlation between the size of *Tamarix aphylla* trees and the salt content beneath the canopy. (After Litwak, 1957.)

vicinity and were comprised of such species as *Mesembryanthemum nodiflorum, Pteranthus dichotomus, Bassia muricata,* and *Chenopodium opulifolium*

Interrelationships between Plants and Spatial Distribution within the Community

Very little is known of the mutual relationships and spatial distribution of individual plants within the halophytic plant community. Such information is of interest, because of the interrelationships between the neighboring halophytic species as well as between halophytic and glycophytic species. Competition exists when two organisms attempt to satisfy their needs of a certain element when the total resources available are less than the sum of their requirements (Donald, 1963). It is generally accepted as axiomatic that glycophytes do not penetrate into saline habitats because of physical environmental factors, such as salt content and higher water stresses, whereas halophytes are unable to survive in nonsaline sites because of competition with glycophytes (cf., Adriani, 1945). Nevertheless, the term "competition" usually refers to any type of allelopathy or mutual inhibition and in a strict sense, its existence is difficult to prove. Most of the work in this subject was done in crop plants and references can be found in the articles of Shinozaki and Kira (1956), Yoda *et al.* (1957), Goodall (1960),

De Wit (1960), and De Wit *et al.* (1966). Information on competition among wild plants is limited, and negligible for wild halophytes.

Competition is mild when plants are sparsely spaced or when their ecological requirements do not overlap. For example, in the salines of Sedom, Israel, plants with a shallow root system (*Suaeda fruticosa, Salicornia herbacea,* etc.) are found side by side with plants with deep root systems (*Nitraria retusa, Atriplex halimus, Prosopis farcta,* etc.). In such cases, the root systems are undermixed (Harper and Litav, 1967), and competition between plants is uncertain.

Analyses of competitive interrelationships were recently made in the author's laboratory by adopting the approaches and techniques developed for the study of enzyme–substrate relationships and for ion uptake processes. Using the transformed equations of Lineweaver and Burk (1934), reciprocals of the rates of a growth parameter ($1/v$) and of plant density ($1/s$) were plotted for monocultures, as well as for mixed cultures of plants.

Such an approach reveals whether an inhibitory effect of one species on the growth of another is of competitive or noncompetitive essence. In cases where the two lines of the $1/v$ on $1/s$ plot have different slopes, but strike the same point on the ordinate, the inhibitory effects of a constant number of plants of species A per unit area decrease with density of plants of species B. At a theoretical infinite density of plants of species B, the effects of plants from species A are nil. Such an inhibition is competitive. However, if the ordinate is intersected in two different places, growth of species B is inhibited by species A, but inhibition is not competitive.

Data presented in Figs. 5.7–5.10 show effects of saline and nonsaline conditions on growth of a glycophyte (*Triticum vulgare* c.v. Florence) and a halophyte (*Hordeum marinum* L). Under nonsaline conditions, growth of the halophytic barley was competitively inhibited by wheat and competition may eventually account for exclusion of halophytes from sites suitable for growth of glycophytes. The inhibitory effects of barley on growth of wheat were negligible. However, under saline conditions, growth of wheat was markedly affected but growth of barley in monocultures was only slightly checked. Under such conditions, the competitive inhibition of growth of barley by wheat was eliminated (Waisel and Schwartzboim, unpublished data).

Very little is known of competitive relationships among plants of halophytic species or among individual plants within one single species (cf., Adriani, 1945). Nevertheless, the spacious dispersion of plants in most salt marshes may imply that such habitats are underpopulated and sufficient nutrients and other growth requirements are available for all plants present. Further evidence for such an assumption can be obtained from analyses of patterns of plant distribution.

Patterns of distribution of individual plants in a community and the exact location of plants in relation to their neighboring plants usually have meaningful ecological implications (Clapham, 1936; Archibald, 1948; Greig-Smith, 1952;

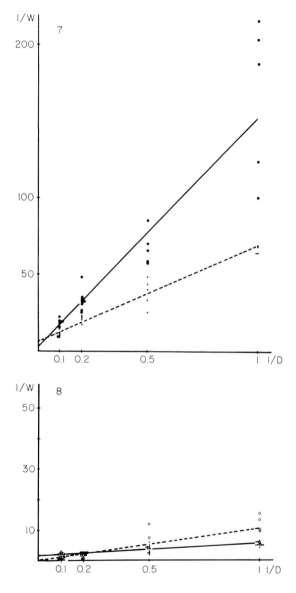

Fig. 5.7–5.10. Competition analysis between *Triticum vulgare* c.v. Florence and *Hordeum marinum* on saline and nonsaline soils. Double-reciprocal plot of plant weight over plant density. Pure culture - - - -, mixed culture ——.

Fig. 5.7. Growth of *Hordeum marinum* under nonsaline conditions; five competing wheat plants.

Fig. 5.8. Growth of *Hordeum marinum* under saline conditions; five competing wheat plants.

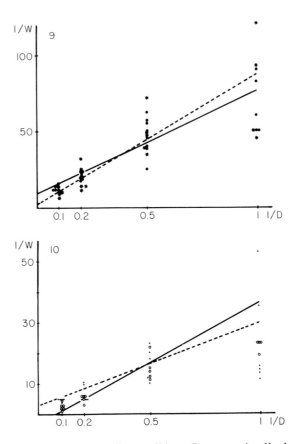

Fig. 5.9. Growth of wheat under nonsaline conditions; five competing *Hordeum* plants.
Fig. 5.10. Growth of wheat under saline conditions; five competing *Hordeum* plants.

Jones, 1955; Pielou, 1960; Kershaw, 1966; Anderson, 1967; Lewis and Taylor, 1967; MacArthur and Conell, 1967; Woodell *et al.,* 1969; Barbour, 1969; Waisel, 1971). Three major patterns are distinguished.

Random Distribution

When resources essential for plant existence are amply available, when distribution and propagation are not limited to short distances, and when no interactions exist between individual plants, the pattern of plant distribution is random. The probability of appearance of randomly distributed plants in various density groups follows the Poisson distribution. Therefore, the variance (S^2) equals the mean (M), and the ratio S^2/M is 1.

Uniform or Regular Distribution

Uniform patterns of distribution are believed to be the result of negative interactions (competition, allelopathy, etc.) between the different individual members of a plant community. Such a pattern of distribution is thus expected in habitats which are poor but still capable of supporting a sparse cover of plants. Usually such habitats are occupied by a single species only. The variance in a pattern analysis of such habitats is nil, and therefore, the ratio S^2/M also tends to approach zero in accordance with the degree of uniformity of plant distribution.

Contagious or Patchy Distribution

Plants appear in groups in a few cases such as: (1) when the physical conditions of the habitat are so poor that their existence is restricted to niches where conditions are somewhat more favorable (excluding the possibility of random distribution of niches and the restriction of one plant per niche); (2) when their propagation is either vegetative and/or restricted to short distances; (3) when the parent plants either supply a suitable germination bed, or cover for the young developing seedlings; (4) when individual plants show mutual attraction.

A contagious pattern of distribution is found in grass communities such as those of *Spartina townsendii* in which the developing clones form a concentric pattern. Such patterns are preserved even in vigorous clones which form a few successive rings (Caldwell, 1957).

In cases where habitat conditions determine the pattern of distribution, it is possible to correlate the existence of plants with one or more of the physical environmental factors. In other cases, where environmental factors are not limiting, although plants may appear in clumps, the distribution patterns of the clumps themselves may well be random. Analyses of habitats with such patterns of plant distribution, using the right quadrat size, yield high variances and the ratio S^2/M is greater than 1, depending on the degree of aggregation.

Patterns of distribution of some perennial halophytic species (*Halocnemum strobilaceum, Arthrocnemum glaucum,* and *Zygophyllum album*) in a coastal saline of northern Sinai are presented in Fig. 5.11. Distribution of all the three species present seems to be contagious (Table 5.2). However, when young and small-sized plants were eliminated from the analysis and only the dominant specimens counted, the S^2/M ratio obtained decreased considerably, hinting at an inclination toward random distribution. The contagious distribution seems thus to result either from short-distance seed dispersion or even more probably from a preference for specific germination beds. The large average distances between plants and the inclination toward random distribution of dominant specimens suggest that no severe intra- or interspecific interactions between plants seem to exist, and plants probably obtain their full supply of

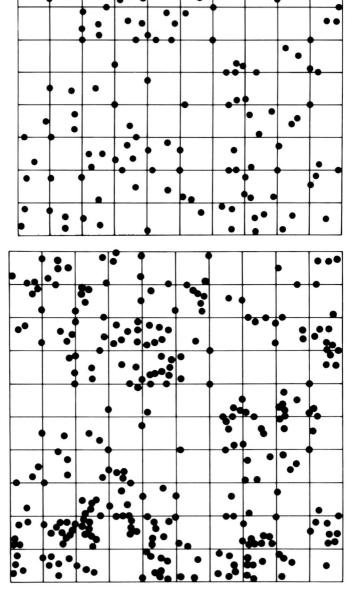

Fig. 5.11. Distribution of *Halocnemum strobilaceum* in a saline in north Sinai, Israel. Left, all the plants; Right, dominant plants only.

Table 5.2

VARIANCE/MEAN RATIO (S^2/M) OF VARIOUS PLANTS IN A COASTAL SALINE
IN NORTH SINAI

Species	Plants examined	Quadrat size		
		4 m²	8 m²	16 m²
Site A				
Halocnemum strobilaceum	All plants	7.34/3.77 = 1.94	19.36/7.52 = 2.57	57.15/15.08 = 3.78
	Adult plants only	1.92/2.18 = 0.88	3.92/4.30 = 0.91	11.76/8.64 = 1.39
Site B				
Halocnemum strobilaceum	All plants	4.92/2.10 = 2.34	12.46/4.20 = 2.96	23.13/8.40 = 2.75
	Adult	1.61/1.13 = 1.42	3.27/2.26 = 1.44	7.73/4.52 = 1.71
Arthrocnemum glaucum	All plants	11.76/1.81 = 6.49	24.90/3.62 = 6.87	56.25/7.24 = 7.76
	Adult	1.66/0.82 = 2.02	3.80/1.64 = 2.31	7.84/3.52 = 2.22
Zygophyllum album	All plants	24.90/5.10 = 4.82	72.76/10.52 = 6.91	222.01/21.04 = 10.55
	Adult	0.90/0.47 = 1.91	2.34/0.94 = 2.48	6.00/1.88 = 3.19
All species present	All plants	42.51/8.89 = 4.78	124.54/17.78 = 7.00	379.47/35.56 = 10.67
	Adult	3.20/2.42 = 1.32	7.84/4.82 = 1.62	16.00/9.64 = 1.65

environmental requirements. Thus, though more specific information on distribution, density, root dispersion, etc., is needed, it seems that once adapted to salinity, halophytes have satisfactory ecological conditions, and are less disturbed than glycophytes by competition.

Life Form, Distribution, and Migration of Halophytes

Despite the fact that different life forms are common in various plant formations and under various climates, only very few habitats have a plant cover which consists of a single life form only. Although in each habitat, one life form is favored and appears in large proportions, nevertheless, other life forms are present as well. The spectrum of life forms, dominating under a certain climate, represents some mode of adaptation of the plants to that specific climate (Raunkiaer, 1934). Some life form spectra represent a response of the organisms to the prevailing climatic or ecological conditions. Others represent residual effects of some historical climatic or biotic conditions on the population of plants.

Diversions from a normal spectrum reflect the specific response of those plants to local habitat conditions.

Differences between life form spectra of the halophytic and of the entire flora of various localities reveal that certain life forms appear in halophytes in higher proportions and apparently have advantage over others. In salt marshes of the temperate region, the hemicryptophytic form dominates the flora and the vegetation is composed mostly of grasses. In the floras of warmer climates and in inland salines, the dominating life forms are various woody chamaephytes and hemicryptophytes (Table 5.3). Moreover, it is well known that only a small fraction of a flora builds up the vegetation, especially under extreme climatic and edaphic conditions (Orshan, 1953). Thus, if only the dominant species are considered for life form analyses, the advantage of the chamaephytic and hemicryptophytic life forms for the halophytic flora of saline habitats becomes even more apparent.

Table 5.3

LIFE FORM DISTRIBUTION OF THE HALOPHYTIC AND OF THE ENTIRE FLORA OF ISRAEL IN COMPARISON TO THE LIFE FORM DISTRIBUTION OF THE NORMAL WORLD FLORA AS COMPILED BY RAUNKIAER

Plant group	Percentage of the group[a]				
	Th	C	H	Ch	Ph
Halophytes	19.4	0	34.5	29.1	17.0
Entire flora of Israel[b]	48.3	11	24.3	9.8	6.6
Normal world flora[c]	13	6	26	9	46

[a] Th, therophytes; C, cryptophytes; H, hemicryptophytes; Ch, chamaephytes; Ph, phanerophytes.
[b] After Eig, 1931–1932.
[c] After Raunkiaer, 1934.

The high proportions of shrubs and perennial grasses are typical of halophytic habitats in most halophytic floras of the world. An arboreal flora is dominant only in wet and salty habitats of warm regions. The relatively high percentage of trees found in certain halophytic floras, such as that of Israel, is due only to the presence of many species of *Tamarix* in wet inland salines of warm regions.

Since halophytes are a special ecological group of plants, their distribution depends more on edaphic than on climatic factors. Thus, it seems only natural that coastal halophytes, typical of climatic regions such as the Mediterranean and the subtropical regions, inhabit also the Atlantic coasts of Europe and North America. However, as may be expected, this phenomenon is not circumglobal. The question which still remains unanswered is why those plants, which are known to comprise typical azonal vegetation units, do not also occupy other saline habitats in other phytogeographical regions of the world. The answer is partly ecological as well as partly historical.

According to Iljin (1946) and Chapman (1960), the halophytic species have developed primarily as seacoast plants. Only at a late stage of their evolution and after changes in coast lines occurred did some species migrate to inland salines. Thus, the present distribution of halophytes which show similarities and close species relationships among some phytogeographical regions, stems back to the era when those taxa originated from one common source on a coast of an ancient sea. Similarities in species composition exhibited by the halophytic species of Europe, North Africa, Middle East, and Central Asia result probably from the fact that those populations have evolved from a few ancestors which have developed along the ancient coast of the Tethys (Iljin, 1946; Gruenberg-Fertig, 1966).

Some species now occupying coastal salt marshes (e.g., *Statice* sp.) have probably migrated to the coast from saline steppe regions (Van der Pijl, 1969). Thus, such species which originated in coastal habitats in the past have now returned to the same habitats in a new evolutionary form.

The small number of taxa from which halophytes have developed and which presently comprise the halophytic floras in different regions is also indicated by data presented in Table 5.4, which compare phytogeographical distribution of plants in the general and halophytic floras of Israel.

Table 5.4
PHYTOGEOGRAPHICAL DISTRIBUTION OF THE HALOPHYTIC FLORA AND OF THE ENTIRE FLORA OF ISRAEL

Flora	Species distribution (%)		
	Uniregional	Biregional	Multiregional
Halophytes	18	25	57
Entire flora	62	22	16

Evidently, although the ultimate numbers of species involved are small, the percentage of species with a multiregional pattern of distribution is much higher in the halophytic than in the glycophytic flora. Only a small fraction of the halophytic flora is endemic, indicating that local speciation is of minor importance. The low degree of speciation in halophytic habitats is further emphasized by comparing the numbers of various taxa in the glycophytic and halophytic floras of Israel (Table 5.5).

The number of families or genera represented in the halophytic flora is low. In addition, the mean number of genera or species within each family is much

Table 5.5

THE NUMBER OF VARIOUS TAXA IN THE GLYCOPHYTIC AND HALOPHYTIC
FLORAS OF ISRAEL

	Families	Genera	Species	Mean no. of genera per family	Mean no. of species per family
Glycophytic flora	112	852	2088	7.6	18.1
Halophytic flora	17	42	76	2.4	4.4

lower in the halophytic than in the glycophytic flora. Some families to which such plants belong (e.g., Chenopodiaceae, Tamaricaceae, Plumbaginaceae, Graminae, and Juncaceae) now dominate halophytic floras throughout the world. Most submerged hydrohalophytes or mangroves also constitute uniform groups and belong to a limited number of families. In other cases, some genera from additional families (e.g., Verbenaceae and Compositae) are also well represented in the halophytic floras. However, in this case their number of subtaxa is also low.

At present, formation of new species may still contribute to the extension of the distribution area of some halophytic genera. For example, in the nineteenth century, a few plants of *Spartina alterniflora* migrated to England from North America. In their new habitats such plants hybridized with the native species, *Spartina maritima*, to yield the hybrid *Spartina townsendii*. The new amphidiploid ($n = 126$) invaded the sparsely populated tidal marshes and *S. townsendii* is presently a successful species along most of the muddy coasts of northern Europe (Teal and Teal, 1969; Van der Pijl, 1969).

————————————————6

Water Relations

General Remarks: Water in Saline Soils

If we regard the chemical potential of pure water at a certain reference state as zero, then the water potential (Ψ) in the soil, atmosphere, or plant has a negative value (Kozlowski, 1964; Hillel, 1965, 1970; Slatyer, 1967). Water potential, an estimate of water stress or water deficit, is expressed in bars (10^6 dynes/cm^2). Differences in water potential account for water movement in a system. Thus, in order to induce an inwardly oriented gradient in water potential and to extract water from their environment, plants must maintain a water potential of their tissues at an even lower level than that of the medium.

Plants obtain most of their water from the soil. However, in soils and especially in saline soils, it is usually difficult to distinguish between the effects exerted on their water relations by the osmotic potential of the salt solutions (Ψ solution) from those exerted by the soil matrix potential (Ψ matrix). According to various investigators (Wadleigh and Ayers, 1945; Slatyer, 1967) both components (total soil moisture stress—Ψ soil) in saline soils have a cumulative effect on plants, and during short-period exposures are inseparable. However, when plants are exposed to a saline environment for prolonged periods, the osmotic component of the total soil moisture stress (TSMS) loses its dominant effect.

Halophytes are subjected throughout their life cycle, or at least during some of its phases, to an environment of low water potentials.

A shortage in water, expected in plants exposed to saline conditions, brought Schimper (1903) to formulate his "physiological drought" theory, which postulated that, under saline conditions, halophytes suffered from dehydration. This theory was favored for many years and low hydration of protoplasm was frequently cited as being responsible for changes in the metabolic activity of plants subjected to low water potentials (Iljin, 1953; Kramer, 1959; Levitt, 1956, 1962; Kozlowski, 1964).

A high ion content apparently affects viscosity of protoplasm, and consequently its streaming velocity and all other functions. High viscosity of protoplasm is known to evolve in certain halophytes, partly because of their high ion content and partly due to their high content of mucilages. The accumulation of osmotically active substances in the protoplasm of halophytes is not evenly distributed throughout the cell. Ions may be located differently in intracellular organelles (Stocker, 1925). Since electrolytes influence hydration of cell proteins in plants, entirely different physiological effects are expected in halophytes which accumulate chlorides than in those which accumulate sulfates. Sulfate is known to decrease protoplasmatic hydration whereas chloride increases it (Berger-Landfeldt, 1933; Walter, 1961).

Availability of Water to Plants

The supply of soil water to plants depends on the moisture retaining capacity of the soil or plant, on water potential differences, and on rates of flow from the soil volume to the region of the roots, from which water is continuously depleted. Moisture gradients in the vicinity of roots are not easily measured and, therefore, separation between soil and plant factors which govern water availability is difficult (Richards and Wadleigh, 1952).

Common terms such as "field capacity" and "wilting point" are used in saline environments in their general connotation only, and their validity under such conditions is doubtful. The most serious objections involve the determination of the wilting point. Wilting point is defined as the water content of the soil at that point in which the water potential of the soil equals that of plant roots. A few methods are employed for wilting point determination. Most commonly, it is determined by the use of a certain variety of sunflowers which are grown to the third leaf pair stage and then left to wilt. When the lower leaves lose their turgor the water content of the soil is said to be at the wilting point (Briggs and Shantz, 1912). However, such a method is unsuitable for saline soils because (1) sunflowers cannot tolerate even moderate soil salinity, and (2) most halophytes do not show apparent wilting. Moreover, the commonly accepted range of soil-water potentials at the wilting point (−15 bars) is far above the water potential which still enables halophytes to absorb water from saline soils. Thus, other methods had to be sought. One of the latest methods (Waisel and Pollak, 1969a) is based on the pioneer work of Breazeale (1930) and Breazeale and Crider (1934). Those investigators emphasized that plants can absorb water from the soil by some of their roots, and transport it to other roots if the portions of a split root system are subjected to different water potentials. Such capacity to transport water is not infinite and water deficits of the main supporting root system were shown to be lower than those of secondary auxiliary roots (Fig. 6.1). According to Breazeale and McGeorge (1949), when the supporting roots of a tomato plant were amply irrigated, i.e., the soil moisture deficit was negligible, the deficits that developed in the soil around a secondary auxiliary

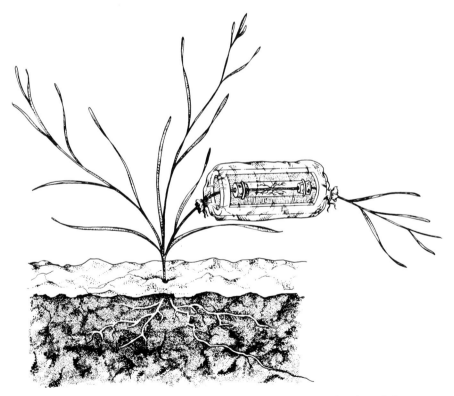

Fig. 6.1. A sleeved branch with an auxiliary root system for estimation of the water potential in the supporting root zone by the double-root system technique. (Modified after Waisel and Pollak, 1969b.)

root system reached the -15 bar range, i.e., the wilting point. Similar results were obtained for the halophytes, *Aeluropus litoralis* and *Suaeda monoica,* in which the auxiliary roots developed inside soil-filled glass sleeves (Waisel and Pollak, 1969a,b).

As may be seen in Table 6.1, even though visible wilting was not observed in *Aeluropus litoralis* or in *Suaeda monoica* plants, the water potential of the sleeved soil equilibrated under nonsaline conditions with that of the plants, at practically the same point as with sunflowers. The equilibration point of tomato plants was at a slightly higher water potential.

Table 6.1

EQUILIBRATION POINTS (PERCENTAGE WATER IN STANDARD SOIL) OF VARIOUS PLANTS UNDER NONSALINE LABORATORY CONDITIONS[a]

Plants	Water (%) at equilibration	
	Mean	Range
Sunflower	17.5	17.4–17.6
Tomato	18.1	12.9–23.1
Aeluropus litoralis	17.0	16.4–17.5
Suaeda monoica	17.0	14.4–20.3

[a] After Waisel and Pollak, 1969a.

When the osmotic potential around the supporting root system was increased by addition of NaCl, the water deficits that developed around the sleeved auxiliary roots increased proportionally (Fig. 6.2).

The double-root system technique is thus a method by which measurements taken from roots that have been developed on above-ground plant organs give indications about the water potentials to which the supporting roots are exposed. Such water deficits are determined by the water potential of roots at the point of equilibration and, therefore, are equal to the wilting point.

Water potentials vary greatly among different plant organs (Catsky, 1962; Campbell *et al.,* 1966; Hoffman and Splinter, 1968). Usually a gradual decrease in potential is observed from the lower roots toward the upper parts of the shoot. However, this may differ in cases where hydrostatic pressures play a role (cf. Scholander *et al.,* 1965). Under steady-state conditions, the water potentials in plant organs will stabilize at different but intercorrelated specific levels.

Correlation between the water deficits around the two splits of the root system may result either from a direct continuous decrease in the water potential from the outer solution across the plant to the sleeved soil, or from lower water potentials of the auxiliary root system following ion uptake and osmotic adaptation. The scattered data available at the moment favor the osmotic adaptation hypothesis.

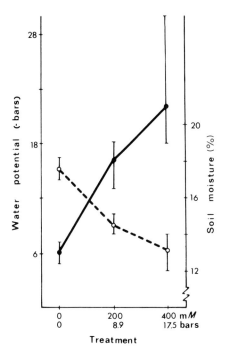

Fig. 6.2. The effect of exposing the supporting roots of *Aeluropus litoralis* to various concentrations of NaCl on the water content (%) (open circles) and the water potential (filled circles) of the sleeved soil. Bars represent the range of the data obtained for various experiments. (After Waisel and Pollak, 1969a.)

When a plant is placed with its supporting root system in pure water, the water content of the sleeved soil equilibrates with the auxiliary roots at a potential that is far below that of the medium. Roots of plants which were subjected to different salt solutions equilibrated with the sleeved soil at a water potential that was respectively lower than that of the wilting point of common plants (c f. Michel and El Sharkawi, 1970). Furthermore, the moisture content in the sleeved soil did not change appreciably even when the plants were placed under conditions favoring high transpiration, i.e., conditions that increased water stress inside the water conducting system. Such a correlation can, therefore, be used for determination of the wilting point for halophytes or other species that undergo osmotic adaptations under saline conditions.

The precise determination of the soil water potential to which plant roots are subjected in native habitats is of great ecological and physiological importance. Such determinations enable comparisons to be made among water potentials of roots located in different layers of a soil profile, among different zones in a single habitat, among water conditions during various seasons of the year, and among species of plants growing together. Nevertheless, because of the extremely high variability of most natural habitats, such determinations by soil sampling are almost impossible.

What makes determinations of soil water potentials most difficult is the fact that variability of a habitat in a vertical direction usually is as high as, or even higher than that on the surface. It is easy to show, in various habitats, that a single vertical root may penetrate through several soil layers which differ in texture, structure, mineral composition and content, nutrient status, and water availability. Analyses of the soil profile in such habitats may give general information on water conditions only. Nevetheless, such analysis does not provide information on actual water deficits in the exact regions which support the plant, i.e., regions where roots are active. At present such information can be obtained only by indirect methods. Consequently, most ecological analyses of natural habitats do not report on the precise water stresses to which plant roots were exposed.

Waisel and Pollak (1969b) used the double-root system technique to estimate water potentials (Ψ) in the active root zone of five halophytic species, *Aeluropus litoralis* (Willd.) Parl., *Arthrocnemum glaucum* Del., *Nitraria retusa* (Forsk.) Aschers., *Suaeda monoica* Forsk., and *Tamarix jordanis* Boiss. growing in a saline of the sebha type in Sedom, Israel.

Glass sleeves were installed on branches of naturally growing established plants. The tubes were filled with a nonsaline, heavy clay soil having a known pF curve, watered, and left untouched for periods of 6–12 weeks. During this period, auxiliary root systems developed inside the sleeves, and water potentials of the roots and soil had equilibrated. As may be seen in Table 6.2, the mean water potential at the equilibration point obtained ranged from -0.3 to about -50 bars during various months, depending on species and their ecological niches. Annual averages of water potentials for the various species also differed and ranged between -10 for *Tamarix* and -32 bars for *Suaeda*.

Some species (*Tamarix jordanis, Arthrocnemum glaucum* and *Aeluropus litoralis*) exhibited marked variations in water potentials between summer and winter (e.g., -48 to -3 bars). During the rainy season, the total soil moisture stresses around the supporting roots of these plants were low. However, since high water stresses existed in those species during summer, the plants should be classified as poikilohydric species. Such behavior may be partly genetically controlled, but in part may also result from limitation of root distribution in specific and different soil layers. During periods of high water deficits in the summer, *Aeluropus* entered a period of inactivity (poikilohydric drought-avoiding species) whereas growth and flowering of *Arthrocnemum* and *Tamarix* continued. It is interesting to note that some growth still occurred in those two species even at equilibration water potentials of -34 (*Tamarix*) and -45 bars (*Arthrocnemum*). These species may thus be classified as poikilohydric drought-enduring species.

Table 6.2

MONTHLY CHANGES IN THE EQUILIBRATION POINT OF A FEW HALOPHYTIC SPECIES GROWING IN THEIR NATURAL HABITAT IN THE SALINE OF SEDOM[a,b]

Month	Aeluropus litoralis			Arthrocnemum glaucum			Suaeda monoica			Tamarix jordanis			Nitraria retusa		
	%W	Ψ	Rep	%W	Ψ	Rep	%W	Ψ	Rep	%W	Ψ	Rep	%W	Ψ	Rep
January	17.8±3.2	8.0	34	20.1±3.4	3.0	4	11.7	32.0	4	26.4±8.0	0.3	7	15.8±2.5	14.0	4
February															
March	9.8±0.5	55.0	14	13.2±1.2	24.0	7	11.2	35.0	1	19.1±2.6	4.5	4	13.0	24.5	1
April	10.9±1.7	38.0	8							12.3	28.0	1	11.5	33.5	1
May															
June	9.8±2.0	55.0	26	10.2±0.8	48.0	30	10.2±1.1	48.0	14	11.3±3.7	34.0	12	12.0±1.5	30.0	22
July															
August							11.7±1.0	32.0	4	13.3	23.5	1			
September															
October				10.3	45.0	1	12.9±1.6	25.0	9				12.3±2.9	28.0	9
November	17.4±2.0	9.5	5												
December	13.9±2.1	21.0	5	12.6±1.1	26.0	3	13.7±1.6	21.5	6				13.1±1.2	24.0	29
Annual mean	13.4	23.0		11.7	31.5		11.7	32.0		16.9	10.5		12.8	25.0	

[a] After Waisel and Pollak, 1969a.

[b] %W, Water percentage in the sleeved soil; Ψ, Water potential (−bars) of the soil inside the sleeve at equilibration point; Rep, no. of replications per experiment.

Growth under such low water potentials apparently is possible also in two other species, *Suaeda monoica* and *Nitraria retusa*. However, those species show only minor annual fluctuations in equilibration potentials (stenohydric drought-enduring species).

The data presented also give additional information on the water relations of different species within a single plant community. For example, *Tamarix jordanis* and *Suaeda monoica* plants are found side by side and are members of one plant association (Zohary, 1962). Nevertheless, it seems that roots of those plants are exploiting different soil horizons. Apparent annual fluctuations in the equilibration of water potentials could be observed in *Tamarix*. Such behavior may imply that roots of these plants are located in coarse textured, highly permeable soil horizons. On the other hand, more or less constant water potentials were observed during the year in *Suaeda*. This might occur if roots of such plants are located either very deep or in heavy clay soil layers which are affected only slightly by seasonal climatic fluctuations.

Suaeda monoica is usually associated with the wet sites, whereas *Nitraria retusa* is located on the apparently drier sites of the Middle East salines. It is interesting to note that despite such segregation of plant distribution among niches of different water conditions, water deficits at the equilibration points of *Suaeda monoica* plants were higher than those in *Nitraria retusa*.

Salinity and Water Flow

Flow of water from the soil into plant roots depends on differences in water potentials, on water conductivity of roots and soils, and on rates of root growth toward wet layers (Kramer and Coile, 1940; Gingrich and Russell, 1957). Water flow toward plant roots is negatively affected by saline conditions because of reduction in water permeability of the soil. Salinity may also adversely affect plant–water relations by reducing water conductivity of roots (O'Leary, 1969). Thus, as a result of reduction in permeability of cell membranes, water movement into such plants is hampered despite the fact that there is an osmotic gradient between the plant and the soil.

Growth rate of bean and barley plants, which had their split root systems immersed in differently salinized nutrient media, was intermediate between that of plants which grew solely in either of the solutions. Under such conditions, resistance to water flow through those roots increased (Shalhevet and Bernstein, 1968; Kirkham *et al.*, 1969). However, if rates of root growth in halophytes remain under saline conditions high enough, they may substitute for reduction in soil water conductivity, and thus keep the plant water supply balanced.

Osmotic Adaptation

Plants growing under saline conditions must maintain a high concentration of osmotically active substances in order to compete successfully with the water-retaining capacity of the surrounding medium. Most plants exposed to such conditions suffer initially from a certain period of "physiological drought" (Schimper, 1903). However, it is now well established that when the osmotic potential (OP) of the solution around plant roots increases, the OP of plant tissues increases proportionally (Tables 6.3a and b). Such osmotic adjustment occurs whether the osmotically active substance in the growth medium is a permeating or nonpermeating one (Eaton, 1927, 1942; Black, 1956, 1960; Bernstein. 1961, 1963; Slatyer, 1961; Lagerwerff and Eagle, 1961; Rufelt, 1963; Jarvis and Jarvis, 1963; Meiri and Poljakoff-Mayber, 1969). In some species, only partial osmotic adjustment occurs (Rufelt, 1963); whereas in others overadjustment was observed (Boyer, 1965). It is probably true that the great majority of the halophytic species belong to the adjustable group, and that their osmotic adjustment occurs rapidly. Recovery from osmotic stress occurs faster in salt-accumulating halophytes than in salt-excluding ones (Greenway, 1968).

Table 6.3a

GROWTH AND OSMOTIC POTENTIALS (OP) OF COTTON PLANTS AT FOUR DIFFERENT LEVELS OF SALINITY (NaCl) IN THE NUTRIENT SOLUTION[a]

Added NaCl (atm)	Top fresh weight (gm)	Root fresh weight (gm)	OP (atm) of expressed saps and OP_a [b] (atm)					
			Leaves		Stems		Unrinsed roots	
			OP	OP_a	OP	OP_a	OP	OP_a
0	268	94	10.2	9.8	10.3	9.9	4.9	4.5
3	185	75	12.5	9.1	11.5	8.1	8.8	5.4
6	165	69	17.9	11.5	13.9	7.5	11.2	4.8
12	107	46	21.3	8.9	22.4	10.0	16.7	4.3

[a] Adapted from Bernstein, 1961.
[b] OP_a, differences between sap and culture solution.

Table 6.3b

OSMOTIC POTENTIALS (OP IN ATM) OBSERVED IN TOMATO SHOOTS 28 HOURS AFTER THE ROOTS WERE IMMERSED IN VARIOUS OSMOTIC SUBSTRATES[a]

Substrate	OP of substrate	OP of expressed sap	Difference	OP of substrate	OP of expressed sap	Difference
Potassium nitrate	5.7	14.9	9.2	10.7	20.8	10.1
Sodium chloride	5.7	16.7	11.0	10.7	21.6	10.9
Sucrose	5.7	15.2	9.5	10.7	19.7	9.0
Control	0.7	10.9	10.2	0.7	10.9	10.2

[a] Adapted from Slatyer, 1961.

According to Walter (1961), halophytes which had been moved into saline media showed osmotic adaptation at an average rate of 1 atm/day. Similar, but somewhat higher rates were reported for glycophytes by Bernstein (1963). Higher rates of adaptation were observed in *Suaeda monoica* too. However, in glycophytes, salts may be continuously taken up until toxic ionic effects result (Walter, 1961). In such a situation it can be expected that growth inhibition by salinity is associated with ion accumulation in the plant, rather than with gross reduction in water availability. However, according to some investigators (Scholander *et al.*, 1962; Greenway and Thomas, 1965), at least some growth inhibition may still be attributed to reduced water availability, when osmotic adjustment is incomplete. In other cases I have suggested that even though entire leaves show osmotic adaptation, the protoplasts of various leaf cells remain only partly adjusted. Growth of such tissues will thus be inhibited.

The increase in osmotic potential of cell sap associated with osmotic adjustment of plants to salinity is partly accomplished by accumulation of ions; under certain conditions it can be achieved by accumulation of large amounts of organic acids (Bernstein, 1961; Jackson *et al.*, 1970).

Sap concentration of *Atriplex confertifolia* may attain osmotic potentials equivalent to 150 atm (Harris *et al.*, 1924). Osmotic potential measurements of sap of plants of three species of mangroves (*Avicennia nitida, Rhizophora mangle,* and *Laguncularia racemosa*) were not as high, but still considerable. In Jamaica and southern Florida (Harris and Lawrence, 1917) plants growing on freshwater had OP's that ranged between 20–30 atm whereas those on saline and dry substrates had OP's up to 50 atm.

Avicennia developed much higher water deficits than did all other genera tested. Osmotic potentials of *Rhizophora* and *Avicennia* leaves reached, according to von Faber (1923), values of 148 and 163 atm, respectively. Such steep osmotic gradients between plants and environment together with the negative pressures in the xylem (Scholander, 1968) are sufficient to draw water even from the sea.

Walter and Steiner (1936) also presented data on osmotic potentials of mangroves, but their values were lower. The osmotic potential of leaves of mangrove plants always exceeded that of seawater by some 7–10 atm. The water potentials of *Avicennia* leaves were higher than those of *Rhizophora* and exceeded those of seawater by 21.6 atm. However, since *Avicennia* is a salt-secreting plant, such data may reflect errors of methodology.

High osmotic potentials are common in halophytes. Steiner (1935) summarized some detailed information on the halophytic vegetation of the eastern United States (Table 6.4).

Generally, differences in the −5 to −15 bars range existed between the osmotic potential of the medium and that of glycophytic roots. For halophytes, the ΔOP yield much higher values, and in certain cases reaches −30 bars or

Table 6.4

OSMOTIC POTENTIALS (ATM) OF SOME HALOPHYTES OF THE
EASTERN COAST OF THE UNITED STATES[a]

Species	OP
Seawater (New Jersey)	**23.2**
Spartina glabra	31.1
S. patens	31.1
S. michauxiana	31.1
Salicornia europaea	31.1
Distichlis spicata	28.8
Limonium carolinianum	28.8
Juncus gerardii	28.8
Baccharis halimifolia	26.1
Atriplex hastata	26.1
Hibiscus moscheutos	12.2

[a] After Steiner, 1935

more. According to Ashby and Beadle (1957) and Black (1956, 1960), the osmotic potential of *Atriplex* leaves was always above the OP of the soil solution with the difference being more or less constant. *Aster tripolium* plants also exhibited osmotic adjustment (Table 6.5) when plants were moved from freshwater to a 0.8 *M* NaCl solution (Benecke, 1930a). Similar results were also reported for *Statice gmelinii* (Stocker, 1933), and several other halophytes (Table 6.6) by Mullan (1932, 1933).

Somewhat contradictory results were reported by Repp (1939). Osmotic potentials of leaf sap of halophytes collected from saline and nonsaline habitats in the Neusiedler See region of Austria (Table 6.7) showed relatively small differences in water potentials between tissues taken from plants of the two types of habitats. As halophytes seem to increase the OP to high levels even in habitats low in salt, Repp (1939) believed that a high ion content in halophytes

Table 6.5

THE OSMOTIC POTENTIAL (ATM) OF ROOTS AND LEAF SAP OF *Aster tripolium*
PLANTS GROWN ON SOLUTIONS WITH A DIFFERENT OSMOTIC POTENTIAL[a]

	OP		ΔOP	
Solution	Roots	Leaves	Roots	Leaves
0	15	30	15	30
32	50	65	18	33
40	55	65	15	25

[a] After Benecke, 1930.

Table 6.6

OSMOTIC POTENTIALS (ATM) OF VARIOUS HALOPHYTES SUBJECTED TO
DIFFERENT SALT SOLUTIONS[a]

Species	OP				
	Tap water	6.04	11.02	16.04	22.13
Triglochin maritimum	10.08	20.16	23.52	26.88	30.24
Aster tripolium	13.44	23.52	–	–	–
Salicornia herbacea	16.80	26.88	30.24	33.60	36.96
Juncus maritimus	6.72	20.16	23.52	26.88	–

[a] After Mullan 1932–1933.

Table 6.7

OSMOTIC POTENTIALS OF LEAF SAP OF PLANTS COLLECTED FROM SALINE
AND NONSALINE HABITATS IN THE NEUSIEDLER SEE REGION, AUSTRIA[a]

Species	Osmotic potential (atm)	
	Salt-free habitat	Saline habitat
Aster pannonicus	18.5	26.4
Scorzonera parviflora	16.6	22.5
Plantago maritima	19	23
Triglochin maritimum	27	28.5

[a] After Repp, 1939, 1963.

was not related to osmotic adaptation. However, this seems to differ from species to species and among populations. The OP of leaf sap extracted from *Aster tripolium* plants reached 30 atm on a saline soil, but only about 15 atm on a nonsaline one. Since most of the sodium in the latter group of plants was substituted by potassium, it seems that the difference in OP between such groups of plants was due to differences in chloride accumulation.

As roots directly contact a saline soil, they must be osmotically adapted. Nevertheless, the OP's of roots are consistently lower than those of leaves of the same plants. Maximal osmotic potentials recorded for roots were 50 atm for plants in the Egyptian deserts (Stocker, 1933), 40 atm for plants in the Hungarian "szik" soils (Stocker, 1928, 1930), and 96 atm for Indian mangroves (von Faber, 1923). The higher values of the OP of leaves probably resulted from their continuous uptake of ions.

Changes in salt content of the soil solution must eventually be reflected by an equivalent change in the content of osmotically active substances of plants as they undergo osmotic adaptation. Such adaptation can be achieved in several ways. The chemical components which contribute to the OP of cell sap differ in various species. According to Knodel (1938), only in leaves of some glycophytic evergreen tree species (e.g., *Picea, Buxus*) did the sugar concentration contribute more than 30–50% of the total OP. Usually, sugars constitute less than 25% of such osmotic potentials. The electrolyte content and consequently the osmotic potential of crop plants may be increased by potassium fertilization (Knodel, 1938).

Osmotic adaptation in plants occurs mostly because of increase of chloride content, and is especially high in halosucculents (Adriani, 1958). Chloride ions have been shown to account for 67–88% of the increase in OP of different species of salt marsh plants (Steiner, 1935) (Fig. 6.3). Increase in OP from 31.9 to 43.4 atm in *Salicornia europaea,* of the east coast of the United States was found by Steiner (1935) when soil salinity increased. Chlorides in these plants

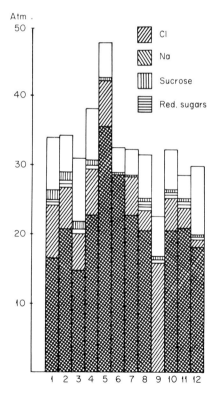

Fig. 6.3. Contribution of various substances to the osmotic potential of various plant species. (After Steiner 1935.) 1. *Spartina glabra*, 2. *S. patens,* 3. *Distichlis spicata*, 4. *Juncus gerardii*, 5. *Salicornia mucronata,* 6. *S. europaea,* 7. *Plantago decipiens,* 8. *A triplex hastata,* 9. *Aster subulatus,* 10. *Limonium carolinianum,* 11. *Suaeda linearis,* 12. *Iva ovaria.*

accounted for changes in OP from 27.8 to 38.3 atm, whereas other osmotically active substances had only a negligible effect. The same applied to *Juncus gerardii*. While the total osmotic potential of the plant sap rose from 22.0 to 33.3 atm, chlorides accounted for changes from 10.5 to 20.2 atm. The contribution of all other substances to this change was not considerable, from 11.5 to 13.1 atm only.

In many species of halophytes, e.g., *Atriplex* sp., organic ions may appreciably increase the cellular osmotic potential (Osmond, 1963). Osmotic adaptation of plants, such as *Artemisia salina* and *Anabasis salsola,* results mostly from accumulation of oxalates (Strogonov, 1964).

Data presented by Önal (1964, 1966) dealt with effects of inorganic and organic ions on the osmotic potential of plant cells. The high contribution of inorganic ions to the OP of halophytes constituted another differential characteristic between halophytes and glycophytes (Table 6.8).

Transport of large quantities of ions and maintenance of their concentration require high input of metabolic energy. Osmotic adaptation may, thus, consume energy which could otherwise be used for growth.

Variations in OP between closely related organs are also frequently observed. The osmotic potential of *Sonneratia* seeds was about 14.4 atm whereas that of other parts of the flower, was almost twice as great (Walter and Steiner,

Table 6.8

CONTRIBUTION OF NaCl TO THE OSMOTIC POTENTIAL OF GLYCOPHYTES AND HALOPHYTES[a]

Species	Osmotic potential of soil solution (atm)	OP of plant sap (atm)		
		OP chloride calculated as NaCl	OP due to other substances	Total OP
Halophytes				
Atriplex portulacoides	27.7	36.4	4.7	41.1
Salicornia fruticosa	20.6	31.7	9.6	41.3
Inula crithmoides	17.0	17.6	7.1	24.7
Statice limonium	10.5	18.5	5.0	23.5
Juncus acutus	9.3	11.9	7.5	19.4
Plantago coronopus	4.0	7.7	4.0	11.7
Glycophytes				
Pistacia lentiscus	[b]	4.5	20.1	24.6
Phillyrea latifolia	[b]	3.4	19.7	23.1
Pinus pinaster	[b]	6.9	15.0	21.9
Quercus ilex	[b]	2.2	24.6	26.8

[a] After Önal, 1964, 1966.
[b] Osmotic potential was not measured but is presumably very low.

1936). The question of how the developing embryos of this species obtain water against an osmotic potential gradient still remains unanswered (cf. Pannier, 1962). A clearer case of water absorption may be found in semiparasitic plants. The xylem sap tension of a *Viscum* plant growing on *Rhizophora* was measured with the pressure bomb. Results indicate that a negative pressure of 1 to 2 atm over that of the host plant existed, thus enabling the mistletoe to obtain water (Scholander, 1968).

Osmotic adaptation differs not only between species but also between organs and even between cells. Root hairs vary greatly in this respect, even in the same individual plant (Hill, 1908). Generally, the OP was higher in young root hairs of *Salicornia herbacea* or *Suaeda maritima* plants than in older ones. When plants were transported from concentrated salt solutions to dilute ones, the tips of young root hairs became swollen as a result of high salt content. Root cap cells had a very low osmotic potential.

Osmotic adaptation of epidermal cells in plants of two closely related species was studied by Fukuda (1937) and Takaoki (1957). The OP of epidermal cells of *Plantago coronopus* increased when plants were subjected to saline media. The relationships between internal and external OP's yielded a concave curve with NaCl, but a convex curve with solutions of other salts. The osmotic potential of the glycophyte, *Plantago major,* yielded convex curves with changes in the OP of the medium. These were independent of ionic composition. Since similar trends were found for other pairs of species as well, these investigators concluded that osmotic adaptation varied in plants of different ecological groups (see also Fig. 11.1).

As osmotic adaptation of plants is not a static phenomenon, periodic changes in OP of plants occur (Bernstein, 1963). Ions are translocated between leaves and roots periodically and, in certain cases, are exuded from roots. Experiments reported by Repp (1939) revealed that the osmotic potential of *Suaeda maritima, Chenopodium glaucum, Lepidium crassifolium,* and *Aster pannonicus* plants decreased by an average of 1 to 3 atm in 12 hours, when high salt plants were placed in distilled water. Such decreases occurred normally in native habitats where diurnal changes in water potentials were encountered. Fluctuations in OP may affect the metabolism of plants in a different manner from those of constant conditions (Vaadia and Waisel, 1967).

Water Transport in Stems

Very little is known of paths of water conduction in halophytic shrubs and trees. Mangroves seem to comprise a distinct group. Osmotic potentials of the expressed sap of leaves and roots of mangrove trees exceed those of seawater, with those of leaves having the highest potential (Walter and Steiner, 1936).

Such a gradient could have explained the entry of water from the sea into the plants. However, measurements made by Scholander *et al.* (1962, 1965) and Scholander (1968) revealed that the xylem sap which is delivered from the roots to the shoots is practically fresh water. Water should then be sucked along potential gradients from the xylem back into the roots, on one side of the stem, and into leaf cells, on its other side. A certain stress must then be expected to develop inside the vascular system, and an equilibration between osmotic and hydrostatic pressures was postulated (Fig. 6.4). In fact, measurements of

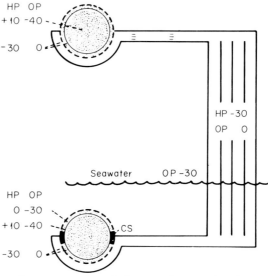

Fig. 6.4. A schematic representation of the osmotic potential (OP) and hydrostatic potentials (HP) in a mangrove plant immersed in seawater. All data in bars. CS, Casparian strip. (After Scholander *et al., Science,* **148,** 339–346, 1965. Copyright 1965 by the AAAS.)

hydrostatic pressure in xylem sap of mangrove trees conducted by Scholander (1968) with a pressure bomb yielded negative values of about 25 atm. The data suggested that water was sucked into the plant because of such negative hydrostatic suction gradients. Since seawater salts are prevented from entering the sap by cell membranes, the pressure differences induce a forced ultrafiltration. Such a process seemed to remain functional even when roots were exposed to carbon monoxide, DNP, or lack of oxygen. It seems, thus, that membrane characteristics leading to ultrafiltration are static–structural rather than flexible–metabolic.

　　Ascent of water in stems of shrubs and trees constitutes another important step in their water supply. A few distinct patterns of upward movement were described: sectorial straight, sectorial winding, sectorial straight turning into ring ascent, interlocked, and spiral ascent. Attempts were made to correlate those patterns either with the plants' water economy (Vité, 1959; Vité and Rudinsky, 1959; Rudinsky and Vité, 1959; Hendrickson and Vité, 1960; Kramer and Kozlowski, 1960; Kozlowski, 1961) or with their growth form (Waisel *et al.,* 1972).

Only scattered data are available on paths of water movement in stems of halophytic shrubs and trees. However, as far as the data permit, it is suggested that halophytes do not differ in this respect from glycophytes. Water ascent (traced either by acid or by basic fuchsin) in *Reaumuria palaestina* showed the sectorial straight pattern typical of shrubs of arid habitats (Waisel *et al.,* 1972). On the other hand, water ascent in stems of *Suaeda monoica, Atriplex halimus,* and *Seidlitzia rosmarinus* showed an interlocked pattern (Fig. 6.5). However,

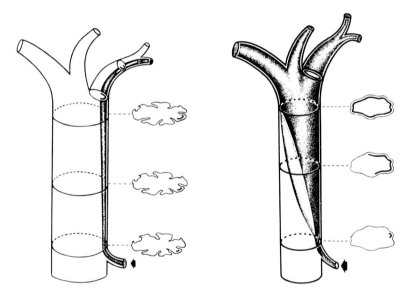

Fig. 6.5 Patterns of water ascent in stems of *Reaumuria hirtella* var. *palaestina* (left) and *Suaeda monoica* (right). (Adapted after Waisel *et al.,* 1972.)

since two other shrubs of the Chenopodiaceae (*Noaea mucronata* and *Chenolea arabica*), which are xerophytes rather than halophytes, exhibited a similar pattern, it is tempting to ascribe that pattern to the structure of their stems, rather than to their ecological group. Stems of the Chenopodiaceae have scattered vascular bundles. Flow of water in them is interlocked.

Capacity of stems to conduct water decreases in plants upon exposure to saline conditions. The number and size of vascular elements were found to be small in tomato and cotton plants grown under saline conditions (Strogonov, 1964). In certain cases, only the size of vessels was smaller, while their number either remained unchanged or even increased. In both cases, the area of the conducting tissues in the stem cross section was smaller under saline conditions (Strogonov, 1964). Similarly, also in halophytic shrubs, which were exposed to highly saline conditions, the number of vascular bundles and the size of the conducting elements were smaller than in plants grown in fresh water (Rumyantseva, 1959; Liphschitz and Waisel, 1970a,b).

In some plants, the capacity to transport water from roots to shoots also varies seasonally and is related to cambial activity and production of more vessels. Waisel (1960) showed that the annual curve of transpiration of *Tamarix aphylla,* and especially the curve for relative transpiration, were characterized by two maxima in early spring and late summer (Fig. 6.6). The peaks were

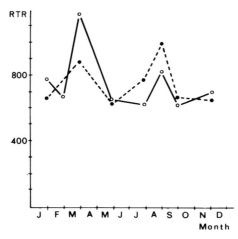

Fig. 6.6. Annual march of relative transpiration rate (RTR) of two *Tamarix aphylla* trees in the Negev, Israel. (After Waisel, 1960a.)

positively correlated with the two seasonal maxima of cambial activity as reported by Fahn (1958), and probably reflect an increase in the number of water-conducting xylem elements.

Transpiration

Low vapor pressure of saline water results in lower rates of evaporation, and in smaller heat loss as the latent heat of vaporization (Penman, 1948; Ferguson, 1952; Bonython, 1958). Consequently, a heat buildup can be observed in saline solutions which are exposed to full solar radiation. Lower rates of evaporation and higher temperature must influence water in plant tissues in the same way they affect water in an evaporating pan. However, no such effect was found in various halophytes, probably due to increasing rates of transpiration. Again great variations are found between species.

High salt *Atriplex* plants grown in sodium chloride solutions had low transpiration rates per unit of dry weight when reaching maturity, i.e., when their highest salt content was obtained (Ashby and Beadle, 1957). Even though many factors presumably contribute to this phenomenon, salt concentration may still be a major factor.

Saline conditions are not essentially "dry" for all species of plants. Under saline conditions, some halophytes transpire more than their neighboring glycophytes (Delf, 1911; Braun-Blanquet, 1931). Such plants seem to be capable of extracting water from saline soils more efficiently than any other group of plants because of a high gradient of osmotic potential between the plants and soil.

Usually sodium and chloride are found in nature together and, in most cases, their effects seem to be inseparable. Nonetheless, their effects on transpiration differ. While sodium did not affect the transpiration coefficient* of young tomato plants, chloride increased it (Arnold, 1955). According to this investigator, chloride ions also influenced other aspects of plant water relations which were unaffected by sodium.

Accurate measurements of transpiration of halophytes are scanty, and some reported data are even contradictory. This can be ascribed primarily to the methods used in transpiration measurements. Data on transpiration rates of halophytes, measured with the rapid weight method or with potted plants, are ample (Keller, 1926, 1929; Seybold, 1930; Braun-Blanquet, 1931; Walter and Steiner, 1936; Evenari and Richter, 1937; Adriani, 1945; Shmueli, 1948). Some of this information was summarized by Adriani (1956) (Table 6.9). Evidently, halophytes are not a homogenous group and include species with high, as well as species with low, transpiration rates (Stocker, 1933). Consequently, separation of halophytes from glycophytes cannot be made on the basis of transpiration measurements. However, transpiration rates of most halophytic species investigated were lower than those of glycophytes, when calculated for comparable conditions. The basis for calculation can also cause differences in the conclusions. According to Schratz (1937), the transpiration rates of coastal halophytes were lower when calculated on a fresh weight than a surface area basis (Table 6.10).

As shown in Table 6.11, fluctuations in salinity affected transpiration rates of halophytes much less than those of glycophytes.

Upon transfer to a saline medium beans suffered from osmotic shock, and their stomata closed. However, no differences in the behavior of *Suaeda* plants were observed. Also Henkel and Shakhov (1945), noticed that most

---

* Transpiration coefficient expresses liters of water transpired during the production of 1 gm dry weight.

Table 6.9
AVERAGE TRANSPIRATION RATES OF VARIOUS HALOPHYTES[a]

Halophyte	Transpiration rate (mg/gm/min)
North Sea	
Spartina stricta	5.4
Salicornia herbacea	1.5–2.3
Suaeda maritima	2.5–2.6
Aster tripolium	2.5–5.0
Triglochin maritimum	3.5
Statice limonium	5.1–7.4
Obione portulacoides	1.1
Artemisia maritima	11.1
Spergularia marginata	3.1
Plantago maritima	2.9
Glaux maritima	7.3
Agrostis stolonifera	7.3
Plantago coronopus	9.1
Mediterranean	
Salicornia herbacea	2.3
Salicornia radicans	2.5–3.8
Salicornia fruticosa	0.9–2.6
Salicornia macrostachya	0.5–1.7
Obione portulacoides	1.5–2.9
Suaeda maritima	3.9
Juncus maritimus	2.7–3.8
Inula crithmoides	2.2
Statice limonium	2.9
Tamarix gallica	11.6
Mangroves	
Sonneratia alba (normal)	2.6
Sonneratia alba (highly succulent leaves)	0.85
Rhizophora mucronata	1.8
Ceriops candoleana	1.9
Avicennia marina	4.1
Lumnitzera racemosa	2.0

[a] After Adriani, 1956.

hydrohalophytes transpire profusely. Such plants may thus develop only on wet sites. No correlation was found between the leaf water content and transpiration of various species of *Suaeda*. Higher rates of transpiration were found in *Suaeda monoica* than those in *Suaeda fruticosa* and *Suaeda palaestina,* although water percentages in the leaves of the latter species were higher (Shmueli, 1948). Among the hydrohalophytes, salt-secreting species had higher rates of transpiration than did halosucculents (Shmueli, 1948).

Table 6.10

TRANSPIRATION RATES OF GLYCOPHYTES AND COASTAL HALOPHYTES
CALCULATED ON FRESH WEIGHT AND ON SURFACE AREA BASIS[a]

Species	Fresh weight basis (mg/gm/min)	Surface area (mg/dm^2/min)
Glycophytes		
Helianthus	16.1	24.8
Fragaria	12.9	14.5
Halophytes		
Glaux maritima	6.4	15.2
Aster tripolium	5.0	22.7
Cakile maritima	4.1	17.0
Salsola kali	2.1	14.0
Salicornia herbacea	1.5	14.6

[a] After Schratz, 1937.

Table 6.11

INFLUENCE OF MOVING A GLYCOPHYTE (*Phaseolus vulgaris*) AND A HALOPHYTE
(*Suaeda monoica*) FROM LOW TO HIGH SALT MEDIA AND VICE VERSA, ON THEIR
RATES OF TRANSPIRATION (PERCENTAGE OF INITIAL RATES)

Species	Treatment				
	Basic nutrient solution		Nutrient solution +100 mM NaCl		Basic nutrient solution
Phaseolus vulgaris	100%	→	−46.8%	→	+24.0%
Suaeda monoica	100%	→	+ 5.2%	→	+ 8.2%

Effects of increasing concentrations of NaCl in the growth medium, on growth and transpiration coefficient of *Salicornia herbacea, Suaeda maritima, Salsola kali,* and *Spergularia salina,* were investigated by Önal (1971). Transpiration coefficients decreased with increasing salinity. Values of 512 ± 4.4 (gm/gm) were obtained for *Salicornia herbacea* plants when grown on a salt-free solution, but 271 ± 7.2 (gm/gm) when only 1% NaCl was added to the growth medium.

Transpiration measurements of a salt-secreting hydrohalophyte, *Tamarix aphylla,* were reported by Waisel (1960a). Daily fluctuations in transpiration during various months are shown in Figs. 6.8–6.10.

Three types of curves can be noted:

(a) Curves with one peak at noon (Fig. 6.8). Such types reflect high soil water availability. In fact, this type of curve was obtained in February, when the soil was wet (Fig. 6.7) and evaporation not too intensive.

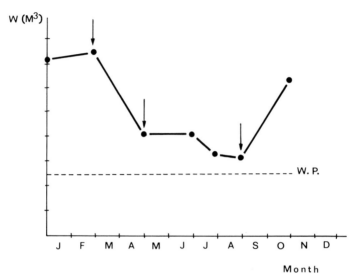

Fig. 6.7. Annual changes in soil water content in a stand of *Tamarix aphylla* trees in the Negev, Israel. Arrows indicate dates of transpiration measurements; W.P., water content at the wilting point, in the root zone. (After Waisel, 1960a.)

(b) Curves with two peaks, one in the morning, and the other in the afternoon (Fig. 6.9), indicating either restricted water supply during midday, or regulation of transpiration by the plants' aerial organs. The midday depression may result from a temporary lack of water in the soil near the roots, or from the inability of the conducting system to transfer regularly the large quantities of water required. A curve of this type was obtained in May.

(c) Curves with one peak in the morning (Fig. 6.10). This type is believed to reflect conditions of lack of water in the soil near the roots (cf. Fig. 6.7). In fact this type of curve was obtained in the late summer.

Very few measurements of transpiration were conducted during recent years with more reliable and modern methods than the cut-leaf method. Nevertheless, from existing data one can learn that transpiration rates of halophytes can be related more to their structure and water sources than to their salt content. Whether such a conclusion is true or whether it only appears so because of inaccurate methods of measurement, still awaits an answer.

A comparative study of photosynthesis, growth, and transpiration was conducted in two species of *Atriplex* by Slatyer (1970). A combination of low leaf mesophyll resistance (r_m) and high stomatal resistance (R_L) enabled a higher efficiency of water use in *Atriplex spongiosa* than in *A. hastata*. The former species showed faster rates of leaf growth and thus seemed to be better adapted to conditions of high water deficits.

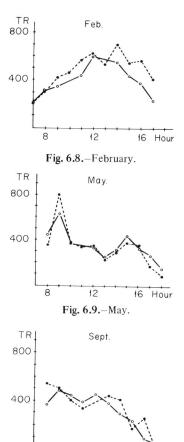

Fig. 6.8.–February.

Fig. 6.9.–May.

Fig. 6.10.–September.

Fig. 6.8–6.10. Daily curves of transpiration rate (TR) of two *Tamarix aphylla* trees in the Negev, Israel. (After Waisel, 1960a.)

The mesophyll cell surface may be subjected to high stresses, during transpiration. Values of up to −90 bars were reported for corn plants (Shimshi, 1963). Measurements of leaf water potentials of the xerohalophyte, *Reaumuria hirtella,* were made by Whiteman and Koller (1964). They measured the water vapor pressures of leaves of intact plants when equilibrated with those of the atmosphere, at the point when transpiration dropped to zero. Leaf water potentials equilibrated at extremely high deficits. The water potentials of such plants, containing about 30% soluble salts in their expressed leaf sap, varied between −180 and −320 bars in different plants. Since such potentials have to be balanced somewhere by osmotic or by hydrostatic pressures, they seem to be extraordinarily high and probably represent an equilibration point also with salt crystals which were previously secreted. Nevertheless, such data still indicate that water stresses which develop in halophytes—especially in xerohalophytes—are high, and that the metabolic activity of those plants must be geared to operate under such conditions.

To a large extent transpiration is controlled by the stomata. Stomatal movement is known to be controlled by a specific and selective light-dependent potassium transport mechanism. Good data on stomatal responses under saline conditions were scarcely reported. However, available data suggest that plants subjected to NaCl solutions behave differently from those subjected to KCl. According to Thomas (1970), stomata of tobacco leaves tended to open in the dark, when immersed in a 10 mM NaCl solution (Fig. 6.11). The same effect of

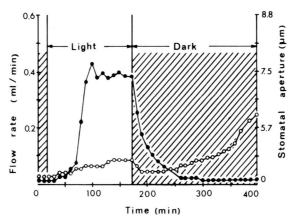

Fig. 6.11. Effect of 10 mM KCl (filled circles) or 10 mM NaCl (open circles) bathing media on tobacco stomata opening in the light and dark. (After Thomas, 1970.)

NaCl was observed also in *Kalanchoe marmorata,* a plant with acid metabolism, in which stomata normally open in the dark and close in the light (Fig. 6.12).

Only little is known of stomatal behavior of halophytes. However, since such plants are exposed to an environment with a high NaCl content and as they have acid metabolism, at least to some extent, it is tempting to assume that reaction to NaCl is similar in halophytes and in *Kalanchoe*. Indirect information on stomatal behavior can be obtained from leaf temperature measurements. Measurements were made on plants subjected either to common nutrient media, or to media containing NaCl. While pepper plants were overheated under conditions of water deficits, temperatures of *Suaeda, Aeluropus,* and *Atriplex* leaves were below ambient. Nevertheless, it should be noted that such heat regulation is effective only when water is readily available. Leaves of *Suaeda* plants which were not immersed in water, became overheated whether the

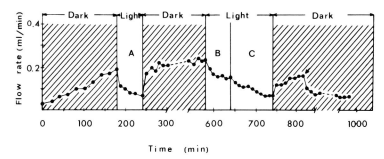

Fig. 6.12. Effects of light on the opening of *Kalanchoe marmorata* stomata bathed in 10 mM NaCl. Light fluxes at A and B were 22.6 mW · cm^{-2} and at C 33.4 mW · cm^{-2}. (After Thomas, 1970.)

medium was saline or not. Similar results were reported for *Nitraria retusa* (Seybold, 1930; Lange, 1959).

Absorption of Atmospheric Moisture

Branches of many salt-secreting species are highly hygroscopic due to their cover with salt crystals; on windless clear nights, such plants may absorb water vapor from the atmosphere. When wilted twigs of various species of *Tamarix* were left during the night under a saturated atmosphere, or sometimes even on a laboratory bench, they became moist and regained their turgor. This is evidence that *Tamarix* plants may absorb water from the surrounding humid atmosphere, even in the field, and that such absorption provides the plant with an additional source of water (Waisel, 1960a). Furthermore, the salt solution, which is formed on the twigs, reduces considerably the extent of transpiration in the morning, and improves the water economy of those plants.

There are many difficulties in a quantitative determination of water uptake by such plants. Results of controlled experiments (Waisel, 1960a) revealed that water absorption by the aerial organs of *T. aphylla* occurred even when plants were covered with salt solutions exceeding 1 *M* NaCl. However, relatively large quantities of water could be absorbed only if plants were undergoing very high saturation deficits, and over long periods of continuous exposure. Under field conditions, the time during which twigs were exposed to moistening was relatively short and saturation deficits were low. Under such conditions, only small quantities of water could be absorbed, thus excluding atmospheric moisture from being an ecological factor of great importance. Nevertheless, since *T. aphylla* is most sensitive to dehydration of its twigs (Waisel, 1959), water absorption by aerial organs may still be of value to the plant in recovery from sublethal water deficits. Similar data, showing water absorption even from a 3% salt solution, were reported for *Salicornia ramosissima* plants by Halket (1911).

Nevertheless, differences between species were found. Vesiculated hairs of *Atriplex halimus* are unable to absorb water vapor from air despite their high salt content (Mozafar and Goodin, 1970).

In regions where fogs occur frequently, it was demonstrated long ago (Means, 1929) that the soil under trees contains higher percentages of moisture than soil under herbaceous vegetation cover.

Fog drops, blown by the wind, collide with the high solid bodies of trees, drop down, and moisten the soil below the retaining body. This phenomenon has attracted many investigators. Bleasdale (1957) who dealt with the various factors affecting afforestation, felt such sources important, taking into account precipitation in the calculations of the water balance of entire catchment areas.

A similar phenomenon was observed with various trees, even in such an arid region as the Negev of Israel (Waisel, 1960b). Such precipitation was found to be more pronounced under the salt-secreting *Tamarix* trees, than under any other tree species in that area. Precipitated water moistened the soil under *Tamarix* trees to a depth of 50 cm. Moistening the soil with precipitated fog water may thus contribute to their water balance during dry summer months.

7

Ion Transport and Mineral Nutrition

General Remarks: Nutrition in a Hostile Environment

Nutrition of plants cannot be considered as one general process and its meaning lies far beyond mere penetration of certain quantities of ions into the whole plant. A balanced nutrition also means persistence of ions in a functional state within the various tissues and cells, i.e., in a state that enables them to fulfill their role in metabolism. Thus, nutritional requirements have to be satisfied for the whole plant as well as for each and every individual cell of an organism. Nevertheless, even though plant cells are omnipotent in many respects, the nutritional requirements of different cells and the problems related to their ion uptake and accumulation are not always similar. Effects of salinity on cell nutrition may thus differ for each type of cell and in accordance with its specific location and function. Root cells differ in their nutritional requirements not only from one another, but also from those of the hypocotyl, stem, or leaf. Thus, understanding the specific behavior of each individual cell within the plant body will eventually provide us with the knowledge of the common, as well as with the specific mechanisms which govern plant nutrition.

Nutritional requirements of plants which grow in the presence of high concentrations of NaCl are of special interest. In this respect, halophytes provide choice material for students of plant nutrition. When cells of *Zostera marina, Rhizophora mangle,* or *Suaeda monoica* plants absorb potassium or nitrate, this is accomplished in the presence of sodium and chloride, and despite the effects of their overwhelming external and internal concentrations. The intriguing question of how such plants manage to absorb nutrients, where other species fail, still remains to be answered.

Salts affect the general nutritional status as well as the metabolism of plants. Nevertheless, these are not one-step effects and long series of sequential processes lead to a certain ionic composition of a plant. Ion uptake is only one of many processes which are affected by sodium and chloride. Consequent transport across roots, secretion into the xylem sap, transport through the hypocotyl and stem tissues, and uptake by the leaf mesophyll and meristem cells are all affected. Depending on species, salinity may either disturb or accelerate each of these processes.

In this chapter, the intricate processes involved in ion uptake and transport into the multicellular plant body (Fig. 7.1) will be followed, with special emphasis on those processes which characterize halophytes.

Salt Movement Toward Roots

The starting point for the long paths of translocation, which ions traverse on their way into a cell, somewhere inside the plant, is usually in the soil. There, ions are present in a few forms. They constitute part of the rock crystals or part of the lattice of soil clay minerals; they may be adsorbed on the surface of such clay minerals or may be dissolved in the soil solution. Usually the three different phases, mineralized, adsorbed, and dissolved, are in a dynamic equilibrium. Nevertheless, rates of transformation from one phase to another are of different magnitudes and differ in accordance with the environmental conditions.

In some cases, ions are readily absorbed by plants from the soil solution, while in other cases the nutritive cations adsorbed on the soil clay minerals are exchanged against hydrogen ions (Jenny and Ayers, 1939; Epstein and Stout, 1952). For example, excised barley roots take up zinc from zinc-saturated bentonite or kaolinite suspensions, almost as readily as from true solutions. Such an exchange and the consequent decrease in pH around absorbing roots was demonstrated by Sachs, as early as 1875, by growing plants on a polished piece of marble.

Movement of ions to root surfaces may take place by gradual exchange reactions (Fig. 7.2). Since this is a relatively slow process, a region of low ion content is gradually formed near the root surface. The radius of the cylinder of

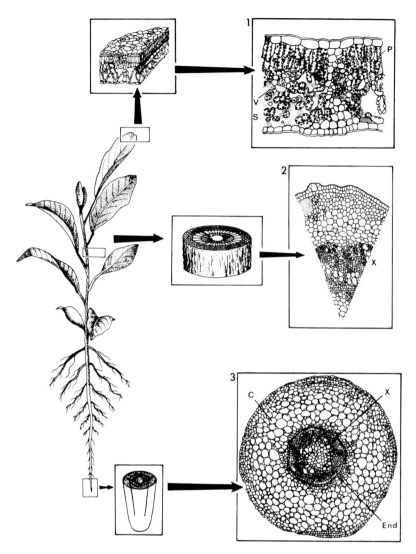

Fig. 7.1. Schematic structure of a terrestrial plant. P. palisade tissue; V, vein; S, spongy tissue; X, xylem; C, cortex; End, endodermis. (After Waisel *et al.,* 1966b.)

depleted soil thus formed approximates the length of fine root hairs. Consequently, further uptake of ions by roots is limited by the rates of their supply from the bulk of the soil to the root surface. Such a supply of ions depends on two processes: ion diffusion and mass flow. Rates of uptake of certain ions were found to be far higher than their rates of diffusion in the soil (cf. Walker and Barber, 1962; Wilkinson *et al.,* 1968). In such cases, the supply

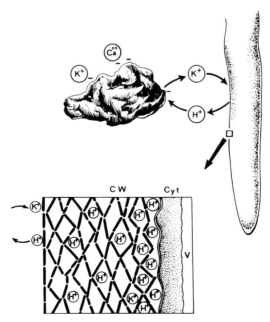

Fig. 7.2. Exchange reactions between root epidermal cells and a soil particle. CW, cell wall; Cyt, cytoplasm; V, vacuole. (Cell wall redrawn after Frey-Wyssling, 1935.)

of nutrients to the plant may depend on mass flow of the soil solution. Calculating transpiration coefficients for corn enabled Walker and Barber (1962) to show that the total supply of calcium to the plants could be explained by mass flow. On the other hand, concentrations of potassium and phosphorus in the soil solutions were too low and additional mechanisms were sought for uptake of those ions. Growth of roots into unexploited volumes of soil appeared to be the best explanation.

Information on mass flow and rates of diffusion of ions in the soil matrix is mostly related to movement of K and Ca in salt-free soils. Since, in halophytes, concentration gradients in the inward direction may be negligible or inverse, diffusion of sodium and chloride in such a case from the bulk of the soil to the root surface seems to be of minor importance.

Roots may exploit only a limited volume of the soil and the uptake of nutrients by them is limited by a few factors, five of which are outlined below:

1. Growth capability of roots into unexploited soil.
2. The time during which a root is active in a certain site.
3. Changes in uptake ability along the root.
4. Rates of ion movement from the bulk of the soil to the root surface.
5. The capability of the soil adsorption complex to retain ions.

The first three factors are dependent on plant characteristics, while the last two depend mostly upon the physical nature of the ions involved, and upon soil characteristics.

Generally, the transport of ions from the soil into the root system can be summarized by the scheme shown in Fig. 7.3. Once ions reach the root surface, they can enter the root via the free (outer) space or osmotic (inner) space. In the

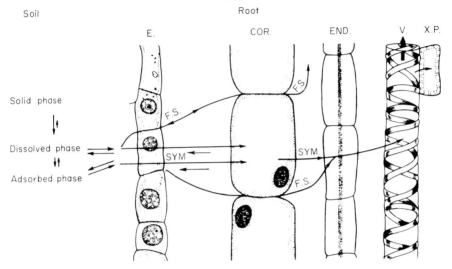

Fig. 7.3. A general scheme of ion movement from the soil into a root. F.S., free space movement; Sym, symplast transport; Cor, cortex; End, endodermis; V, vessel; X.P., xylem parenchyma; E, epidermis.

latter case, they can be taken up into epidermal, cortical, or endodermal cells and retained or transported inside the symplast further into the xylem. In such cases, transport may be accomplished either by passive or by active movement, the latter requiring expenditure of metabolic energy.

Metabolic accumulation is probably the fastest process in the described chain of reactions leading to ion uptake. Thus, the rate of the overall process of ion uptake depends on rate of metabolism. Nevertheless, besides metabolism, uptake of ions still responds noticeably also to many other factors that open physical "bottlenecks" in the process; for example, an increase in flow of solutes by the transpiration stream or to increases in the activity of ions in the external solution.

Salt Uptake and Uptake Mechanisms

Upon contacting a root surface, ions penetrate plant tissues. Entry of an element into a plant is dependent on many external and internal conditions, which were most commonly summarized as permeability. This is defined as the rate of entry in a unit time, through a unit area, under conditions of a unit activity difference. Usually permeability differs for movement of ions inward and outward, the movement into the cell being higher than out of it. (cf. Dainty, 1963).

For the sake of simplicity, the processes involved are divided into nonmetabolic (passive) and metabolic (active) ones.

When ions are transported into cells as a result of electrochemical potential differences, the process is called passive or nonmetabolic uptake. Rates of ion uptake by such a process are linearly correlated with the external concentration of ions, and are only very little affected by temperature changes.

Nonmetabolic uptake starts with the entry of an ion into the free space of a root. The free space is defined as the volume of tissues which is available for free diffusion (cf. Briggs and Robertson, 1957). Cell walls are considered the major component of the free space; they are negatively charged and adsorb cations. Winter (1961) showed that about 96% of the capacity of *Vallisneria* tissues to adsorb cations was concentrated in the cell walls. The remaining sites of adsorption were probably on the plasma membranes. Similar conditions also prevail in roots. In some species most ion uptake is into the free space. Entry of phosphate, bromide, and sulfate into about 80% of the volume of yeast cells is by passive diffusion. Additional accumulation occurs only into the remaining 20% of the cell volume (Leggett and Olsen, 1964).

Movement of ions across cell walls is not a simple process because its kinetics are not of undisturbed diffusion but rather of diffusion across charged surfaces. Ions are adsorbed and discharged along their way inward, and rates of movements are thus much slower. The nature of the adsorbing surfaces of plant tissues still seems to be obscure, although pectates can account for most of them.

Questions were raised whether ions are first adsorbed on the cell walls and only later accumulated, or whether adsorption does not constitute a prerequisite for ion uptake into the osmotic space. In *Vallisneria* leaves, ions move directly from the free space into the osmotic space without intermediate adsorption (Winter, 1961). At this stage, discrimination between ions may occur. Electrical potential differences which exist between the osmotic space and the medium (approximately -100 mV) certainly accelerate movement of cations and inhibit the movement of anions (cf. Higinbotham, 1968, 1970; Higinbotham *et al.*, 1962).

As ions penetrate plant cells beyond the walls, they must pass through a membrane which comprises a barrier to free diffusion. Without the existence of such a barrier, plants would have constituted one large free space and ions would have moved in and out freely. Such a membrane is selective and its properties determine the quality and quantity of the moving ions. The question of whether the limiting membrane is the plasmalemma or the tonoplast is in dispute. Evidently, both membranes exhibit low permeability. Under certain conditions the plasmalemma is limiting and only small quantities of ions will be accumulated or transported across it. On the other hand, when the tonoplast is limiting, accumulation in the vacuoles will be hampered, but ions will continue to move in the symplast. Limitation of ion transport by a selective barrier is critical for plants which usually are exposed to high salt concentrations. Such limitation may take place at the surface of epidermal cells, cortical cells, or at the endodermis.

The epidermis was thought to constitute a barrier to ion movement by various investigators. Using di-*n*-amylacetic acid for stripping the epidermis off wheat roots, Sandström (1950) succeeded in increasing the rates of inward movement of salts. Nevertheless, data reported subsequently indicate that such effects are not universal. Furthermore, it seems that in certain species transport into the roots may be affected by cortical cells more than by the epidermis.

Ions move across the cortex either along the free space in the cell walls or across the plasmalemma inside the cytoplasm (Fig. 7.3). In the symplast, ions move with the cytoplasmic flow, are assimilated, or are further transported into vacuoles (Arisz, 1963, 1964).

Uptake of ions into the osmotic space against an electrochemical potential gradient requires an expenditure of metabolic energy and, therefore, a normal supply of oxygen and metabolites.

Halophytes may exhibit an exceptional behavior. According to Jennings (1968a), excess sodium crosses the plasmalemma in cells of succulent halophytes without any need for energy expenditure. Moreover, the overall process of sodium transport into the vacuole is postulated to result in ATP production. Such ATP is later used for additional growth, thus explaining the growth stimulation caused by NaCl in halophytes.

The rates of such active uptake are generally temperature-dependent, although the uptake of cations is affected less by temperature than that of anions. It is generally assumed that a carrier mechanism is involved in metabolic uptake. According to this view, a temporary association of ions with certain compounds constitutes a step in the metabolic uptake.

The kinetics of enzyme reactions were adapted to analyses of ion uptake processes by Epstein and Hagen (1952), Epstein and Leggett (1954), and others. Attempts were made to use such an analysis to distinguish between competition and other antagonistic effects between ions. Although the general approach serves as a convenient working hypothesis, such kinetic analyses have strict limitations. They overlook many critical phases in the uptake process, e.g., do not provide satisfactory explanations to cation–anion interactions, do not explain effects of calcium, changes in response to temperature, limitation of uptake by diffusion, etc. (Briggs *et al.,* 1961; Waisel, 1962a; Noggle *et al.,* 1964). For these reasons interpretations based on kinetic analyses must be regarded with caution.

Sometimes metabolic energy is expended for exporting ions rather than for importing them. Experiments involving the insertion of microelectrodes into higher plant cells, i.e., pea roots and epicotyls, oat roots and coleoptiles (Etherton and Higinbotham, 1960; Etherton, 1963; Higinbotham et al., 1967; Etherton, 1967) and leaf cells of Hookeria lucens (Sinclair, 1969) indicated that sodium is probably being pumped out of the cell against an electrochemical potential gradient. Also information about electrochemical potential gradients of sodium between the external medium and the bleeding sap of Ricinus communis and Helianthus annuus (Bowling and Spanswick, 1964; Bowling et al., 1966; Bowling, 1966) was interpreted in terms of an outwardly directed sodium pump. No information of this kind is available for halophytes.

Very little is known of ionic interrelationships in plant cells and of the nature of the substances that bind them specifically. Sodium was shown, in certain tissues, to be bound to phosphatidic acid. Specific binding of chloride by certain proteins of higher plants was also reported (Strogonov, 1964). However, whether this really occurs is still uncertain. Specific proteins with a high affinity for sulfate and potassium were isolated from bacteria (Pardee and Prestidge, 1966; Damadian, 1969); but none is known for halophytic angiosperms.

Specific accumulation of ions by intracellular organelles is also controlled by similar active mechanisms. This is of special interest because of the possibility to tie the transport mechanism to a certain metabolic process. Thus, it was assumed that the initial stage in ion uptake involved accumulation in mitochondria or in chloroplasts because the cytochrome system is mostly located on such organelles. Only subsequently were ions released into the cytoplasm or the vacuole.

As mentioned, metabolic energy is directly involved in uptake processes, but is also essential for the ability of the cells to maintain a selective ion metabolism. The sites where metabolic energy is expended to keep selectivity is in dispute. A few possible sites have been suggested: (a) at the membrane, maintaining its selective permeability; (b) at the membrane, operating the metabolic transport mechanism; (c) beyond the membrane, operating the metabolic transport mechanism; and (d) beyond the membrane, building the cell's electric potential.

Only the necessity of metabolic energy for the maintenance of permeability is agreed on among investigators of this field. As to the other possibilities, it is unknown whether either one of them may explain selectivity alone or that all three operate together. When cells are poisoned or die, their selective activity is lost and their ionic content equals that of the bathing solution. Only in Avicennia roots the selective mechanism retains its properties even when treated with carbon monoxide (CO) or 2, 4-dinitrophenol (DNP) (Scholander, 1968).

When plotted against time, ion uptake in most plant species investigated so far, is expressed by a saturation type of curve. Such a curve indicates that the uptake capacity of plants is limited. Limitation may result in three cases: (a) saturation of a limited number of intracellular stationary binding sites, and formation of an exchange equilibrium; (b) saturation of the attachment sites on a dynamic transport pump; and (c) reduction of the uptake rates.

For many years, the second approach was favored by investigators of ion uptake, but recently more emphasis has been placed on the other approaches as well. Based on experiments with barley roots, Pitman (1970) concluded that salt uptake from concentrations above 0.5 mM was linked to release of H$^+$. Although he assumed that the H$^+$ efflux was conducted by an active pump and balanced the efflux of HCO$_3^-$, it still might contribute to an exchange between cellular H$^+$ and absorbed cations. Also sodium uptake by *Atriplex* leaf disks was largely balanced by a loss of potassium from the tissue (Osmond, 1968). It becomes evident that information on ion uptake processes can be explained by both approaches and that information obtained by each of these approaches is true to a certain extent only.

Various investigators reported a dual or multiple uptake isotherm when plants were subjected to a wide range of salt concentrations. This was interpreted as involving two mechanisms that operate in the absorption process for a given ion (Epstein *et al.*, 1963; Epstein, 1966; Lüttge and Laties, 1966, 1967; Rains and Epstein, 1967). The first mechanism (system I) operates at low salt concentrations (below 0.5 mM) and its affinity for potassium, rubidium, and cesium is high. The other mechanism (system II) has a much lower affinity for ions and does not operate except at a high salt concentration, i.e., 1 mM and above. System II absorbs most of the sodium ions.

System I appears to be located in the plasmalemma, but there is a dispute whether system II works in the plasmalemma (Epstein, 1966, 1969; Rains and Epstein, 1967; Welch and Epstein, 1968) or in the tonoplast (Torrii and Laties, 1966; Lüttge and Laties, 1966, 1967; Osmond and Laties, 1968). A statistical evaluation suggests that more than two mechanisms may be present, some localized in the plasmalemma and some in the tonoplast (P. Nissen, personal communication).

Changes in rates of ion uptake may also occur as a result of changes in permeability (Schaedle and Jacobson, 1966, 1967) or changes in organic or amino acid production (Hiatt and Lowe, 1967).

The mechanisms of sodium uptake were investigated by Ganmore-Neumann and Waisel (1972). In barley roots, two parallel mechanisms which participate in the uptake process were postulated; a nonmetabolic mechanism (P) and a temperature-dependent metabolic mechanism (M). The two mechanisms operate behind a diffusion barrier (D) which ions must cross, in order to get into the cell (Fig. 7.4). At low concentrations, diffusion is mostly limiting, whereas at high concentrations, mechanisms P and M dominate the process. In barley roots Rb is preferred by the P mechanism whereas sodium is preferred by the M mechanism.

The uptake of cations by *Atriplex vesicaria* was also believed to take place by

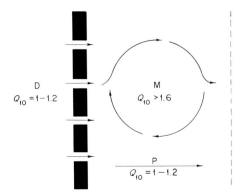

Fig. 7.4. A scheme of ion uptake mechanisms in barley roots. (See text for explanation. After Ganmore-Neumann, 1970.)

two mechanisms, a sodium mechanism with a low affinity for potassium, and a specific potassium mechanism (Black, 1960). A similar interpretation was given for ion uptake by *Avicennia* leaves (Rains and Epstein, 1967).

Various environmental factors affect the processes of ion uptake by plant tissues. Some of these factors influence uptake directly while others do so indirectly. The following factors are of special importance.

TEMPERATURE

Movement of ions from the bulk of the soil into plant roots depends mainly on mass flow or diffusion and is only negligibly affected by temperature. This also is true for certain processes of ion uptake within plant tissues.

Physical processes such as diffusion are practically unaffected by temperature and estimates of Q_{10} values for diffusion across biological membranes do not exceed 1.2 (Giese, 1957; Briggs *et al.*, 1961). Nevertheless, certain diffusion processes require high activation energy and, therefore, are affected by temperature to a greater degree (Danielli, 1943). Chemical processes have a far higher response to temperature increases and reported Q_{10} values for metabolic processes are closer to 2. High Q_{10} values obtained for a transport process can thus mean different things, while for similar uptake rates, low values can only be interpreted as diffusion. Intermediate responses to temperature are obtained when the physical and metabolic processes operate in parallel (Waisel, 1962a; Waisel *et al.*, 1966a, 1970b; Ganmore-Neumann, 1970).

AERATION

A sufficient supply of oxygen is generally a basic requirement for plant respiration and Lundegardh's "ion respiration" was mostly based on the quantitative relationship between anion uptake and respiration (Lundegardh and Burström, 1935). A restricted supply may change the entire energy transfer so essential for membrane and organelle integrity. It alters the production of high molecular organic acids, the amount of indiffusible anions and, consequently, the Donnan equilibrium. It may change rates of assimilation of elements such as phosphorus and sulfate (cf., Robertson, 1958). Thus, under conditions which impair respiration, plants cannot maintain a normal composition of ions.

Regarding uptake processes, aeration affects membrane permeability as well as the metabolic uptake mechanism. Thus, lack of O_2 and a high CO_2 content in the root medium causes roots to be leaky. Various mineral deficiencies may then develop in plants.

Periods of low O_2 pressure may cause a temporary change in metabolic activity, or even permanent damage—depending on plant species, ions involved, and various other environmental factors. Data reported by Leggett and Stolzy (1961) revealed that short periods of anaerobiosis impaired the accumulation of sodium in barley roots as well as its transport to the shoot. However, uptake returned to normal rates and rates of transport to the shoot even increased, upon returning the plants to aerobic conditions. Such conditions may affect swamp and marsh halophytes. Nevertheless, these plants are mostly tolerant to short periods of anaerobiosis, and some species even prefer habitats with such conditions.

LIGHT

Uptake of chloride into green tissues is affected by many environmental factors. According to Arisz and Sol (1956), light affected chloride uptake by *Vallisneria* leaves, but uptake was independent of the synthesis of metabolites. Such a conclusion had been drawn, first, because the effects of light were observed even in the absence of CO_2, and, second, because additions of sugar to the medium accelerated the rate of uptake even under conditions of light saturation. Thus, it seems that among other influences, light affects permeability of cell membranes, and consequently, enables higher rates of chloride permeation into the symplast.

Light had a significant stimulative effect on chloride uptake by leaves of species possessing the C_4 pathway of photosynthesis (e.g. *Atriplex spongiosa*). In species with the C_3 pathway (e.g. *Atriplex hastata*), effects of light on chloride accumulation varied with the physiological conditions of the leaves (Lüttge *et al.*, 1971). Light also stimulated the uptake of phosphate by leaves of the submerged marine angiosperm, *Zostera marina* (McRoy and Barsdate, 1970). In this plant, phosphate was loosely held by the leaves and some of it leaked out.

EFFECTS OF WATER STATUS

Changes in the soil water content affect salt uptake by plants directly as well as indirectly. A low water content in the medium increases the concentration of the soil solution, thus enabling higher rates of uptake of some ions. However, under such conditions, precipitation of less soluble ions occurs, enabling plants to take up more of the soluble ones. An increased concentration of the soil solution also lowers the permeability of plant roots and causes a slower flow of water and solutes into the plant (O'Leary, 1969). Low water potentials within roots affect the absorption and translocation of different ions in various ways. Transport of sodium, bromide, and phosphate to the shoots was lower when the water potential around tomato roots was reduced by -5.4 bars. Such treatment had no effect on rates of accumulation by roots or even increased it (Greenway *et al.*, 1969). Treatments with low water potentials had a delayed effect and caused lower transport of ions even after stresses were discontinued.

According to Richards and Wadleigh (1952), high water deficits caused an increase in nitrogen and a decrease in potassium content of most plants. Effects on other ions varied in different species. In each of these cases, it is still unknown whether such effects were due to changes in ion uptake or to differences in their assimilation.

Only scanty data are available on effects of water stresses on the uptake processes (Kylin, 1960a). Low osmotic potentials of the medium (-1 bar) stimulated accumulation of phosphate by roots, but higher potentials (-4 bars) reduced its uptake (Linser and Herwig, 1963). Under extreme water deficits, when external concentrations induced plasmolysis in root cells, a marked decrease in P uptake was observed (Ingelsten, 1966). Under such conditions, transport of sulfate to the shoot was entirely dependent on mass flow, indicating that damage might have been caused to the endodermis. Besides, under such conditions of high salinity, no direct relationships between rates of transpiration and rates of ion transport were found, either for soluble and fast moving ions (K^+, Na^+) or for slowly moving ones (Ca^{2+}, SO_4^{2-}). Nevertheless, more monovalent than divalent ions were taken up by plants when transpiration was prevented.

Water surplus in the root region also affects ion uptake, but in such a case, the effect is indirect. Inundation reduces the amount of oxygen available to the roots and increases the concentration of CO_2.

IONIC COMPOSITION OF THE MEDIUM

Ions may affect uptake and consequently the metabolic role of one another in different ways. Some ionic effects are positive (*synergistic*), i.e., when one ion is taken up faster or functions more efficiently in the presence of another ion. Synergistic effects are known to be exerted by certain concentrations of di- or polyvalent ions, on the uptake of potassium or cesium. Synergism is not essentially mutual; calcium ions accelerate the uptake of potassium ions but uptake of calcium remains unaffected by potassium.

A more common case of interaction between ions is when one ion exerts a general negative effect (*antagonistic*) on uptake or metabolism of a second ion. For example, such relationships exist between calcium and sodium. A special case of antagonism arises when two ions function on precisely the same site. When this can be quantitatively proved, ions are said to be in *competition*. Competitive relationships are known to exist between H^+ and Na^+ or K^+, between K^+ and Rb^+ or between Ca^{2+} and Sr^{2+} (Epstein and Hagen, 1952; Epstein and Leggett, 1954; Fried and Shapiro, 1961). Contradictory opinions were expressed regarding competition between sodium and potassium. Epstein and Hagen (1952) suggested that those two ions did not compete on the same attachment sites, whereas Waisel (1962a) and Waisel *et al.* (1966a) were of the opinion that sodium and potassium competed on the same sites of a metabolic uptake mechanism.

When ions are taken up from a certain mixture in different proportions, the tissue is said to be *selective for* those absorbed more and *selective against* those taken up less. Several physical properties of ions effect their selectivity by a live cell such as their size, degree of hydration, charge, and mobility. Nevertheless, even when ions have inferior physical characteristics, i.e., when some ions are diffusing slower than others, or when their hydration shell is thicker, they may still be selected for, in high proportions. Such a preferential absorption seems to be the case of sodium accumulation in halophytes.

EFFECTS OF pH

The pH of saline soils is usually close to neutrality (pH 7–8.5), that of saline-sodic soils is close to pH 8.5, whereas sodic soils have a high pH (>9.0). Saline soils of the temperate region which have been leached (solod type soils) tend to be acidic (pH 3.5). Such extreme pH ranges, which many halophytic species are able to withstand, certainly are outstanding.

The pH of the growth medium affects the nutrition of plants in several ways: pH changes the ion species present in the growth medium. Uptake of phosphorus by plants is far higher from slightly acid solutions than from alkaline ones, because under such conditions, phosphorus is available as the monovalent $H_2PO_4^-$ ion and not as di- or trivalent ones (Van den Honert, 1937).

Low pH conditions tend to decrease dissociation of organic acids within plant cells and thus affect the Donnan equilibrium. Under such conditions, less cations, but more chloride, borate, and other mobile anions are transported into plants. Uptake rates at different pH levels may help in estimating the state in which ions are taken up. Molybdenum, for example, shows higher uptake rates at low than at high pH because at low pH competition with OH^- is less severe (Gauch and Dugger, 1954). Molybdenum is, thus, taken up mostly as molybdate.

Concentrations of hydrogen ions in root media have marked effects on the mechanisms involved in accumulation of ions. Usually, interference of pH in the accumulation of cations and anions is least between pH 5 and 7.

Effects of pH on uptake of ions depend very much on temperature and ion concentration (Jacobson *et al.*, 1957; Waisel, 1962a). At low pH and low ion concentrations, uptake of rubidium and sodium was practically independent of temperature changes. Under such conditions the metabolic uptake mechanism seems to be saturated by H^+ and uptake is mostly nonmetabolic (Waisel, 1962a; Waisel *et al.*, 1966a).

Extreme pH values may induce temporary alterations in the nature of the outer surface of the protoplast—the plasmalemma (Wedding and Erickson, 1957; Kavanau, 1966). Such changes may reduce the observed net accumulation of cations in two ways: either by decreasing permeation of cations through the outer surface into the cells, or by allowing an increased leakage of accumulated cations outward. Clear indications as to the type of mechanism which is affected by pH can be drawn from the temperature responses of the mechanism involved. Data reported by Jacobson *et al.* (1957) indicate that at low temperature conditions pH had no effect on the uptake of Br^- or of K^+. The effects became noticeable only at higher temperatures which enable operation of the metabolic mechanisms.

According to Higinbotham *et al.* (1964), lowering the pH of the absorption solution either from pH 7 to 5.6 or from pH 5.5 to 3.6, caused a small but apparently significant increase in the transmembrane electropotential of *Avena* coleoptile cells. As explained by those investigators, such a change in potential could be expected only if hydrogen ions are actively transported, i.e., when they are directly linked to a metabolic temperature-dependent transport mechanism.

Hydrogen ions exerted their effects on uptake only when concentrations of the substrate ions were low. Temperature coefficients for rubidium and sodium accumulation from 0.5 mM solutions at pH 3.5 were very close to 1. However, when the substrate ion concentration was increased, a parallel increase in the response to temperature was observed (Waisel *et al.*, 1966a). When concentrations of rubidium or sodium in the medium reached the range of saturation of the uptake mechanisms of barley roots, neither further effects of pH nor effects of other competitive ions could be seen.

EFFECTS OF COUNTER IONS

Despite various suggestions that cation uptake follows the uptake of anions (cf. Lundegardh and Burström, 1935; Robertson, 1958), it seems that the uptake mechanisms for cations and for anions are different, and to a great extent, independent. For barley roots, it was shown that the two major components of salinity, i.e., chloride and sodium behave differently. Sodium was retained by root cells while a steady-state equilibrium between uptake and leakage was established for chloride (Ganmore-Neumann, 1970).

Accumulation of sodium, potassium, rubidium, or calcium was not correlated with uptake of the counter ion with which they were originally supplied. Similar uptake rates of those cations were observed when the accompanying anions were NO_3^-, Cl^-, or SO_4^{2-}, i.e., anions which exhibit different rates of uptake (Hiatt, 1970; Ganmore-Neumann, 1970).

Correlations between uptake of anions and cations seem, in certain species, inconceivable since the behavior of such ions in response to changes in various environmental factors is completely different. Moreover, due to production of negatively charged macromolecules and due to the fact that such anions are unable to diffuse outward through the membranes, a Donnan equilibrium is established favoring an excess uptake of cations over anions. Donnan phenomena can account for cation accumulation only. Anions enter cells either by exchange for OH^- or by a metabolic transport system. Usually it is believed that uptake of cations is independent of anions in plants with a high capacity for organic acid production. In plants with a low organic acid content, the uptake of cations is dependent on the counter anion. Exchange phenomena can explain such a behavior. Not many halophytes were investigated for that issue. Sodium uptake by *Suaeda monoica* was found to be dependent on the accompanying anion. Higher rates of uptake were obtained when sodium was given to those plants as NaCl rather than Na_2SO_4 (Ovadia, 1969). Sodium content of young *Suaeda monoica* shoots was 12.5% of their dry weight, when plants were grown on a NaCl solution, whereas that of plants grown on an equivalent Na_2SO_4 solution was only 5.3%.

Uptake of chloride by terrestrial halophytes seems, in most cases, to be unaffected by the accompanying cation. However, presence of other anions, such as phosphate or bromide in the medium affected its uptake rates considerably.

Difference in Ion Uptake along Roots

Although roots are organized multicellular structures comprising several differentiated tissues, mechanisms of ion uptake by roots were in most cases calculated on the whole root weight basis. Such an approach raised a number of questions:

1. Do tissues which are located in different places along the root behave uniformly?

2. Do different ions enter roots in the same places, or do they enter on different sites along and across the roots?

Since some of the existing theories depend very much on precise and definite answers to these questions, the problems raised deserve special attention.

It is well known that plant tissues undergo continuous changes as they grow and mature. Such changes involve structural and physiological modifications that eventually determine the capability of each cell along the root to absorb or transport ions (Scott and Martin, 1962; Brown, 1963; Bowen and Rovira, 1967, 1968).

Several investigators tackled the question of ion distribution, and a few patterns of uptake along roots were reported (Steward *et al.*, 1942; Kramer and Wiebe, 1952; Canning and Kramer, 1958; Bowen and Rovira, 1967; Bowen, 1968; Rovira and Bowen, 1968; Kramer, 1969; Weavind and Hodgson, 1971). Practically all data reported were for roots of glycophytes. It was generally accepted that the highest capacity of roots to absorb ions was near their apex. Nevertheless, very little attention was paid to the mechanisms involved. Only during the last 20 years, it was noticed that most of the uptake in the undifferentiated apical segments of corn roots is nonmetabolic and ions are easily exchangeable (Handley *et al.*, 1960, 1963, 1965).

In older segments of corn roots (2nd–8th cm), very clear patterns were shown for sodium uptake. Moreover, evidence was obtained that mechanisms of accumulation differed in various segments along corn roots (Eshel, 1971; Eshel and Waisel, 1972).

Similar data were also obtained in preliminary experiments with secondary unbranched young roots of *Suaeda monoica* Forsk. plants. Apical 8-cm long portions were immersed for short uptake periods in an aerated 1 mM ^{22}NaCl solution at 30°C. Roots were then washed and desorbed. The apical 6 cm of each root were then cut by a guillotine into 2-mm segments, and their sodium content was traced. Results of such an experiment are presented in Fig. 7.5. Similar patterns were observed for most roots. Nevertheless, the absolute values varied to a certain degree between individuals, probably due to different weights per unit length.

A high and sharp peak was found to exist in the 2–6 mm behind the root tip. Uptake of sodium into this portion almost doubled the average uptake of the entire root. Such a peak means a high rate of sodium accumulation into the nonexchangeable fraction of the cells and occurs in the zone of elongation and xylem differentiation. A second peak of sodium uptake was observed in the 35–45 mm portion, following a segment of low uptake capability in the 10–20 mm region.

Attempts to correlate ion uptake with changes in enzyme activity, cell size, protein content, etc. along roots were made by Brown and Broadbent (1950), Robinson and Brown (1952), and Sutcliffe and Sexton (1968). No evidence was

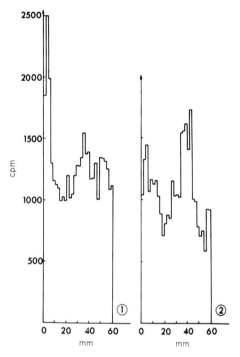

Fig. 7.5. Patterns of sodium distribution along roots of *Suaeda monoica*. (1, fresh roots; 2, aged roots.)

obtained as to any connection between the measured parameters and uptake rates. Neither could differences in uptake capability along *Suaeda* roots be correlated with anatomical characteristics. The zone of cell division is well inside the first 2-mm segment. Vacuolated cells appear already in the second millimeter and a gradual increase in cell length was observed from that point and up. The first differentiated xylem elements appeared already in the fourth millimeter. Thus, it seems more likely that physiological rather than anatomical variations are responsible for the different accumulation patterns.

Roots are also not homogeneous with regard to uptake of chloride. Accumulation of chloride by onion root cells and its transport to the shoot differed along the roots, being most intensive in the distal 3 cm. Nevertheless, it is interesting to note that the region which was most active in water transport was not as efficient for chloride transport, being located mainly in the 6–9 cm section (Hodges and Vaadia, 1964). Thus, osmotic flow of water in the xylem does not seem to be a "one site phenomenon" and probably sums up the contribution of different segments, along the root.

Activity of microorganisms around the plant roots also affects the patterns of ion uptake and metabolism. Experiments reported by Bowen and Rovira (1966) and Rovira and Bowen (1970) revealed that more phosphate was found in inoculated roots. Furthermore, under such conditions, rates of phosphate transported to the shoots were at least four times higher.

Ion Distribution Throughout Plants

TRANSLOCATION OF SODIUM AND CHLORIDE

Plant growth is dependent upon a continuous supply of nutrients. Thus, only a small part of the ions which reach root surfaces are accumulated and utilized *in situ*. Most of these ions are transported to the shoot.

Further transport of ions from the free space, or from root cortex cells into the xylem sap depends very much upon root characteristics as well as on their water and ionic environment (House and Findlay, 1966; Anderson *et al.*, 1970; Anderson and Allen, 1970). Once across the plasmalemma of any of the root cells, ions can move relatively free in the symplast. Movement in the symplast across tissues occurs either inside the cytoplasm or inside the interconnected vacuoles (cf. Helder and Boerma, 1969).

A certain portion of ions which enter a root can move unbound inside the free space across the root cortex until they reach the endodermis. This layer constitutes a barrier to free ion movement, due to the existence in the radial walls of special lignified strips with low permeability to which the cytoplasm is attached (Van Fleet, 1961). Thus, ions cross the endodermis mostly through the cytoplasm and the quality and quantity of most ions which reach the central vascular cylinder is controlled by live cytoplasmic membranes.

Two mechanisms, i.e., a mechanism which accumulates ions in the vacuoles and a mechanism which transports ions into the xylem, evidently compete with each other for ions which cross the root. Although such mechanisms were thoroughly investigated (cf., Hodges and Vaadia, 1964; Lüttge and Laties, 1966) the interrelationships between them and their location remain obscure.

The fact that the endodermis constitutes a barrier to free transport of ions across roots was implied also from data exhibiting the pattern of sulfate distribution. Lüttge and Weigl (1962) used microautoradiographical techniques to show that an apparent accumulation of sulfate occurred outside the endodermis. This was explained as excess accumulation, at the site of the barrier to free SO_4^{2-} movement into the xylem. No evidences of this type are known for halophytes, though endodermal control of ion movement probably plays an important role in such plants. The ecological significance of the Casparian strips in desert plants was stressed by Fahn (1964) and Ginzburg (1964). Those investigators found wide and thick Casparian strips in the endodermis of primary roots of various halophytic species (Fig. 7.6). It was thus suggested that in habitats in which an easy penetration of salts into plant tops may endanger their normal existence, the endodermal barrier appears in a highly developed form. The cortex in primary roots of such plants is comprised of a few cell layers only. However, the ecological significance of a narrow cortical layer is still obscure. (See also Chapter 13.)

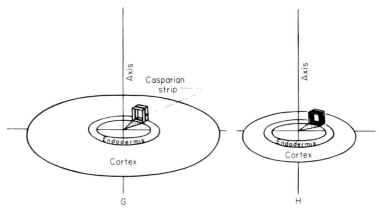

Fig. 7.6. Location and structure of the endodermis in primary roots of a glycophyte, G and a halophyte, H.

The major part of ion transport into the shoot occurs via the xylem. Such transport is believed to be metabolically controlled and located behind a protoplasmic barrier. This is implied from the fact that otherwise the negative electrical potential of the xylem sap and the negative charges of the cells across the roots would have greatly reduced the ion content in the sap (Bowling *et al.,* 1966).

A few hypotheses regarding the location of the transport mechanism were established. Some investigators located it in the epidermis. Sandström (1950) obtained an increase in ion transport by wheat roots from which the epidermis was chemically ruptured. On the other hand, Anderson and Reilly (1968) observed either no changes, or a decrease in ion concentrations of root exudates, following mechanical peeling of the epidermis.

According to Crafts and Broyer (1938), Arisz (1956), and Lüttge and Laties (1967), the active mechanism is located in the cortex. In such a case, energy is spent on accumulation of ions in parenchyma cells of the cortex whereas subsequent transport of ions occurs passively in the symplast.

In another hypothesis, the active transport mechanism resides either in the endodermis or in the xylem parenchyma. Accordingly, movement of ions up to the endodermis can be either in the free or in the osmotic space, respectively (Van Andel, 1953; Steward and Sutcliffe, 1959; Weigel and Lüttge, 1962).

House and Findlay (1966) and Anderson and House (1967) located the active transport mechanism inside the immature xylem elements.

Similar transport mechanisms were suggested for glycophytes and halophytes. However, rates of ion uptake and ion transport into the shoots seem to vary greatly within each group, being rapid in many species of terrestrial halophytes (Black, 1956; Adriani, 1958).

There is some information on ion content of xylem sap of halophytes (Scholander *et al.*, 1962, 1966; Scholander, 1968; Atkinson *et al.*, 1967). Generally such information concerns concentration of chloride in the sap, but there are some data for sodium and potassium content and there is an indication from freezing point determinations that other compounds may be present as well.

In some halophytes, ion accumulation in shoots is high even in plants which grow in media with low content of sodium chloride (Black, 1956; Adriani, 1958). On the other hand, salt concentration in the xylem sap of most halophytes is low. Sap of roots of *Spartina* or of various mangrove species is almost fresh water.

Data obtained for the bleeding sap of *Aeluropus litoralis* and *Atriplex halimus* roots (Table 7.1) indicate that while potassium content in xylem sap of *Atriplex* remained constant, it differed for *Aeluropus*. Addition of NaCl increased the content of sodium in the sap of both species. Nevertheless, sodium concentration in sap did not exceed 6% of the sodium content in the bathing solution.

In many species slow ion transport comprises a protection mechanism against excess salt accumulation and damage to their shoots. For mangroves, and a few coastal halophytes, it was shown that rates of salt entry into the xylem and consequently ion concentrations in the sap are extremely low (Scholander,

Table 7.1

SODIUM AND POTASSIUM CONTENT IN THE BLEEDING SAP OF
Aeluropus litoralis AND *Atriplex halimus* ROOTS[a]

Species	Medium	Ion content (mM)	
		Na	K
Aeluropus litoralis	Hoagland's solution	1.6	5.6
	Hoagland's solution + 100 mM NaCl	5.7	5.4
Atriplex halimus	Hoagland's solution	2.6	12.0
	Hoagland's solution + 100 mM NaCl	5.3	9.1

[a] Twenty-four hour experiments at 30°C, pH 5.5.

1968). This is achieved by an efficient mechanism of ultrafiltration of the solution, which keeps salt content in the xylem sap at only 1 : 10,000 of that of the environment. Similar information is also available for *Spartina* plants. Accumulation of salt in the tops of such plants is evidently very slow (see also Chapter 6).

Chloride ions which penetrate into the xylem sap originate from two sources. At high external salt concentrations, most of the chloride which reaches the xylem sap originates in the external solution. However, when chloride is removed from the growth medium, cortex cells serve as a reservoir and release some of their chloride content into the xylem and the shoot (Hodges and Vaadia, 1964).

When ion activity in the xylem sap equilibrates with that of the external solution, only a fast removal of sap followed by its replenishment with fresh water may enable further ion uptake. Thus, transpiration streams affect ion movement to the root surface as well as its uptake and transport by the root.

Effects of transpiration rates or of external NaCl concentration on sodium transport into the shoot are much greater—at least double—than their effects on the transport of chloride. It seems that the passive component in the total uptake of sodium is greater than that of chloride, and that relatively more chloride is transported via the symplast than via the apoplast.

In this respect it is worth noticing that ion concentration in the xylem sap is not constant. Changes in relative concentrations of ions in the xylem sap may result, among other things, from the fact that different root zones react to environmental water stresses differently (Greenway and Klepper, 1969).

An increase in the external salt concentration results usually in higher rates of salt transport into the shoot. However, at a certain concentration the osmotic gradient becomes so low that transpiration stream is greatly reduced. Consequently, transport of ions by the xylem sap is limited and salt content of the shoots is kept low.

Ions in the xylem sap do not remain untouched, and uptake of ions out of the flowing sap by shoot cells is a known phenomenon. Special emphasis was placed on selective accumulation of sodium out of the sap and the consequent reduction in sodium content of leaves and upper parts of the shoot. This phenomenon was reported for several glycophytic and halophytic species such as beans, stone fruits, and *Prosopis farcta* (Bernstein *et al.,* 1956; Jacoby, 1964, 1965; Eshel, 1966; Waisel *et al.,* 1970b) (Fig. 7.7.). As far as we know today, it is the xylem parenchyma cells which actively and selectively absorb sodium, but not rubidium, out of the flowing sap. The rate of such accumulation may be so high that it keeps the sap which reaches the upper parts of the shoot practically free of sodium (Bernstein *et al.,* 1956). Such a control system may operate in glycophytic and halophytic legumes for long periods and sodium is retained in such tissues until their death. When almond trees age, and their sapwood turns

Fig. 7.7. Dried specimens (a) and autoradiograms (b) of *Prosopis farcta* plants subjected to ^{22}NaCl (1) or to ^{86}RbCl solutions (2); (24-hour experiment; 100 mM salt solutions).

into heartwood, sodium is released from the xylem parenchyma, moves into the leaves, and causes sodium toxicity.

Transport of nutrients in plants is usually polar and despite their great mobility, lateral movement is mostly limited. A *Coleus* plant, which received nutrients only from one side of a split stem, grew considerably better on that side (Caldwell, 1961). The same phenomenon was also observed in the halophyte *Suaeda monoica.* Single organs which were individually supplied with NaCl grew faster and were more succulent than adjacent organs which were kept on a salt-free diet (Ovadia, 1969; H. Wilcox and Y. Waisel, unpublished data). Thus, despite the high mobility of Na^+ and Cl^-, their transport upward or downward along the axis is basically restricted to sections of the stem or to individual organs only. An intensive competition for such ions probably occurs between different organs, tissues, or cells.

Concurrent uptake and transport of ions by roots were measured by Greenway *et al.* (1966) in a salt-sensitive species, *Phaseolus vulgaris,* and in the halophyte, *Atriplex hastata.* Plants with high sodium chloride content exported from older leaves only a small percentage of previously absorbed ions. However, in plants with low chloride content, retranslocation of chloride from older leaves was appreciable. Similar results were obtained by the same authors also for a salt-tolerant nonhalophyte (*Hordeum vulgare*). Thus, it seems that retransportation of ions is not a special characteristic of halophytes.

Ion concentrations in leaves of halophytes such as *Atriplex* are well regulated. However, the limited content of salts seems to be primarily related to continued growth and increased succulence, and not due to salt export. Cessation of growth of halophytes under saline conditions would result in an increase in their ion concentrations unless uptake stopped or export became appreciable (Greenway and Thomas, 1965). In some species, salts are disposed of by means of salt glands, while in others, phloem export occurs (cf. Adriani, 1958; Black, 1956, 1960; Cooil *et al.,* 1965).

Transport of salt downward in the phloem was reported for a few cases, e.g., by Barbier and Chabannes (1951) and Cooil *et al.* (1965) in *Cucurbita pepo* and by Wallace and Hemaiden (1963) in beans. Such export may play an important role in salt evading characteristic of certain species.

Inorganic salts tend to leave maturing or dying leaves and migrate into live tissues (LeClerc and Breazeale, 1908). Such transport occurs mostly upward via the xylem, but also downward in the phloem, and includes sodium, chloride, and sulfate (cf. Cooil *et al.,* 1965). Rates of flow in the xylem depend on the transpiration stream. Transport in the phloem is less dependent on environmental factors. As a matter of fact, phloem transport seems to be the only explanation for downward movement of salts from the top to lower leaves or to the roots.

Cycling of sodium among organs and tissues was found in the shoots of various halophytes investigated. When NaCl labeled with [22]Na was applied to single leaves of *Atriplex halimus* or *Suaeda monoica* plants, sodium could later be traced in various organs both above and below the points of application. No great differences in rates of transport were observed between low salt and high salt plants, although slightly higher rates were still observed in the latter. The general trend of sodium movement was from old leaves into apical or axillary buds. Such movement is similar to what is known of the patterns of movement of various essential nutrients in plants, and certainly is not similar to transport patterns of waste or end plant products. Sodium, seems thus to behave in halophytes as an essential nutrient.

Similar results were obtained in *Suaeda monoica* when plants were flash-labeled by subjecting roots to a [22]NaCl solution (Table 7.2). In this case, [22]Na tended to move from the roots into the shoot. After 14 days, almost half of the sodium which was originally retained by the roots moved into the shoot. Moreover, newly formed branches received [22]Na from older tissues, Exactly as in *Atriplex,* also in *Suaeda* sodium tended to concentrate in buds and young leaves.

Table 7.2

TRANSPORT OF SODIUM INTO VARIOUS ORGANS OF *Suadea monoica* PLANTS[a]

	Time after labeling	
Organ	7th day	21st day
Roots	5.2	2.2
Hypocotyl	3.0	3.0
Stem	5.8	5.0
Leaves	2.1	5.1
Buds	2.02	–
Young leaves (less than 14 days old)	–	13.5
Young buds (less than 14 days old)	–	31.5

[a] Roots subjected for 24 hours to a [22]NaCl solution and later transferred into Hoagland's nutrient solution to which 5 mM NaCl were added. Data denotes μmoles Na/100 mg dry weight of the specific tissue. (After Ovadia, 1969.)

Rates of sodium retransportation in *Atriplex halimus* leaves were rather low (Fig. 7.8). Evidently, most of the sodium applied ([22]Na) was retained at the site of application. A small percentage was distributed among other leaves and buds, but none reached the roots.

PATTERNS OF DISTRIBUTION AND ACCUMULATION OF SALTS AT THE ORGAN AND TISSUE LEVEL

The fact that different substances are accumulated by plants in different quantities was already noted by de Saussure (1804) and Sachs (1875). The latter

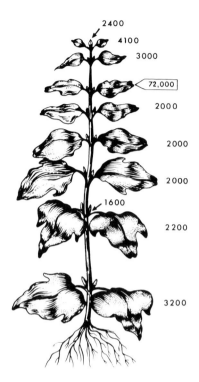

Fig. 7.8. Distribution of ^{22}Na in various leaves of an *Atriplex halimus* plant labeled with ^{22}NaCl on one leaf. Figures denote cpm.

ascribed this phenomenon either to the rates of ion entry into the plants or to differences in their utilization. Even closely related species or closely related ecotypes vary greatly in their capability to select ions. In the sea alga, *Valonia,* the content of potassium exceeds that of sodium by a factor of 40. On the other hand, in a closely related alga, *Halicystis,* more sodium is accumulated than potassium. As both plants are submerged in the same seawater, the differences in contents must be ascribed to plant properties. Differences among terrestrial species in content of sodium, potassium, and rubidium were shown by Collander (1941). The twenty-one species investigated (Fig. 7.9) showed wide variations in their content of sodium, but had very similar contents of potassium or rubidium. Virtual exclusion of sodium was found in *Fagopyrum* while the halophytic species *Plantago* and *Atriplex* had the highest accumulation of sodium among those species investigated.

Ion content and composition differ also greatly between organs. Shoots generally deplete their respective roots of ions. However, the extent and rates of such processes differ among species. For example, the relative concentration of

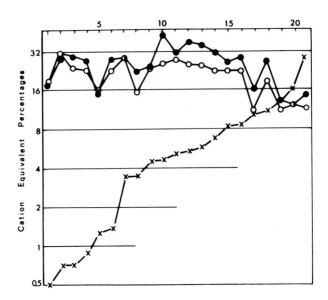

Fig. 7.9. Content of sodium (x – x), potassium (○ – ○), and rubidium (● – ●) in various species of plants. (After Collander, 1941.) Species left to right: *Fagopyrum; Zea; Helianthus; Chenopodium; Salsola; Pisum; Nicotiana; Solanum; Spinacia; Avena; Aster; Papaver; Lactuca; Plantago lanceolata; Melilotus; Vicia; Atriplex litorale; Sinapis; Salicornia; Plantago maritima; Atriplex hortense.*

bromide in *Cucurbita* plants is in the descending order of leaves >stem >root. In *Narcissus*, this order is root >stem >leaves. The order differs also for various ions. Magnesium is selected over calcium by shoot tissues while calcium is selected by root tissues (Lazaroff and Pitman, 1966).

Bean hypocotyls absorb sodium selectively while leaves of the same plants are selective against it. Thus, total analyses of entire plants do not always give a good representation of their overall ion metabolism.

Ions may also be differently balanced in different organs. In the leaves of *Atriplex hastata,* sodium is balanced by organic acids, while in roots of the same plants sodium is balanced by chloride (Black, 1956). Sodium is balanced by chloride in stems of *Suaeda monoica,* but not in the leaves (Table 7.3).

DISTRIBUTION OF SODIUM AND CHLORIDE

Ion content of tops of nonsecreting species may be regulated by a limited intake, by the dilution due to an increase in cell volume (e.g., increased succulence), or by both. The high content of chloride and sodium, even in shoots of halophytes grown on soils low in salts (Adriani, 1958), suggests that halophytes often have a higher capability for salt uptake than nonhalophytes.

Table 7.3

MINERAL CONTENT[a] OF *Suaeda monoica* PLANTS COLLECTED IN THE SALINE OF SEDOM[b]

Organ examined	Minerals					Total % measured	% Ash of dry weight
	Na$^+$	K$^+$	Ca^{2+}	Cl$^-$	SO$_4^{2-}$		
Leaves	15.6	3.4	0.8	6.8	4.4	31.0	38
Stems	5.6	5.3	0.6	5.4	Traces	16.9	19

[a] Values in percentage of dry weight.
[b] Sample taken in the spring.

Halophytic species usually exhibit a clear preference for sodium over potassium and rates of sodium uptake are usually high. However, sites of accumulation vary greatly in different plants. In *Atriplex hastata*, high chloride and sodium concentrations were found in buds and very young leaves (Black, 1956). The quantities of sodium taken up by leaves of *Suaeda monoica* can amount to 20% of their weight. Sodium tends to accumulate in the young leaves and in buds of those plants, while its content in their roots is low. Such patterns of accumulation are in marked contrast with the accumulation of salts in nonhalophytes such as *Hordeum vulgare* and *Phaseolus vulgaris* (Greenway and Thomas, 1965; Greenway *et al.,* 1966). No definite explanations can be offered at present for such phenomena. Tracer experiments revealed that most of the chloride that reached the buds did not come directly from the soil solution, but was previously accumulated in older leaves.

Distribution among plant organs differed for other ions as well. Calcium tended to concentrate in the stems of *Suaeda monoica.* The content of potassium in leaves of this species was low and changed only slightly beyond a minimal limit of NaCl content in the medium. This is also true for leaves of *Atriplex halimus* and shoots of *Arthrocnemum glaucum* where potassium content remained constant even in media containing fiftyfold more sodium than potassium.

Sodium and chloride are taken up by roots of halophytes in nonequivalent quantities. Higher content of sodium was usually reported for many halophytic algae (Collander, 1930; Steward and Martin, 1937), for *Salicornia herbacea* plants (Van Eijk, 1939), and for leaves of *Suaeda monoica* (Waisel and Eshel, 1970). Sodium and chloride differ also in the patterns of their distribution within plant organs. For example, overaccumulation and toxicity of chloride occur at tips and edges of leaves while sodium is accumulated mostly in the mesophyll cells which surround the veins. A higher uptake of sodium over chloride was noticed in halophytes even under such conditions where the chloride content of the growth media is far higher than that of sodium. Under such conditions, glycophytes usually absorb much more chloride than sodium.

The uptake of sodium has a certain autocatalytic characteristic to it (Wybenga, 1957). Beet plants grown on media low in sodium absorbed more sodium when transferred to a sodium-rich medium than plants which were previously grown on a sodium-free medium.

Sodium content in the roots, cotyledons, leaves, and stems of *Suaeda vulgaris* plants increased with age (Binet, 1963). For *Prosopis farcta* plants, such an increase was found to be more or less linear with age. It was practically unaffected by the presence of other ions (Eshel and Waisel, 1965). Sodium content of *Suaeda vulgaris* leaves could reach at maturity, approximately 20% of their dry weight, and is thus very similar to the values reported for *Suaeda fruticosa* in Pakistan and *Suaeda monoica* in Israel (Chaudhri *et al.,* 1964; Ovadia, 1969). A high content of potassium was also observed under certain conditions in *Suaeda* plants, but the ultimate content of potassium in the leaves was far below that of sodium. Moreover, since rates of sodium uptake exceeded those of potassium by approximately a factor of 10, the ratio of sodium/ potassium in such plants dropped as plants aged (Binet, 1963).

The content of sodium and chloride in various organs of *Salicornia* and *Suaeda* plants increased from the root upward. In young seedlings, highest salt content was found in the upper parts of the hypocotyl and in the cotyledons (Hill, 1908).

Uptake of anions was also found to vary among different species. The chloride content of *Atriplex* leaves is positively correlated with its concentration in the outer solutions. While in plants which grew on low-salt media, chloride constituted only about 1% of the plants' dry weight, the content of chloride increased to 3.3 and 5.8% in treatments containing 50 mM and 100 mM NaCl respectively. The affinity of halophytes, and especially succulent halophytes for chloride is higher than for any other anion. Even in sulfate-rich soils of the Namib Desert, Southwest Africa (Walter, 1937) and in the sodic soils of the Neusiedler See region (Zellner. 1926) succulent plants accumulated more chloride than sulfate. On the salty soils of the Namib Desert, succulent and less succulent halophytes grow together. According to Walter (1955), a ratio of chlorides : sulfates in the cell sap of succulent halophytes was about 300 : 1; that of less succulent halophytes varied between 30 : 1 to 11 : 1 only. In a few nonsucculent halophytes, the sulfate proportion even surpassed that of the chlorides. In contrast to chloride, sulfate ions exert a strongly dehydrating action on proteins, and lack of succulence may be attributed to specific effects of sulfates.

Accumulation of chloride in *Citrus* and avocado (*Persea* sp.) leaves occurs mostly in leaf edges and results, in due time, in their early death. Nevertheless, no general relationships between levels of accumulation and the degree of damage caused to a tissue by chloride can be established, and high variations are found between species. Accumulation of chloride varies in different grafts of stocks and scions. Thus, under identical environmental conditions, and identical stocks, the content of chloride in plant tops varies according to the scion (Altman, 1969).

Uptake of chloride by algae is highly affected by temperature. It was found to be higher from KCl than from NaCl solutions (Nielsen, 1963). Chloride transport in various Characeae, seems to be directly related to pH gradients across the plasmalemma, decreasing with increasing pH of the medium. It is believed that chloride uptake is directly linked to efflux of OH^- ions and that the metabolic energy for ion uptake is linked to H^+ or OH^- pumps (Smith, 1970; Pitman, 1970).

Uptake of ions via the free space occurs also in leaves (Kylin, 1960b). Solutes entering leaves with the xylem sap move out of the dead xylem vessels into the live cells. The path of such movement involves the leaf's free space. In certain halophytes, e.g., *Aeluropus litoralis,* the free space of the leaves may constitute a considerable part of the ion transport system.

Leaf parenchyma cells have different characteristics than similar root cells. Permeability to ions of root cortex cells of *Statice gmelinii* is lower than that of the leaf parenchyma cells of the same plants (Ruhland, 1915).

Excess accumulation of chloride reduces uptake of phosphate and nitrate by *Salicornia* and *Suaeda* plants. Such a reduction results in a reddish color in those plants.

Accumulation of salts in tissues of some halophytic species may reach extremely high values. Ash content of leaves of three species of *Atriplex (A. vesicaria; A. nummularia,* and *A. inflata)* was found to be as high as 24–38%. The ash consisted mostly of sodium and potassium. It was balanced in certain tissues by Cl^-, while in others, by organic acids (Beadle *et al.,* 1957). Chaudri *et al.* (1964) calculated that about 2.5 tons/ha of sodium chloride could be accumulated annually by a dense stand of *Suaeda fruticosa* in Pakistan. Even higher values (4.5 tons/ha) were obtained by the present author for *Suaeda monoica* in Israel. Such a considerable accumulation of salts suggested that halophytes may contribute to desalinization processes in some saline habitats. However, for fast growth, such plants have to be irrigated, and as in arid zones irrigation water may add NaCl in equal amounts to those which can be removed with the halophyte plant material, such an assumption seems to be somewhat premature (Waisel and Ovadia, 1972).

UPTAKE AND DISTRIBUTION OF SULFATE

Uptake and accumulation of sulfate excesses were investigated mostly in glycophytes (Kylin, 1953; Pettersson, 1960, 1961; Sutcliffe, 1962). Accumulation of this ion is achieved by metabolically controlled mechanisms. However, being a divalent ion, rates of sulfate uptake are low.

Accumulation of sulfate by excised barley roots was stimulated by calcium ions. Uptake of sulfate was unaffected by monovalent anions but was inhibited by di- and trivalent anions. Sulfate and selenate seem to compete on the same sites of attachment on the metabolic carriers (Leggett and Epstein, 1956). Chloride, phosphate, or nitrate are presumably attached to different sites on the carrier system and are thus unaffected by the presence of sulfate (Epstein and Leggett, 1954). More sulfate is usually taken up by plant tissues at low pH conditions than at high ones.

Most of the sulfate which is absorbed by plant cells is rapidly assimilated into organic compounds. According to Kylin (1953), 15% of the sulfate taken up by wheat roots and 26% of the sulfate taken up by the shoots were assimilated within 1 hour.

Very little information is available on sulfate uptake and transport in halophytes. Nevertheless, from unpublished data obtained in this laboratory, we know that the few halophytic species investigated behave like glycophytes, i.e., their rates of sulfate uptake are low and the amounts of sulfate absorbed are not more than one tenth of those of chloride.

UPTAKE AND DISTRIBUTION OF CALCIUM

Very little is known of calcium being a toxic constituent of salinity. On the contrary, calcium is known to be an essential component of a balanced nutrient medium, and one of the elements which enables plants to survive in hostile sodic environments (Hoagland, 1948; Epstein, 1961). According to Montfort (1926, 1927) and Montfort and Brandrup (1928), many organisms that are unable to live in a pure solution of NaCl survive in the balanced solution of seawater. The same holds also for terrestrial halophytes.

Calcium is absorbed by plants in small quantities and at low rates (cf. Loneragan and Snowball, 1969). In certain plant species, mostly in grasses, calcium is absorbed by a nonmetabolic mechanism and is practically unaffected by changes in temperature. In other species, it is taken up metabolically and increasing the temperature from $20°$ to $30°C$ almost doubles its uptake. In both types of plants, the nature of the uptake mechanisms remains obscure. Long and Levitt (1952), Honda and Robertson (1956), and Grunwald (1966) used fractionation techniques to show that in certain species, calcium was, to a great extent, associated with mitochondria. Grunwald (1966) found at least two fractions containing calcium in potato mitochondria. One fraction was easily extracted and another, presumably a Ca–RNA complex, was only slowly released of its calcium ions and was believed to be a calcium carrier system.

On the whole plant level, calcium was found, in most cases, to move passively in the transpiration stream. However, no correlation was found between calcium accumulation by bean leaves and either rate of transpiration or calcium concentration in the medium. Regulation of calcium content in bean leaves was probably achieved either by retransportation to the roots or by direct leaching. Limitation of calcium uptake is lessened in older leaves and calcium content is positively correlated with leaf age.

Calcium ions have multiple effects on plant behavior and function (Wyn-Jones and Lunt, 1967). Regarding monovalent ion nutrition, calcium is better known for its stimulative effects on the uptake of potassium and inhibitive effects on transport of sodium, lithium, and hydrogen ions. Nevertheless, in some halophytes $CaCl_2$ has different effects. Growth of *Atriplex nummularia* was negatively affected by $CaCl_2$. Similar behavior was observed for *Salicornia herbacea* (van Eijk, 1939; Osmond, 1965; Greenway, 1968).

Effects of calcium on potassium uptake are dependent on the Ca : K ratio in the medium, more than on its ultimate concentration (Waisel, 1962a; Hiatt, 1970). Moreover, since the effects of calcium can be replaced, to a certain extent, by those of strontium, magnesium, barium, and aluminium, it seems to be at least partly a general polyvalent cation effect—and presumably a structural one.

Calcium content in various plants is low and usually does not exceed 0.5% of their dry weight. The content is affected by the presence of NaCl in the medium. Data presented in Fig. 7.10 show that in leaves of *Atriplex halimus* as well as in

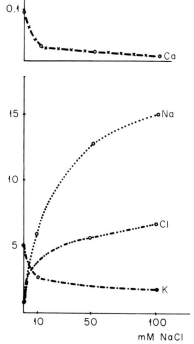

Fig. 7.10. Effects of various concentrations of NaCl on content of potassium, calcium, sodium, and chloride in *Atriplex halimus* leaves. Ordinate, content in μmoles/gm fresh weight.

some other halophytes investigated (e.g., *Suaeda monoica* and *Aeluropus litoralis*), the content of calcium dropped to a basic level following a rise in NaCl concentration of the medium; that level was retained even when salinity of the medium was further increased. Sodium chloride also caused a decrease in the calcium content in *Prosopis farcta* plants. Calcium uptake in this species was not affected by direct competition with sodium and inhibition was not mutual (Eshel, 1966). Similar conclusions were reached also for *Suaeda monoica* (Ovadia, 1969).

In certain hydrohalophytes, uptake of calcium is extremely slow and precipitates of calcium carbonate are found on leaf surfaces of plants which grow in alkaline water. Submerged marine halophytes are no exception in this respect and leaves of *Zostera* sp. and *Cymodocea* sp., etc., are encrusted with such precipitates.

UPTAKE AND DISTRIBUTION OF BORON

Boron is another element which may locally constitute an important component of salinity. Despite being an essential element, high concentrations of boron are not tolerated by most plants. It is one of the elements in which the span between essentiality and toxicity is extremely narrow.

The essentiality of boron for higher plants was established almost 50 years ago (Warington, 1923). Nevertheless, very little is known of boron metabolism in plants. Boron deficiencies have an effect on ribonuclease activity (Shertsnev and Ragomova, 1965), distribution of calcium (Gavalas and Demetriades, 1964), nitrogen metabolism (Sherill, 1966), sugar transport (Gauch and Dugger, 1954), cell division (Yih *et al.,* 1966), incorporation of phosphorus into nucleic acids (Cory and Finch, 1967), etc. It is also essential for pollination, and pollen tubes do not grow unless supplied with boron. Brown and Ambler (1969) observed degeneration of the apical root meristem and formation of a root rosette in boron-deficient soybean plants. Most of the boron is presumably absorbed by plants as borates $(H_2BO_3^-)$, but at least some of it is taken up as the undissociated H_3BO_3. Under high pH conditions, less borate is taken up because of competition with OH^-. Under such conditions, plants tend to show boron deficiency unless its concentration in the soil is high (Gauch and Dugger, 1954).

Depending on plant species or organ, boron concentrations of 1–50 mg/liter are toxic. Only few species, including halophytes such as *Suaeda monoica* and *Atriplex halimus,* can survive under such concentrations of boron. Calcium antagonizes, to a certain degree, such toxic effects (de Broyn, 1966).

Localization of Ions in Subcellular Compartments

The lack of information on the exact location of ions inside cells constitutes a major obstacle in understanding of plant nutrition and plant adaptation to saline environments. Different ions may be separated from one another even in the same tissue, by being located in different cells, but also in different compartments within one cell. In such a case, speaking of interactions between ions, of mutual effects, or of mutual morphogenetic effects, may be erroneous.

It was generally accepted that ions which were absorbed in excess by plant cells, had been accumulated inside their vacuoles. Halophytes seemed to be no exception in this respect, and a considerable accumulation of chloride and sodium in their vacuoles was frequently suggested (Stocker, 1928; Walter and Steiner, 1936; Steiner, 1939; Adriani, 1958; Jennings, 1968a,). However, the techniques used for such determinations were crude, and distinction between ions which were accumulated inside the vacuole and those retained by the cytoplasm remained, for many years, a matter of speculation. It was also suggested that two compartments exist for chloride accumulation; the cytoplasm from which ions may easily leak out and the vacuole which retains ions rather firmly (Hodges and Vaadia, 1964). Support for such assumption was obtained in *Vallisneria* leaves also by treatment with sugar. Addition of sugar affected movement of chloride from the cytoplasm into the vacuole but not inside the cytoplasm. Such an effect was observed even when sugar was added to the absorbing site via another segment of the leaf (Arisz and Sol, 1956). Compartmentation of sodium in roots is still in dispute. Several investigators (Epstein *et al.,* 1963; Epstein, 1966; Rains and Epstein, 1967; Osmond and Laties, 1968; Laties, 1969) suggested the existence of two compartments. Others think of the osmotic space as being one only.

Destination of a nutritive ion in a plant is, in most cases, also the site of its activity. Nevertheless, when ions are accumulated within tissues above the required quantities for nutrition, excesses must be transported into some kind of sink and inactivated there. In some cases, ions are pumped outward by sodium pumps. Such pumps are well known in animal tissues, in sea algae, in freshwater algae (Kylin, 1964, 1966), in salt glands, and were also postulated for pea epicotyl tissues (cf. Higinbotham, 1968).

The chemical state which ions have in the vacuole is also obscure. The ease of extraction of monovalent ions from plant tissues and especially from tissues of halophytes hints that they are free in the vacuole. Nevertheless, in certain cases, it was claimed that potassium or chloride was bound to a protein fraction (Strogonov, 1964). However, even in such cases, it seems that only a small fraction of the entire ion content of the cells was bound. Very little is known regarding the localization of chloride within tissues or within cells. According to Arnold (1955), the cytoplasm of cells of halophytic species is rich in chloride where it is maintained unbound as free ions. An unbound state of chloride was also suggested for *Vallisneria* cells (Arisz, 1964). Autoradiographical data brought Osmond *et al.* (1969) to the conclusion that in bladder cells of the salt hairs of *Atriplex,* chloride is concentrated mostly in the cytoplasm. On the other hand, concentrations of chloride in the mesophyll cells of *Suaeda monoica* were low (Waisel and Eshel, 1970), and in those cells, chloride was similarly distributed between the cytoplasm and vacuole.

Besides chloride, also other ions may remain in a free state inside various compartments within the cell. Walter (1968) claimed that nitrate ions remain unchanged inside the cytoplasm of halophytic plants. Nitrate, in certain plants (e.g., beets) also may substitute for chloride in the nutrient solution (Schmalfuss, 1945) and cause an increase in succulence in halophytes (Van Eijk, 1939). Free ions of chloride and nitrate seem, thus, to affect similar sites, and at least some of them are localized in the cytoplasm.

The exact location of sodium is also uncertain and was investigated in very few species indeed. Jennings (1967, 1968a) tended to believe that sodium is located inside the vacuoles, for two reasons. First, the expanded vacuoles of succulent halophytes seem to attract salts due to their ability to dilute them. Second, on the basis of various considerations connected with ATP metabolism, it was suggested that most of the sodium accumulated in cells of halophytic plants is probably pumped inward into the vacuole. This is presumably done by a metabolically operated pump located at the tonoplast. Such a pump thus serves as a mechanism for diminishing the effects of high intracytoplasmic salt contents (Arnold, 1955; Jennings, 1967). Consequently, the conclusions reached by Jennings were that a relatively low sodium content must be found in the cytoplasm of succulent halophytes.

The techniques used for such analyses of ion localization were usually fractionation techniques. Such techniques are based on the assumption that nothing had changed its location inside the cell during preparation; thus, they suffer from severe drawbacks.

Localization of ions inside cells became much simpler with the development of modern nondestructive microanalytical methods. Nevertheless, such methods have been used for only a limited number of plants (Läuchli and Lüttge, 1968; Waisel et al., 1970a; Rasmussen, 1970). To the best of our knowledge, only one paper on element distribution in halophytes is available.

Using an X-ray microanalyzer, Waisel and Eshel (1970) were able to scan leaf mesophyll cells of *Suaeda monoica* and to follow the distribution of elements such as sodium, chloride, potassium, and phosphorus within those cells (Figs. 7.11 and 7.12). Determination of the exact localization of such highly mobile elements is rather tricky. Still, data obtained for frozen dry leaf cross sections of *Suaeda monoica* revealed that the various mineral constituents differed in their distribution. Such differences depended on the growth conditions to which plants were previously subjected. Sodium was located mainly in the cytoplasm of mesophyll cells, in plants grown on a saline nutrient solution. A similar tendency was found for potassium. Relatively, only a small portion of the sodium and potassium contents of those cells were found inside their vacuoles (Fig. 7.11). The content of phosphorus inside the vacuoles of the

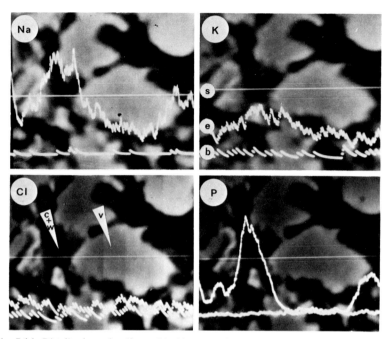

Fig. 7.11. Distribution of sodium, chloride, potassium, and phosphorus in the mesophyll cells of *Suaeda monoica* plants grown on half strength Hoagland's solution + 50 mM NaCl. (After Waisel and Eshel, 1970.) Absorbed electron image. Magnification ×900. Range of determinations (cpm): Na, 3×10^3; K, 10^3; Cl, 10^3; P, 10^4. s, scanning line. e, element distribution; b, background $(2\theta - 1^\circ)$; c+w, cytoplasm and cell wall; v, vacuole.

very same leaf cells was negligible, and practically all of it was concentrated in the cytoplasm. It is interesting to note that the total content of chloride in *Suaeda* mesophyll cells seemed to be low, though the plants were treated with NaCl. Distribution of chloride and sodium did not coincide; while relatively more sodium was concentrated in the cytoplasm, chloride was more or less evenly distributed throughout the cell sections examined.

Distribution of ions inside leaf cells of plants grown on a NaCl-deficient medium was different (Fig. 7.12). Apparently, the signals observed for sodium and chloride were background levels only. In such sodium-deficient cells, potassium was amply found inside the vacuoles, with its content in the vacuole exceeding that of the cytoplasm. Thus, it seems that in sodium-deficient media, potassium may substitute for sodium in cell vacuoles. Phosphorus in such cells was again mostly located in the cytoplasm, though small quantities were also observed in the vacuoles.

It is still difficult to evaluate exactly such data, because the method used (X-ray microanalysis) is not yet refined enough. Nevertheless, it is already an

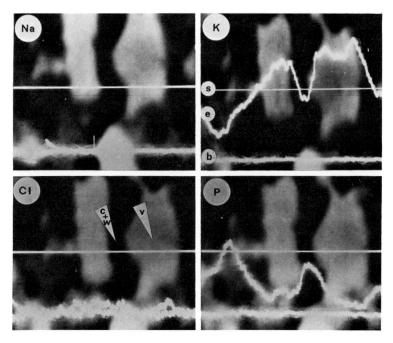

Fig. 7.12. Distribution of sodium, chloride, potassium, and phosphorus in the mesophyll cells of *Suaeda monoica* plants grown on NaCl-deficient solution. (After Waisel and Eshel, 1970.) Absorbed electron image. Magnification ×900. Range of determinations (cpm): Na, 10^3; K, 10^4; Cl, 10^3; P, 10^4. s, scanning line; e, element distribution; b, background $(2\theta - 1°)$; c+w, cytoplasm and cell wall; v, vacuole.

achievement that a gross direct distinction was obtained between the distribution of ions in the cytoplasm and in the vacuole. Conclusions based on such analyses imply that the cytoplasm of the mesophyll cells of *Suaeda monoica* not only absorbed, but also retained, high quantities of sodium. On the other hand, neither sodium nor chloride was excessively accumulated and inactivated in large masses inside vacuoles of cells of *Suaeda monoica* plants grown under saline conditions. Moreover, satisfactory growth of various halophytic plants such as *Suaeda monoica* depends on the presence of high concentrations of sodium ions in the growth medium (Ovadia, 1969). Thus, it seems only reasonable to assume that sodium participates in significant physiological processes and is, therefore, expected to occupy sites which actually affect growth, i.e., sites inside the cytoplasm.

Nutritional Requirements for Sodium and Chloride

It is well established that ten elements are required for plant nutrition in considerable amounts whereas several others are essential in minor quantities.

Many additional elements may also be required by plants in minute quantities, although the question of their necessity is still open. For most plant species sodium is one of the toxic elements when present at high concentrations. In some species, although sodium is of beneficial value under some conditions of stress, it cannot be defined as an essential element. However, for many other plants, and especially for halophytes, sodium is an essential element.

During the last century, scientists have argued over the need of sodium and chloride by higher plants. Only early in this century has the necessity of sodium been commonly accepted, and sodium was incorporated as a micronutrient in many of the nutrient solutions (e.g., Hoagland's A–Z solution). Modern and precise purification techniques are enabling the establishment of the essentiality of sodium for an increasing number of plant species.

Sodium is known to be essential for several wild plants, both cryptogams and phanerogams. It was shown to be an essential element for *Anabaena cylindrica* of the Cyanophyceae (Allen and Arnon, 1955). In *Anabaena*, sodium could not be substituted by lithium, rubidium, or cesium. Brownell and Wood (1957) showed that sodium is essential for growth of *Atriplex vesicaria*, Ovadia (1969) for *Suaeda monoica,* and Keren (1970) for *Nitraria retusa. Halogeton glomeratus* also required sodium for normal growth. The requirements for sodium in some twenty-three species of higher plants were investigated by Brownell (1968). Most of the species of *Atriplex* investigated (*A. nummularia* Lindl; *A. paludosa* R. Br.; *A. quinii* F.v.M.; *A. lindley* Moq.; *A. semibaccata* R. Br., *A. inflata* F.v.M.; *A. leptocarpa* F.v.M.; *A. spongiosa* F.v.M.; *A. semilunalaris* Aellen and *A. hortensis* L.) responded favorably and increased in dry weight when 0.1 mEq/liter Na_2SO_4 was added to their culture solution.

When plants of *Atriplex vesicaria* or *Nitraria retusa* were kept in a sodium-free medium, their leaves became chlorotic, and necrotic patches between veins and on leaf tips appeared (Brownell, 1965; Keren, 1970). *Nitraria* plants became infected, under such conditions, by diseases and most of them died after a short period. Sodium could not be substituted in *Nitraria* by Li, K, or Rb, but could be substituted in *Atriplex* by potassium or magnesium (Greenway, 1968). Incorporation of amino acids into proteins of the nuclei is stimulated by sodium. Potassium increased incorporation into the cytoplasm.

Leaves of *Suaeda monoica* plants grown on a salt-free medium are thin and wrinkled. However, addition of 100 m*M* NaCl to the growth medium accelerated growth of such leaves and straightened them up within a week (Fig. 7.13).

Chloride also was shown to be an essential nutrient for various plants such as tomato and sugar beet (Broyer *et al.,* 1954; Ulrich and Ohki, 1956). Nevertheless, though essential, chloride in tomato may be substituted, to a certain extent by bromide, fluoride, and iodide. Chloride also seems to be essential for *Fagopyrum,* though this species has one of the lowest known contents of sodium (Collander, 1941).

Fig. 7.13. Branches of *Suaeda monoica* plants grown on a salt-free medium (left) and appearance of branches after plants had been exposed to 100 m*M* NaCl for a week (right). (Courtesy of Dr. A. Shomer-Ilan.)

Increases in yield were obtained in sugar beet and in *Aster tripolium* following treatments with 50 m*M* NaCl. However, no increase was induced by minute concentrations of Na_2SO_4 (Brownell, 1968). It seems, thus, that either high concentrations of sodium are required for those plants or that plants respond to NaCl better than to Na_2SO_4. Woolley (1957) also showed that an addition of 1 m*M* NaCl to a highly purified nutrient solution increased growth of tomato plants by about 12% (cf. Harmer and Benne, 1945).

Growth of *Atriplex halimus* was also strongly stimulated by addition of small quantities of NaCl to the growth medium. This was evidently an ionic and not an osmotic effect, because growth was unaffected or even adversely affected by equiosmotic solutions of other salts (Rosenblum and Waisel, 1969). Still, better growth rates were obtained when a mixed NaCl + KCl solution rather than a pure NaCl treatment was given to *A. halimus* plants (Mozafar *et al.,* 1970a, b). Shoots of *Atriplex halimus* grown under mixed salt treatments were about 50% larger than controls, though growth of roots remained practically unchanged. Moreover, plants grown on a NaCl + KCl medium exhibited optimal top growth at osmotic potentials which inhibited growth of the entire plant when composed of either of the participating salts.

The positive effects of sodium on growth of plants can frequently be seen better in species with a high requirement for potassium (Harmer *et al.,* 1953).

Due to the high rates of exploitation, a temporary deficiency of potassium may occur in such species. Under such conditions, substitution of potassium by sodium results in the observed stimulation of growth (Lehr, 1942).

Positive effects of sodium are typically characteristic for wild halophytes, and, therefore, other species which exhibit a similar behavior are also considered as having a halophytic nature.

Essentiality of an element can be expressed in any phase of plant growth. Sodium chloride was shown to be essential for *Salicornia olivieri* at the flowering stage. This species exhibited vegetative growth, but did not flower under salt-free conditions. Only in soils containing 3.4–5% NaCl did *S. olivieri* plants flower abundantly (Halket, 1911). Capacity to inhibit plant reproduction under salt-free conditions makes sodium and chloride essential elements.

Essentiality was not defined quantitatively in any of those reports. It seems that out of the large quantities of NaCl accumulated by the cells only a minor fraction is actually essential.

Effects of Salinity on Metabolism of the Major Nutritive Cations and Anions

An unbalanced nutrient medium leads to an unbalanced uptake of ions and finally to far-reaching changes in metabolism and in ion composition of plants. Disturbances within plant tissues may arise from an excessive salt uptake from the medium. Unbalanced nutrition results from excess accumulation of sodium, calcium, magnesium, potassium, chloride, sulfate, bicarbonate, or nitrate ions and has been well recorded in various species. An excessive accumulation of some of those ions may even cause deficiencies in others. For example, a high content of magnesium in plant tissues may induce calcium or potassium deficiencies. High potassium content may give rise to iron chlorosis or to magnesium deficiency. Low growth rates of *Atriplex nummularia* plants grown under high $CaCl_2$ treatments (Greenway, 1968) may also be indirect and result from deficiencies in other elements. However, halophytes seem to be less susceptible to deficiencies, and to maintain a basic level of nutrients even in the presence of high NaCl concentrations (cf. Heiman and Ratner, 1945).

Content of potassium in *Atriplex halimus* leaves shows, under saline conditions, a typical trend. The content decreased rapidly when the external concentration of NaCl increased from traces up to 10 m*M*, but remained at a constant level even after consequent increases in external NaCl concentrations. Such a level probably constitutes the basic and minimal content of potassium in leaf tissues and remained unchanged in healthy plants, even when the NaCl concentration in the growth medium was increased tenfold. Very similar results were obtained in *A. halimus* leaves also for the content of calcium (Fig. 7.10).

Metabolism of potassium and of calcium in *Suaeda monoica* plants also showed similar trends (Fig. 7.14). Content of potassium and calcium dropped to

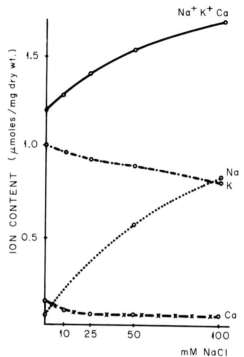

Fig. 7.14. Effects of different concentrations of NaCl on content of sodium, potassium, calcium, and chloride in *Suaeda monoica* plants.

a minimal level in *S. monoica* in the presence of 10 m*M* NaCl and was maintained there even in presence of overwhelming external concentrations of NaCl. On the other hand, when *Suaeda* plants were grown on a KCl solution, potassium substituted for sodium. However, under such conditions, potassium content reached only 9.65% of the plant dry weight, as compared to about 15% sodium in NaCl treated plants.

Rates of sodium uptake by *Suaeda* plants exceeded those of potassium by a factor of 10. Consequently, the ratio Na : K dropped as plants aged (Binet, 1963). However, halophytic species differ greatly in this respect. For example, the ability of *Lepidium crassifolium* to absorb potassium is low and, unlike other halophytes, potassium cannot substitute for sodium in various growth processes (Weissenbock, 1969). Content of potassium in *Prosopis farcta* leaves remained more or less, at a constant level and was unaffected by the presence of NaCl. Even when plants were apparently damaged by Na toxicity, their inner levels of K remained the same (Eshel, 1966).

Atriplex halimus is outstanding with regard to its cation selectivity and ion uptake mechanisms. While in the leaves of other halophytes, e.g., *Avicennia marina,* uptake of sodium and potassium was reported to be intercorrelated (Rains and Epstein, 1967), uptake of those two cations by *Atriplex halimus* seems to be partly independent (Mozafar *et al.,* 1970b).

Sodium chloride caused a decrease in calcium content of the shoots of *Prosopis farcta* but did not interfere with the content of potassium. Calcium content of roots, leaves, and stems of *Prosopis* plants treated with NaCl stayed at a low but steady level. It showed a tendency toward increasing with age. On the contrary, $CaCl_2$ in the medium affected neither the sodium nor the potassium content of most organs of *Prosopis farcta.* The only exception within such organs was found in the hypocotyls where stimulation of potassium content was caused by calcium. Such effect was observed even in the presence of an overwhelmingly high NaCl concentration (Eshel, 1966).

Growth of *Aeluropus litoralis* plants was unaffected by the addition of NaCl even in concentrations of 100 mM NaCl. However, under such conditions, the contents of sodium and chloride in the plant tissues were apparently increased, while the contents of potassium and calcium decreased (Pollak, 1967).

Salt marshes get some of their nitrogenous compounds from external sources. Moreover, under marsh conditions, nitrogen is also fixed *in situ* by nitrogen-fixing blue-green algae. Nevertheless, when grown on saline media, under high light intensity, certain halophytes (*Suaeda* sp., *Salicornia* sp.) take on a purple color. Such color does not show up in plants which received a luxurious supply of nitrogen and phosphorus, and it seems that under such conditions, deficiencies may develop. Thus, despite the fact that salt marshes can be richer in various elements than their surrounding habitats, halophytes may still have there an insufficient supply of certain nutrients.

Specific Problems of Plant Nutrition in Salt Marshes and Saline Springs

Besides the dominating effects of high NaCl concentrations, also other environmental conditions which prevail under salt marsh conditions are hardly endured by most plants. The most important factors include: composition of the soil solution, concentration of specific ions (including pH), temperature, light, water status, partial pressure of oxygen, carbon dioxide, and other factors controlling metabolic activity of the plant.

In many cases, effects are involved with micronutrient nutrition. It was frequently reported that salt marsh plants are highly sensitive to chlorosis. Such plants cannot survive in waterlogged habitats unless more iron becomes available due to the activity of anaerobic iron-reducing bacteria (Bromefield, 1954). Anaerobic conditions affect mostly higher plants.

Unless precipitated as sulfides, iron and manganese are maintained in marshes in their reduced and more soluble divalent form. This might have suggested a higher content of those elements in salt marsh plants. However, analyses made of a number of such plants (*Silene maritima, Limonium binervosum, L. vulgare, L. humile, Aster tripolium, Salicornia perennis,* and *S. stricta*) revealed that although their ash content was higher, the relative content of iron and manganese was far lower than in woodland or in freshwater swamp plants (Gorham and Gorham, 1955). It is possible that a high NaCl content of the medium antagonizes uptake and reduces assimilation of iron and manganese by such plants. However, such an assumption does not seem to be general, and at least one group of halophytes has adapted its nutritional requirement to swamp, marsh, and bog conditions. In experiments conducted in the author's laboratory with *Arthrocnemum glaucum* under well drained and inundated conditions, it was shown that *A. glaucum* grew luxuriously in saline media, even when iron was given at only $^1/_{50}$ of the normal iron content of Hoagland's nutrient solution. Since under such conditions, other plants exhibited iron chlorosis, it seems that the requirements of *Arthrocnemum glaucum* for iron are very low. In fact such plants may withstand salt marsh conditions where most of the iron or other micronutrients was precipitated as sulfides. Similar results seem to hold true also for *Salicornia herbacea.*

On the other hand, *Spartina alterniflora* which may grow also under reducing conditions, has a high requirement for iron. In case of *Spartina,* oxygen diffusing out of plant roots oxidizes some of the precipitated iron sulfide to ferrous sulfate and makes it available again to plants.

Growth of halophytic plants in hot springs with a high sulfuric acid content constitutes another nutritional problem. Such plants not only withstand high temperature, salinity, and acidity, but also the presence of high quantities of aluminium. Phosphate content of such plants is usually lower than normal, while their content of Al_2O_3 is high (von Faber, 1925).

Under acid conditions ($<$pH 4) the high content of Al^{3+} causes toxic effects to most plants. Increases, in pH, that occur in saline soils due to their sodification cause further nutritional problems. In such cases, aluminium may appear in the form of aluminate (AlO_2^-) though its toxicity remains similar.

————————————8

Regulation of Salt Content of Shoots

Salts are continuously transported into plant shoots via the transpiration stream. In plants growing in saline habitats, salt accumulation may eventually reach a hazardous level, and survival of plants may depend on reduction of the salt content of the shoot.

Secretion of ions by special salt glands is the best known mechanism for regulating mineral content of plant shoots. However, this is only one of many means by which salts are released from plant shoots. Salts are released through the cuticle, or in the guttation fluid. They are retransported via the phloem back to the roots and soil, or they are concentrated in special leaf hairs (Stenlid, 1956).

Secretion of Salts

Salt glands were already known and described in the middle of the nineteenth century. Originally, these structures, described as special hydathodes or even as chalk glands, were looked upon as a rarity. However, since Marloth's (1887) description of salt glands of the Tamaricaceae and the publications of Volkens

(1884), Schtscherback (1910), and Ruhland (1915), regarding the Plumbaginaceae, increasing attention was given to the structure, function, and ecological significance of the salt glands.

Transport of substances out of plants was divided by Frey-Wyssling (1935) into the following three categories: *exkrete,* removal outward of metabolic end products; *sekrete,* the transport outward of metabolic products that function outside the organ or plant, and *rekrete,* the transport outward of substances taken up by the plant, but not assimilated. Out of these three categories, only one—"rekrete" can be used for salt secretion. However, as the term "secretion" has been so commonly used for ion transport out of plant tissues, it will also be used here with the same connotation.

Salt secretion is a common phenomenon in various halophytic plant genera. It is found in the following genera of terrestrial halophytes: Convolvulaceae— *Cressa;* Frankeniaceae—*Frankenia;* Gramineae—*Spartina, Chloris, Aeluropus;* Plumbaginaceae—*Statice, Limonium, Plumbago, Armeria;* Primulaceae—*Glaux;* Tamaricaceae—*Tamarix, Reaumuria.* It is also found in the mangrove species: Acanthaceae—*Acanthus;* Avicenniaceae—*Avicennia;* Combretaceae—*Laguncularia;* Myrsinaceae—*Aegiceras;* Plumbaginaceae—*Aegialitis;* Rhizophoraceae— *Ceriops, Bruguiera;* Sonneratiaceae—*Sonneratia.*

However, in spite of the large choice of plants, the process of salt secretion was extensively studied in only a few genera of the Plumbaginaceae (*Aegialitis, Limonium, Statice*), in *Tamarix aphylla,* and in *Avicennia marina.* Investigations were both physiological (Ruhland, 1915; Arisz *et al.,* 1955; Helder, 1956; Waisel, 1961; Scholander *et al.,* 1962; Atkinson *et al.,* 1967; Berry and Thomson, 1967; Pollak and Waisel, 1970), as well as anatomical and ultrastructural (Ruhland, 1915; Walter and Steiner, 1936; Ziegler and Lüttge, 1966, 1967; Atkinson *et al.,* 1967; Thomson and Liu, 1967; Shimony and Fahn, 1968, Thomson *et al.,* 1969). Data obtained from the above investigations posed a few hypotheses as to the mechanisms of salt secretion. Three fundamental features determine the effectiveness of salt glands in removing salt excesses: (a) their structure, location, and abundance; (b) their mechanism; and (c) their physiological and ecological significance.

STRUCTURE OF SALT GLANDS IN VARIOUS PLANT SPECIES

As stated previously, salt glands are common in several families of plants. Most species with salt glands are able to tolerate a high salinity and as a rule their distribution is limited to salines of various types.

In all known cases, salt is secreted from well-defined salt glands located on or sunken into the epidermis. Although the structure of the salt glands varies greatly in different plants, it is strikingly similar in plants of the same genus or even within one family.

The simplest type is the two-celled hairlike gland found in *Spartina* (Fig. 8.1) or in *Aeluropus* (Fig. 8.2). A three-celled gland is found in *Chloris gayana.* In Avicenniaceae, glands are comprised of an indefinite number of cells (usually between five and nine) arranged in a group of four or more cells located on top of one stalk and two–four collecting cells (Fig. 8.3). A similar gland is also found in *Acanthus ilicifolius.* The Tamaricaceae have a more complicated gland composed of eight cells arranged in four pairs (Fig. 8.4).

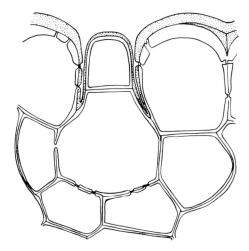

Fig. 8.1. Salt gland of *Spartina townsendii.* (After Skelding and Winterbotham, 1939.)

Multicellular glands comprised of sixteen cells are found in the Plumbaginaceae. In some species of this family (on the stems of *Statice pruinosa*), the glands are located on the top of special structures (Fig. 8.5).

Glands are found on almost every aerial part of those plants, but most abundantly on leaves (Table 8.1). Only in *Statice pruinosa* does the number of glands per unit area on the stem exceed their number on the leaves. Usually, there are fewer glands than stomata, and according to Ruhland (1915), the number of glands in *Statice gmelini* is about one-tenth the number of the stomata.

In all cases investigated, the glands have been initiated at an early stage in the development of the leaf, and their differentiation was completed much earlier than the differentiation of various other leaf tissues. This early development stresses the significance of their special function for the developing organ (Helder, 1956).

Cells of salt glands differ in many respects from the surrounding epidermal or parenchymatous cells. Salt gland cells lack a central vacuole and the number of mitochondria and other organelles is extremely high. This suggests that gland

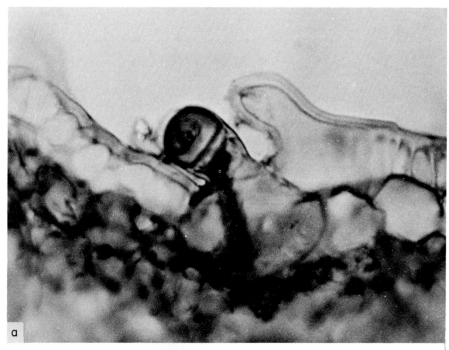

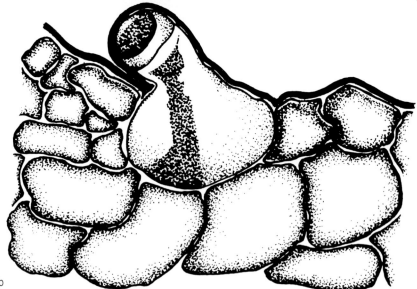

Fig. 8.2. Salt gland of *Aeluropus litoralis.* a, Photograph; b, schematic drawing.

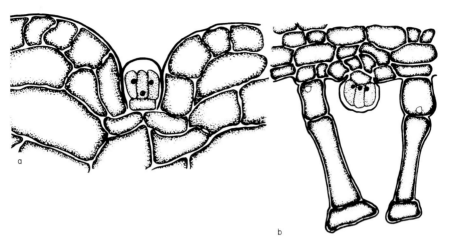

Fig. 8.3. Segments of a leaf cross section of *Avicennia marina,* showing sunken glands in the upper epidermis (a) and elevated glands on the lower one (b).

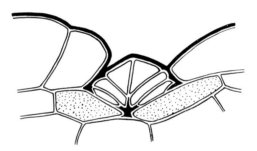

Fig. 8.4. Salt gland of *Tamarix aphylla.* (After Shimony and Fahn, 1968.) Published by permission of the Linnean Society of London, from *J. Linn. Soc. (Bot.),* **60,** 1968.

cells do not act primarily as accumulating organs, but rather as transit cells. Salts are transported outward by specific mechanisms, which consume energy originating on the mitochondrial surface. Nevertheless, concentrations of ions in gland cells may, in certain cases, still be high in comparison to surrounding cells (Ziegler and Lüttge, 1966).

Undoubtedly, there is much similarity between active transport processes operating in salt glands and transport processes in other tissues and organs. Similarity is especially striking between salt secretion, on the one hand, and secretion of nectar in nectaries and secretion of ions into xylem elements of roots, on the other hand. There are also similarities in structural and ultrastructural features.

Structures that control free transport, and are analogous to the Casparian strips of the root endodermis are found in salt-secreting glands. Some of the

Fig. 8.5. Salt gland on a stem of *Statice pruinosa.*

Table 8.1

DISTRIBUTION OF SALT GLANDS IN VARIOUS SPECIES OF PLANTS

Family	Species	Number of glands/cm^2				References
		Leaf epidermis				
		Upper	Mean	Lower	Stem	
Plumbaginaceae	*Statice*[a] *gmelini*	722		644		Ruhland, 1915
	S. bellidifolia	960		830		de Fraine, 1916
	S. sinuata	700		1200		[b]
	S. binervosa	750		1240		de Fraine, 1916
	S. graeca	1100		900		[b]
	S. pruinosa	1900		1300	4000	[b]
	Limonium latifolium	3300		2900		[b]
	Plumbago capensis	2900		2100	300	[b]
	P. europaea	1500				[b]
Gramineae	*Spartina* sp.		1400			Skelding and Winterbotham, 1939
	Aeluropus litoralis		4800			[b]
Tamaricaceae	*Tamarix aphylla*		2300			[b]
	T. jordanis		2200			[b]
	T. tetragyna		1900			[b]
	Reaumuria palaestina		2300			[b]
Avicenniaceae	*Avicennia marina*	2100				[b]

[a] The generic name *Statice,* since used in relevant literature, has not been changed in this book.

[b] Original data.

gland cells are characterized by cutinized and suberized cell walls, especially the walls bordering normal parenchyma cells. In certain places, the cytoplasm is tightly connected to the strips, exactly as in the Casparian strips. Thus, ions can be transported into the gland cells only through the symplast. Although this is true also for water-absorbing hairs (e.g., hairs of the leaf bases of *Tillandsia*) and nectaries (e.g., in the flowers of *Syringa*), it is especially clear in salt glands. Thus, the basal cells of the salt gland of *Avicennia,* the cuticular layers in the glands of *Aegialitis, Limonium,* and *Tamarix,* as well as the stalk cells of the salt-accumulating hairs in *Atriplex,* might be compared to endodermal cells and their Casparian strips. The impermeable walls of the endodermal cells as well as the impermeable cuticular layers in salt glands would serve, in this case, as a barrier to leakage of the secreted salt solution back to the tissue.

The gland of *Limonium* is made up of a complex of sixteen cells. There are four secretory cells arranged in a circle, each occupying one quarter. Each cell is accompanied at its outer side by a smaller adjoining cell. Both, the secretory cells and adjoining cells are surrounded by two cup-shaped cell layers, each comprised of four flat cells, arranged similarly to the secretory cells (Ruhland, 1915; de Fraine, 1916; Ziegler and Lüttge, 1966) (Fig. 8.6a).

The upper sides of the glands and the neighboring epidermal cells are covered by a thick cuticle. The outer walls of the outermost cup-shaped layer of gland cells are also strongly cutinized, thus partially isolating the gland. This process of cutinization is not restricted to these walls only, but also extends, to a slight degree, to the adjacent contiguous walls. In this way, a rigid structure is formed below the leaf epidermis in which the gland is firmly inserted (Ruhland, 1915; Helder, 1956).

The walls of the gland cells are pitted in a few places. One single small pore, approximately 1μ in diameter, is found in the cuticle covering the top of each secretory cell (Fig. 8.6b). The total amount of secretion fluid produced by the gland thus passes through these four small pores. The presence of such pores does not mean that the cytoplasm of the secretory cells is directly exposed to the open atmosphere, because it is still protected by the cell wall. In fact, the chemical composition of the cell wall cellulose near the pores is different from that of the bulk of the cellulose (Helder, 1956).

Large pits are also found in the walls of gland cells adjacent to the assimilating tissue of the leaves. Through these pits, contact is provided with four extraglandular large cells, called collecting cells. Each of the collecting cells usually contacts a few normal mesophyll cells. Presumably, the main function of these cells is to transport ions from the mesophyll into the glands.

Gland cells differ from normal mesophyll cells in shape and arrangement. They possess a characteristic dark granular cytoplasm with a large nucleus and thin membranes. Instead of a central vacuole, a large number of very small vacuoles and organelles may be observed in gland cells. This applies particularly to the four secretory cells. Nuclei are often found near pits, i.e., near places where the path of the transported ions can be expected. Chloroplasts are lacking in both secretory and in collecting cells. In this respect they are similar to epidermal cells, and can easily be distinguished from the bulk of the ground tissue.

Glands of other members of the Plumbaginaceae are extremely similar (De Bary, 1877; Volkens, 1884). It is worthwhile to mention that salt glands located on stems of *Statice pruinosa* are not sunken into the epidermis. Rather those glands are located individually with each one on the top of a special elevated cortical structure (Fig. 8.5).

It is interesting to note that the salt gland cells of *Limonium* do not serve solely as transport organs, but also accumulate large quantities of chloride ions (Lüttge, 1966; Ziegler and Lüttge, 1967). Since these cells lack a central vacuole, the accumulated ions seem to be evenly dispersed over the entire cell volume.

Salt glands of *Spartina* are epidermal structures very close to, or even in contact with the assimilating tissue, but not with the bundles. They are much

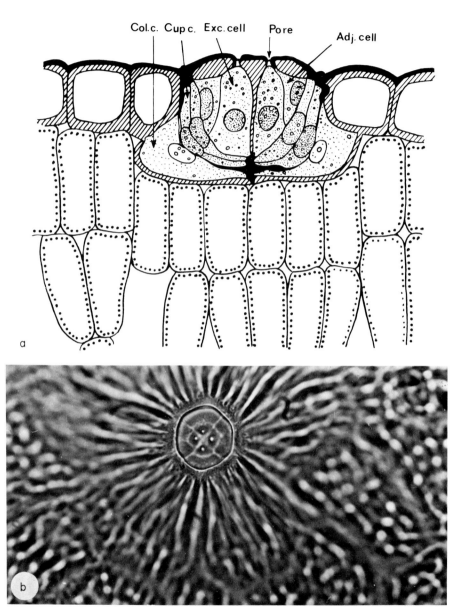

Fig. 8.6. (a) Cross section of a salt gland of *Statice gmelini*. Col. c., collecting cell; Cup. c., cup cell; Exc. cell, excreting cell; Adj. cell, adjacent cell. (After Ruhland, 1915.) (b) Duco print of a salt gland of *Limonium latifolium* showing pores on each of the secreting cells.

simpler in structure than those of *Statice* (Skelding and Winterbotham, 1939; Helder, 1956). Instead of the sixteen-cell complex, there are only two typical cells: a large basal cell and a small cap cell (Fig. 8.1) whose wall is cutinized and not pitted. The basal cell is connected by pits to surrounding cells. On one side, it is connected to cells of assimilating tissue and, on the other side, to the four epidermal cells which form the well-like opening of the gland. The walls of these epidermal cells are perforated by numerous pits. Hence, salt solutions are believed to be forced from the basal gland cell through adjacent epidermal cells outward to the well-like dip in the epidermis. An open well is essential for normal secretion of brine.

The structure of salt glands in *Spartina* is similar to that described for the salt-accumulating hairs of *Atriplex* (Osmond, *et al.,* 1969). In both cases, the basal stalk cell seems to be the functioning one. However, while in *Spartina* the salt is forced outward, it remains trapped in the bladder cells of *Atriplex*.

Glands of *Aeluropus* also belong to the simple two-celled type. They consist of a large basal cell and a small heavily cutinized cap cell. Unlike *Spartina,* the gland is not sunken into the epidermis, but is located erect above it. It resembles more closely a hair than a typical structure of a gland (Fig. 8.2). The role of each of the cells in secretion is unknown.

The salt gland of *Chloris gayana* consists of three cells, a large collecting cell, a stalk cell, and an upper cell, that presumably functions as a secreting cell.

Salt glands of the genus *Tamarix,* and especially of *Tamarix aphylla,* have been frequently described (Brunner, 1909; Decker, 1961; Campbell and Strong, 1964; Thomson and Liu, 1967; Fahn, 1967; Waisel *et al.,* 1966b; Shimony and Fahn, 1968). The gland of *T. aphylla* consists of two highly vacuolated extraglandular collecting cells and six small secretory cells with dense protoplasm (Fig. 8.4). The secretory cells are coated with a sheath of cuticle except for few patches on the walls serving as connections with the collecting cells. Many plasmodesmata pass through these patches and connect the protoplasts of the two groups of cells.

Large numbers of mitochondria and proplastids are observed in all the secretory cells but especially in the upper two pairs (Thomson and Liu, 1967; Shimony and Fahn, 1968). In the secretory cells, as in those of *Limonium* and *Spartina,* many small vacuoles may be seen especially along the periplasm. Such vacuoles contain an electron-dense material, probably a pectic substance and appear dark in an electron microscope image. The same type of material may also be seen on top of the gland. Shimony and Fahn (1968) emphasized that the secretion of such pectic substances was greatly accelerated when NaCl was present in the growth medium, and suggested that pectic substances may serve as carriers for outward transport of salts. According to these investigators, secretory cells of *Tamarix* plants grown under saline conditions contain more ribosomes and polysomes than gland cells of plants grown under nonsaline conditions. Those organelles are presumably involved in the higher rates of secretion observed under saline conditions.

Sites of Salt Pumps

Sites of chloride accumulation within gland cells and the pathway of chloride transport through leaves and salt glands of *Limonium vulgarae* were investigated by Ziegler and Lüttge (1967). According to these investigators, chloride in the leaf was transported both through the symplast as well as through the free space of cell walls. Using autoradiographical and precipitation methods, they showed that chloride was accumulated mostly in the protoplasm of epidermal or mesophyll cells. The concentration of chloride in the cytoplasm of the gland and of parenchyma cells of the mesophyll did not differ greatly. Therefore, the relatively higher activity of ^{36}Cl, seen in microautoradiographs of the gland cells, was linked primarily to a higher content and a denser cytoplasm and was not due to a higher concentration of Cl$^-$ in the cytoplasm (Lüttge, 1969).

A high concentration of chloride was also reported to be inside chloroplasts and nuclei of mesophyll cells. However, since the observations of Ziegler and Lüttge were made only on slowly fixed sections, the exact localization and the path of transport of the ions remain open.

Much of the transport of ions through leaf tissues was found to occur through free space. On the other hand, transport through the symplast has a great advantage, because of its selectivity and independence of ion transport from the rate and direction of the transpiration stream. Plasmodesmata thus seem to play an important role in transport of ions both into gland cells as well as within the gland between the peripheral and central secreting cells.

In spite of the precise structural and ultrastructural descriptions available for most salt glands, very little is known of the connections between their structure and function. Only in very few glands is the exact site of secretion known. Practically in no one of these do we know about the role played by each of the gland cells.

SECRETION MECHANISMS

The need for oxygen, the inhibitory effect of various metabolic inhibitors, and the effects of low temperature on the secretion mechanism suggest that as a whole, this mechanism is an active one, i.e., a mechanism that depends on an undisturbed metabolism for normal operation. It also seems that the secretion mechanism must be located inside the salt gland itself, because brine secretion occurs not only in intact plants, but also in detached leaves and even in leaf disks with nearly isolated glands. Of course, this does not exclude the possible contribution of metabolic energy from other tissues as well to control of secretion.

Concentration Effects

Salt glands are assumed to transport ions against a concentration gradient. This capability was shown by Scholander *et al.* (1962) who found that NaCl concentration in the secreted solution collected from various mangrove species (e.g., *Aegialitis, Aegiceras,* and *Avicennia*) exceeded the NaCl concentration in the root medium. A similar case was also reported for *Aeluropus litoralis* (Pollak and Waisel, 1970), where the concentrations of the secreted fluid (up to 1 M NaCl) were always higher than the concentration of the treatment solutions. On the other hand, according to Ruhland (1915), leaf disks of *Statice gmelini* floated over a salt solution, secreted out of their salt glands a fluid of the same concentration of salts. For *Limonium latifolium,* the concentration of the secreted fluid exceeded that of the external salt solution only at lower values of the latter (Arisz *et al.,* 1955). Also the brine secreted from *Spartina* plants irrigated with a solution, whose concentration equalled that of seawater, gave values of approximately 0.5 M NaCl.

Once brine has been secreted, there is always the risk of measuring subsequent changes in its concentration and osmotic potential. Vapor pressure differences between the secreted droplets and the leaf mesophyll cells, or the environment will tend to equilibrate. However, in short-term experiments, and after certain corrections are made for changes induced in the water content of the secreted droplets, the values obtained seem to give a close estimate of the true concentration (Pollak and Waisel, 1970).

Concentration of the secreted salts is often so high that under dry atmospheric conditions, salt crystallizes, and covers the plants with either salty scales (*Limonium; Spartina*) or salt whiskers (*Aeluropus; Tamarix*) (Fig. 8.7).

This is especially noticeable in plants growing in highly saline habitats or in detached leaves that are exposed to salty solutions. The fine texture and the cylindrical shape of the salt crystals indicate that salts are added in minute quantities at the base of each whisker (Decker, 1961). However, as "growth" of whiskers is not regular, it was assumed that the rate of secretion is not constant. Furthermore, as the whiskers are erect and remain attached, in spite of the added liquids to their base, it is obvious that the secreted brine is viscous and concentrated. The secreted salt whiskers are eventually dissolved and recrystallize into large salt cubes.

The ability to secrete ions against a concentration gradient is not unique to halophytes. Tissue segments of *Nepenthes*, for instance, may transport chloride against a concentration gradient one hundred times higher than their own (Lüttge, 1966). However, the significance of such a transport will evidently be much higher for plants growing in saline media.

The amounts of secreted fluid may be considerable. Under certain conditions, leaf disks of *Limonium latifolium* secreted fluid up to half their weight in a 24-hour period (Arisz *et al.,* 1955). Values of 0.05 mg $\cdot$ mm^{-2} $\cdot$ hr^{-1} of sodium

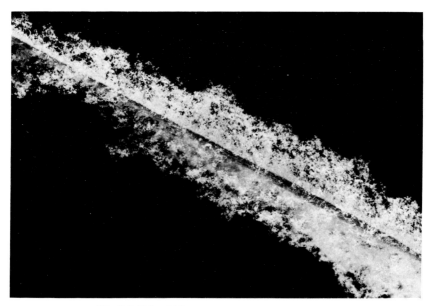

Fig. 8.7. Salt whiskers on a leaf of *Aeluropus litoralis.*

secretion were reported by Ruhland (1915) for *Statice.* Higher values were obtained by Atkinson *et al.* (1967) for chloride secretion from *Aegialitis* leaves. *Aeluropus litoralis* and *Tamarix aphylla* also secrete large amounts of fluids depending on environmental conditions. Very little secretion can be detected in *Aeluropus* under nonsaline conditions. Nevertheless, the addition of even minute quantities of NaCl, e.g., 5 m*M,* into the growth medium resulted in a marked increase in the volume and amount of sodium secreted. When calculated on a gland weight or gland volume basis, rates of sodium secretion by salt glands of *Aeluropus litoralis* are among the highest known in living organisms (Pollak and Waisel, 1970).

Attempts at assessing the influence of the osmotic constituents of the salt solution on the secretion process were reported by Arisz *et al.* (1955). These investigators used *Limonium* leaf disks exposed to solutions with a constant chloride concentration, but with varying osmotic potentials, produced with sucrose. Small increases in osmotic potential mainly affected the amount of liquid secreted and resulted in higher chloride concentrations. Exposure to higher osmotic potentials (-10 to -20 bars) reduced both the volume and concentration of chloride secreted.

Increase in NaCl concentration of the irrigation solution was shown to be positively correlated with both the concentration of secreted brine as well as with salt content of the leaves. Nevertheless, it is apparent that in media with high NaCl contents, secretion of sodium from *Aeluropus* leaves is about three to

Table 8.2

EFFECT OF NaCl CONCENTRATION IN THE ROOT MEDIUM ON SODIUM
SECRETION AND SODIUM CONTENT OF *Aeluropus litoralis* LEAVES[a]

Na$^+$ concentration (M)	Na$^+$ secreted (μmoles/mg dry wt./3 days)	Na$^+$ content in leaves (μmoles/mg dry wt.)	Na$^+$ secreted/ Na$^+$ content
0	0.10	0.17	0.6
0.005	0.25	0.18	1.4
0.05	0.58	0.25	2.3
0.10	1.46	0.43	3.4
0.20	2.55	0.51	5.0
0.30	2.99	0.88	3.4

[a] Light intensity−640 lm/ft^2. Values of contents were taken at the same time, when the secreted Na$^+$ was rinsed from the leaves (3 days after the beginning of the experiment). (After Pollak and Waisel, 1970.)

five times higher than its retention by the leaves (Table 8.2). Evidently, most of the sodium which is absorbed by roots and transported to the shoots, is secreted by the leaves.

Similar results were obtained for chloride secretion from *Limonium* leaf disks exposed to solutions of various NaCl concentrations. High concentrations of NaCl in the medium increased the rate of secretion as well as the concentration of the secreted fluid. However, increase in concentration of the external solution above a certain level reduced chloride content of the leaf as well as the amount of secretion. Despite such inhibition, secretion continued at a low rate and a highly concentrated fluid was produced (Arisz *et al.*, 1955; Helder, 1956).

A concentration rise of the external solution to 0.3 M NaCl still increased the sodium concentration of the secreted fluid of *Aeluropus,* but no increase in amount of solution secreted was observed. Such results point to the possibility that two separate mechanisms are involved in the secretion process: (1) an osmotic mechanism which is involved in secretion of fluid from glands, and (2) an ion transport mechanism which affects salt content and composition of the fluid. Both mechanisms apparently are affected differently by the concentration of the external salt solutions (Pollak and Waisel, 1970). Similar conclusions may be drawn from data presented by Arisz *et al.* (1955) for chloride secretion. They showed that the fluid secreted by *Limonium* leaves, but not its composition, was unaffected either by high or by low salt levels of the leaf. Such assumptions of existence of two different mechanisms were further supported by data of Pollak and Waisel (1970) which showed a higher sodium secretion from plants treated with 0.1 M KCl or 0.05 M CaCl$_2$, than from control plants given the basic nutrient solution alone. It seems that the fluid secretion mechanism must be activated by a critical ionic strength of the outer solution, and that such activation is relatively independent of ionic composition of the activating external solution.

Apart from the positive correlation between concentration of external solution and amount of sodium secreted, a correlation was found between concentration of the medium and sodium content of the leaf. However, since amounts of sodium in the leaves are much lower than the amounts of sodium secreted, it may be assumed that transport of sodium from the roots via the stems and out of the leaves to the secreted solution is very fast. Pulse labeling of roots of *Aeluropus litoralis* plants reveals that the time elapsed between uptake of sodium and its secretion does not exceed 1 hour. The amounts of sodium secreted daily (μmoles $\cdot$ gm^{-1} $\cdot$ day^{-1}) almost equal the sodium content of the leaves. Thus, it is apparent that most of the sodium ions reaching the leaves of *Aeluropus* through the roots, are secreted and rates of accumulation in leaves are low. Two interpretations for this phenomenon are feasible: first, that there are two successive mechanisms operating. Under this hypothesis accumulation of sodium in the leaf cells constitutes an intermediate stage between uptake and secretion, while the period during which sodium is retained in the leaf cells is very short. The second interpretation is based on the existence of two parallel mechanisms: an accumulation mechanism in the leaf mesophyll, and an independent secretory mechanism. The latter usually seems to exhibit a higher efficiency for the uptake of sodium from the xylem sap reaching the leaves.

The second interpretation seems to be favored at this stage.

Specificity of Ions

The brine found on top of the glands is mostly constituted of NaCl. However, many other inorganic substances, in particular, ions such as potassium, magnesium, calcium, nitrate, sulfate, and even phosphate, may be present in the fluid (Schtscherback, 1910; Helder, 1956; Waisel, 1961). Chloride content of the secreted fluid of *Limonium* can account for only about one-half of the total osmotic value. Therefore, a fairly high concentration of other inorganic and organic substances must be present (Ruhland, 1915; Helder, 1956).

Sodium and chloride were located in different compartments of *Limonium vulgare* salt glands. Transport of labeled ions out of leaf disks of *Limonium* plants was characterized by exponential functions with single time constants that differed for each ion (Hill, 1967, 1970a).

Dominance of sodium in the secreted brine may be controlled by several factors located at different sites along the plant axis. The salt glands may comprise just one of those sites.

A preponderant content of sodium in the brine over the content of potassium, calcium, and other ions may occur in a few cases: (a) When the salt glands are selective. Large quantities of sodium are selectively secreted out of an inflowing sap which contains equal concentrations of sodium and potassium. (b)

When the salt glands are not selective and equally secrete both sodium and potassium. However, due to reabsorption of potassium, the remaining brine is comprised mainly of NaCl. (c) When the salt glands are not selective, and secrete both sodium and potassium in accordance with their relative concentration in the inflowing sap. In such a case, if the xylem sap contains higher concentrations of sodium due to a preference in sodium uptake by the root, the secretion of sodium from the glands will dominate.

The first alternative seems to be the case of *Aeluropus litoralis*. In this plant, the xylem sap contains approximately equal concentrations of potassium and sodium ions (12–15 mEq/liter). Furthermore, no considerable reabsorption of potassium seems to take place at the outer surface of the glands, and in KCl treatments potassium is even secreted abundantly.

Secretion of salts was occasionally described (Lüttge, 1962, 1969; Ziegler, 1965) as a possible nonselective process accompanied by a selective reabsorption of utilizable materials. Ion activity gradients for metabolically operated uptake processes are much more favorable this way. However, in view of the continuity of the observed secretion process, and in view of the equal content of proteins, amino acids, potassium, and many other substances in freshly secreted as well as in "aged" droplets, this hypothesis may be dropped, at least as a general hypothesis.

Additional support for the suggestion that selectivity occurs in the glands, is obtained from the behavior of various leaf cells in response to lithium. When lithium is included in the growth medium of *Aeluropus* plants, its rates of uptake and accumulation in leaves are very similar to those of sodium. However, no lithium can be found in the secreted brine. Thus, it seems that the secreting gland itself comprises the main site for ion selectivity of the secretion mechanism (Pollak, personal communication).

The ability to select ions varies from species to species. Scholander *et al.* (1962) found that the fluid secreted from leaves of various mangrove species (e.g., *Aegiceras corniculatum*) contained ions in the same proportions that persist in seawater. Nevertheless, in other species the situation is different. An apparent order of preference of certain ions was found in the secretion fluid of *Aeluropus litoralis*, with sodium being first and potassium and calcium being second and third, respectively. A similar, though somewhat different order was found for *Tamarix aphylla* (Waisel, 1961), where the sequence of preference for ion secretion was found to be in the following order: sodium > calcium > potassium. In *Tamarix aphylla,* the nature of the irrigation solution had a marked effect on salt secretion (Waisel, 1961).

Usually more sodium is secreted than any other cation examined. An optimum type of curve was found in *Tamarix* and in *Aeluropus* plants correlating the amounts of sodium secreted with the concentrations of NaCl in the irrigation solution. Highest secretion was found in *Tamarix* plants irrigated with the 0.2 *M* NaCl solution (Table 8.3). Higher concentrations of the irrigation solutions (e.g., 0.4 *M* NaCl) caused growth reduction, possibly due to osmotic or toxic effects of the NaCl treatments, which may also have affected some essential mechanisms of salt secretion. This may have led to the observed decrease in sodium secretion as well as chloride and potassium secretion.

Table 8.3

EFFECT OF COMPOSITION OF IRRIGATION SOLUTION ON SALT SECRETION BY SAPLINGS OF *Tamarix aphylla*[a]

Irrigation solution (diluted in tap water)	Salt secretion (mg/100 gm fresh plant material per 3 days)			
	Ca	K	Na	Cl
Control (tap water)	43.8	17.6	262.7	54.9
0.1 *M* NaCl	46.1	17.1	461.6	126.2
0.2 *M* NaCl	40.9	16.7	754.3	215.2
0.3 *M* NaCl	42.3	17.1	669.0	128.3
0.4 *M* NaCl	39.6	14.6	401.3	130.7
0.1 *M* KCl	41.8	23.1	155.0	191.0
0.1 *M* MgCl$_2$	42.6	9.0	158.0	106.0
0.1 *M* CaCl$_2$	44.7	15.0	113.7	56.5
0.2 *M* CaCl$_2$	40.0	15.3	77.9	52.5
0.3 *M* CaCl$_2$	42.3	13.0	71.6	55.7
LSD $P = 0.01$	NS[b]	3.68	49.70	24.79
LSD $P = 0.05$	NS[b]	2.68	36.24	18.08

[a] After Waisel, 1961.
[b] No significant difference between treatments.

The presence of potassium, magnesium, and calcium ions in the irrigation solutions depressed the secretion of sodium. In the case of CaCl$_2$ treatments, the decrease in sodium secretion was positively correlated with concentration of CaCl$_2$ in the irrigation solution.

Potassium was secreted in approximately equal amounts by *Tamarix* plants watered with tap water and by plants treated with the three lower concentrations of NaCl. Thus, unless present at toxic concentrations, an excess of sodium in the external solution did not depress the amount of potassium secreted. On the other hand, as mentioned above, an excess of potassium in the irrigation solution caused a decrease in the amounts of sodium secreted (Table 8.3).

Tamarix plants irrigated with CaCl$_2$ and MgCl$_2$ solutions exhibited a decrease in the amount of potassium secreted. This effect was more noticeable in the case of MgCl$_2$ than in any of the other treatments. Besides their toxicity, magnesium ions also functioned as interfering ions in the transport processes. As seen above, toxic concentrations of sodium had a similar effect.

The amounts of calcium secreted by plants such as *Tamarix* sp. did not differ significantly under various treatments and were not affected either by concentration of the external solution or by its composition. Thus, a well organized and efficient mechanism seems to control the secretion of calcium (Waisel, 1961). Different opinions regarding secretion of divalent ions were reported by Berry (1970).

A similar behavior was observed in *Aeluropus litoralis* as well since secretion of sodium was partly (40%) inhibited by potassium (Pollak and Waisel, 1968, 1970). Calcium also inhibited sodium secretion, but only when the ratio Ca : Na was relatively high (Figs. 8.8–8.11). Similar effects of KCl or $CaCl_2$ on the sodium content of the leaves were observed.

Potassium was also secreted by leaves of *Aeluropus litoralis*. However, unlike in *Tamarix*, secretion of potassium in *Aeluropus* was markedly inhibited by sodium.

The total sums of the three cations (sodium + potassium + calcium) contained in the leaves (Fig. 8.11) remained more or less on the same level under various treatments. Thus, when more sodium was secreted out of the leaves, potassium or any other cation was taken up to replace it.

When sodium and potassium were given in equal amounts, much more sodium was secreted. Potassium was preferentially retained by the leaf tissues. Such a

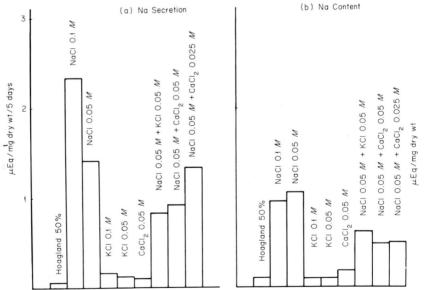

Fig. 8.8. Effects of treatments with various salts and salt combinations on (a) secretion and (b) content of sodium ions in leaves of *Aeluropus litoralis*.

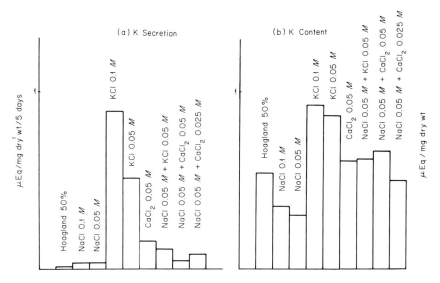

Fig. 8.9. Effects of treatments with various salts and salt combinations on (a) secretion and (b) content of potassium ions in leaves of *Aeluropus litoralis.*

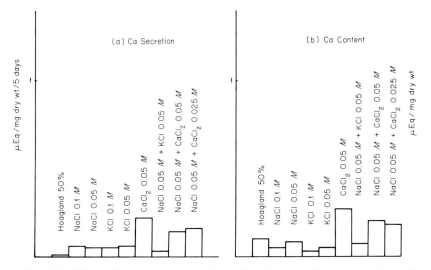

Fig. 8.10. Effects of treatments with various salts and salt combinations on (a) secretion and (b) content of calcium ions in leaves of *Aeluropus litoralis.*

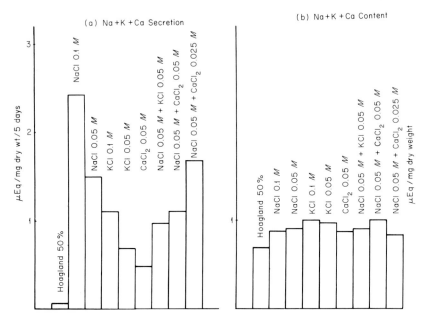

Fig. 8.11. Effects of treatments with various salts and salt combinations on (a) secretion and (b) content of sodium + potassium + calcium in leaves of *Aeluropus litoralis*.

difference, between secretion and accumulation processes, supports the hypothesis that these processes are independent (Pollak and Waisel, 1970). This assumption is also supported by data of Atkinson *et al.* (1967), who observed that chloride can be transported through the leaves directly to the salt glands of *Aegialitis* without an intermediary accumulation in the leaf mesophyll. Further support is also obtained from the fact that while the total sum of ion secretion by *Aeluropus* depends to a great extent upon the concentration and composition of the external solution, the ion content and composition inside the leaves remains more or less constant.

Stimulative effects of calcium on transport of potassium, the so-called "Viets effect" (Viets, 1944) was also observed in the secretion mechanism of *Aeluropus litoralis*. However, this effect was observed only in intact plants and not in detached leaves and was probably exerted on the roots. The effect was abolished in solutions containing sodium.

Potassium content of *Aeluropus* leaves in most treatments was relatively high (about 0.5 mEq/mg dry weight). Potassium content of leaves was higher in plants treated with KCl than in plants treated with NaCl. Potassium content was approximately ten times higher than the content of sodium.

Secretion of calcium by the salt glands and its content in the leaves of *Aeluropus litoralis* were very low.

Apart from the allelopathic relationships between potassium and sodium, such relationships also exist between sodium and calcium. However, the inhibitory effect of calcium on sodium secretion is overwhelmingly higher than the effect of sodium on calcium secretion. Thus, such interrelationships are not competitive.

There is always some calcium bicarbonate present in the secretion fluid, which accounts for the observed incrustations of chalk, causing some investigators to regard the salt glands as chalk glands (Volkens, 1884; Marloth, 1887). Although the concentration of calcium carbonate is mostly low, it is possible that this concept is correct as far as the glands of *Armeria vulgaris* or of certain species of *Tamarix* are concerned (cf. Berry, 1970). However, it is certainly not true for most species of salt-secreting halophytes growing under saline conditions.

Not all ions are secreted by salt glands. Minor elements are strongly retained in leaves, and no secretion of iron could be found in *Tamarix*. As to other elements, only a rare secretion of magnesium was encountered. However, secretion of various minor elements was reported by Berry (1970) in *Tamarix aphylla*.

Secretion of anions was very much affected by both the presence of additional anions, as well as by the composition of cations. A marked increase in the amount of chloride secreted by *Tamarix aphylla* saplings was found in plants treated with NaCl, KCl, and even $MgCl_2$ solutions of increasing concentration (Table 8.3). On the other hand, saplings irrigated with $CaCl_2$ solutions secreted very little chloride even in comparison to control plants.

Selectivity of anions seems to be less definite than that of cations. Secretion of metabolites such as nitrate, phosphate, and sulfate took place whenever *Tamarix* plants were amply supplied with those ions. Secretion of phosphate and sulfate was reported also for *Statice gmelini* (Schtscherback, 1910).

Contents of Organic Material

Fluid secreted by various species of plants contains sugars, amino acids, amines, and proteins. Pollak and Waisel (1970) reported that the protein content in $10\,\mu l$ of the secreted fluid of *Aeluropus* was roughly equivalent to approximately $10\,\mu g$ albumen. Reducing sugars were also detected, but only after hydrolysis. The presence of such organic substances, both with a low and a high molecular weight in secreted fluids, suggested that rupture of membranes and leakage of some protoplasmic contents must occur during the secretion process. Nevertheless, although at first glance it may appear extraordinary that proteins and high molecular hydrocarbons are moving out of salt gland cells, this is not so surprising if it is recognized that such secretion is normal for glands of carnivorous plants. In such glands, enzymes are secreted outward and proteins and amino acids digested from the butt are absorbed inward (Lüttge, 1966).

Effects of Temperature

Influences of temperature on the secretion process were investigated in *Limonium* (Arisz *et al.*, 1955) and *Aeluropus* (Pollak and Waisel, 1970). Increasing the temperature from 5° to 25°C accelerated the rates of salt secretion in *Limonium* by approximately eight times. Nevertheless, although secretion increased, the concentration of chloride in the secreted fluid remained more or less constant under the different temperature conditions. Changes observed in secretion were mostly associated with changes in volume.

Leaves of *Aeluropus* also responded to temperature but this response differed when secreted salts came from internal rather than from external sources. High salt leaves showed little differences between the secretion rates under 10° or 20°C, when placed on distilled water, but exhibited a marked increase in secretion under 30°C. On the other hand, when leaves low in salt content were placed in a salt solution, little secretion was noted at 10°C, but high and similar rates were obtained at 20° and 30°C (Pollak, personal communication).

Effects of Metabolic Inhibitors

Toxic and narcotic substances may either stimulate or inhibit the secretion process (Ruhland, 1915). Experiments conducted by Arisz *et al.* (1955) showed that chloride secretion was slightly stimulated at low concentrations of various inhibitors including potassium cyanide, sodium azide, sodium fluoride, and arsenite, whereas a marked inhibition of the secretion occurred at concentrations above $5 \times 10^{-3} M$. Dinitrophenol had a similar effect. In all cases, the effects of inhibitors were on the volume of brine secreted, and not on its composition or concentration.

Carbonylcyanide 3-chlorophenylhydrazone (CCCP) greatly inhibited the secretion process in *Aegialitis*. When applied to detached leaves through the petioles, CCCP did not affect secretion immediately, the response being noticeable only after 2 hours (Atkinson *et al.*, 1967). No stimulation of secretion was observed in treatments of low CCCP concentrations. Although information is available on the effect of inhibitors, the chain of events which leads to their final inhibition and the exact sites of their effects are still obscure.

Effects of Oxygen Pressure

Effects of oxygen pressure were reported only by Arisz *et al.* (1955) for *Limonium* leaves. Evidently, since the secretion mechanism seems to be an active one, anaerobic conditions had a considerable effect, and inhibited the secretion process altogether.

Effects of Light

Arisz *et al.* (1955) first reported that salt secretion in *Limonium* leaves was light-dependent. Transferring leaf disks from light to the dark, resulted in a sharp decrease in the rate of secretion. However, light dependency of salt secretion by other species occurred only under specific conditions.

While salt secretion by intact plants of *Aeluropus litoralis* showed a positive response to light (Pollak and Waisel, 1970), detached leaves secreted similarly in the light and in the dark. Similar behavior was reported by Atkinson *et al.* (1967) also for salt secretion by *Aegialitis*.

Diurnal variations in salt secretion have been reported for *Aegialitis* and *Aegiceras*, with highest rates at noon (Scholander *et al.*, 1962). Secretion by *Aegialitis* was affected much more by light than was the secretion by *Aegiceras* (Fig. 8.12).

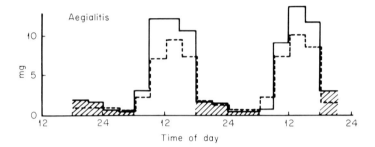

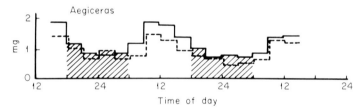

Fig. 8.12. Diurnal variation in NaCl secretion from leaves of two species of mangroves, each represented by determinations in two leaves. Shaded areas–night; unshaded areas–day. (After Scholander *et al.,* 1962.)

Effects of light on salt secretion seem to be dual. A direct effect may couple light-induced electron transport to ion transport. Light may also affect secretion indirectly by changing transpiration rates and the rate of flow of xylem sap. The salt status of the leaves is consequently changed.

THE NATURE OF SECRETION MECHANISMS

Exact sites where secretion takes place, and the role played by each of the cells in the salt glands are still to be investigated. Nevertheless, some information regarding the general nature of the mechanisms involved has been accumulated.

A transport mechanism is considered as being active when ions are transported against an electrochemical potential gradient and nonelectrolytes are transported against a concentration gradient. Transport capacity of such a mechanism is limited and absorption isotherms are thus represented by a saturation curve. Active transport mechanisms depend on an unhampered supply of metabolic energy and thus are sensitive to a low partial pressure of oxygen and to low temperatures (below $5°C$). Very little information regarding such effects on salt secretion mechanisms, is available. Nevertheless, although the exact nature of this mechanism is still obscure, undoubtedly it is an active one and depends on metabolic energy for its operation. Several hypotheses have been formulated to explain its form of operation.

Shachar-Hill and Hill (1970) suggested that buildup of transport capacity by salt glands of *Limonium vulgare* involved RNA and protein synthesis. Glandular transport in *Limonium* leaf disks yielded a sigmoid type of curve. A lag period which was observed, corresponded with the time required for the induction of ion pumps (Hill, 1970b).

Arisz *et al.* (1955) suggested an osmotic mechanism for the secretion process. According to their theory, ions are actively accumulated inside the gland cells. This increase in osmotic potential results in a consequent buildup of pressure. After having reached a peak level inside the gland cell, pressure is released by a periodic outward ejection of droplets. The fact that droplets are ejected under pressure was shown already by Ruhland (1915) who obtained secretion from *Limonium* glands against an external pressure of 22 cm mercury. Droplets were ejected through special pores in the cell walls and cuticle, and caused on their way, rupture of plasma membranes. However, after excess liquid was removed, the membranes healed rapidly, thereby enabling continuation of the process.

This hypothesis was strongly criticized by Sutcliffe (1962) because of inconsistent relationships between the secretion fluid and its concentration. If it was osmotic pressure which caused fluid ejection, then pressure—at the moment of ejection—should always be of the same magnitude, whether rates of secretion are fast or slow. However, in each of the investigated species, the concentrations of the secreted fluid, i.e., their osmotic potentials, were found to differ under various environmental conditions.

Sutcliffe (1962) suggested that the mechanism of salt transport out of salt glands may be an active transport process similar to what is known in practically every mature plant tissue. According to this hypothesis, ions are actively transported outward across the membranes near the pore surface of gland cells. This process is followed by an osmotic withdrawal of water from the tissue to the pore region and formation of brine droplets. According to Sutcliffe's theory, salts and water which make up the brine are transported by different mechanisms. This view was not fully accepted by Atkinson *et al.* (1967), who claimed that a concomitant movement of water and salts occurs in *Aegialitis annulata.*

The most recent hypothesis for explanation of salt secretion was presented by Ziegler and Lüttge (1967) and by Shimony and Fahn (1968). These investigators proposed that secretion of salts out of salt glands is an inverse pynocytosis. According to their hypotheses, presumably secretion is carried out by the discharge at outer cell surfaces of a solution of salts that were accumulated previously in small cytoplasmic vesicles. It may also happen that at a certain stage, the secretion of salts out of salt glands involves a temporary change in structure of membranes and leakage of cytoplasmic constituents. Shimony and Fahn (1968) also raised the possibility that certain secreted materials, presumably pectic substances, may serve as carriers for such an ion transport.

It is too early to accept one of these theories as being the most applicable. The physiological basis is similar for all of them, i.e., an active metabolism-dependent transport mechanism. The only difference in this regard is that Arisz *et al.* (1955) located the site of operation of this mechanism beyond the membrane surfaces while Ziegler and Lüttge (1967) and Shimony and Fahn (1968) located it on the surfaces of the cytoplasmic vesicles and Sutcliffe (1962) on the outer cell surface.

Similar differences were also found with regard to water transport. Ruhland (1915) suggested that, since salt concentration of the secreted fluid was equal to its concentration in other cells of the tissue, there was no need for glands to perform osmotic work. On the contrary, Arisz *et al.* (1955) among others, found that concentration of the secreted fluid was higher than the mean salt concentration in leaves, and accordingly stressed the need for osmotic work. Most investigators assume that water for brine droplets is taken up inside the vacuoles or vesicles. Only Sutcliffe (1962) suggested that osmotic transport of water takes place outside the glands, i.e., after salts have been actively secreted outward. If the vesicles or minute vacuoles are viewed as carriers or as the potential constituents of the active mechanism transporting ions outward, all pseudodiscrepancies in approach may actually be nullified and the hypothesis unified into one scheme. Such a unified scheme is well supported by structural evidence and eventually may explain all the physiological data available at present.

Salt Leaching

Besides salt secretion, other means are known in plants for removal of salts out of the plant shoots. Soluble substances move by the transpiration stream from the interior of leaves and accumulate on their surfaces. Such substances may subsequently be washed from leaves through ectodesmata or cracks in the cuticle, resulting in pseudosecretion (cf. Klepper and Barrs, 1968). Leaching of salts is caused by rain drizzle in humid regions, and in arid regions by heavy fogs. Sodium seems to be the most leachable among the cations and chloride among the anions (Tukey et al., 1958; Tukey and Morgan, 1962).

Leaching of electrolytes depends very much upon water conditions of plants. Leaves that have lost their turgor retain a lower proportion of their ion content, although the wilting is only temporary (Maximov, 1929).

Only scanty information is available on leaching from halophytic species. Salts leached by rains from leaves of *Atriplex* plants, may, under certain conditions, amount to almost half of the salt content in those leaves. Similarly leaching of salts in tropical mangrove species and in other coastal halophytes serves as a major desalinization process.

Guttation

Salt glands are not the only structures through which salts are removed from the plant shoot. Water-secreting structures, e.g., hydathodes, may also serve this purpose. As a matter of fact, the abundance of these structures even caused some of the early investigators to name the secretion of salt glands "guttation fluid." Guttation out of leaves is a widely occurring phenomenon and is most common in young leaves. Guttation fluid is not a direct secretion of the xylem sap, i.e., its nutritive ion content is much lower than that of the bleeding sap obtained from the same or from similar cut organs (Schardakov, 1928). A few examples are given in Table 8.4.

Thus, ions which are transported into the shoot by the xylem sap are, to a great extent, retained on their way. In most plants (e.g., *Triticum, Zea, Papaver* sp.) screening is for nutritive ions. In other species (beans, almonds, *Prosopis farcta*, etc.), sodium is prevented, in this way, from reaching the leaves in large quantities. Furthermore, it seems that even hydathodes function as selective glands which do not permit transport of essential nutrients but allow movement of other ions. This is supported by findings that even certain ferns, e.g., *Polypodium* sp., selectively secrete salts (Frey-Wyssling, 1935). Such secretion results mainly from guttation, and the fluid contains mostly calcium, carbonate, sodium, and silicate ions.

Hydathodes, may thus function in certain species of plants as salt glands and participate in selective removal of specific ions. Secretion of salts in this manner may be quite important in young halophytic plants growing under humid conditions.

Removal of Salt-Saturated Organs

This is another mechanism for removal of large salt quantities from halophytic plants. In certain species, e.g., *Juncus maritimus* or *J. gerardii,* leaves are shed after being loaded with undesirable ions.

Table 8.4

COMPOSITION OF ROOT EXUDATE, LEAF BASE EXUDATE, AND GUTTATION
FLUID IN SOME SPECIES OF PLANTS[a]

(A)

Species	Source of liquid	Ion concentration (mEq/liter)		
		K^+	Ca^{2+}	$H_2PO_4^-$
Papaver somniferum	Root exudate	9.9	17.3	9.3
	Guttation liquid	0.4	7.4	0.5
Brassica oleracea	Leaf base exudate	3.4	49.5	10.0
	Guttation liquid	0.3	6.2	0.6

(B)

Species	Source of liquid	Treatment			
		Hoagland's solution		Hoagland's solution +50 mM NaCl	
		K^+	Na^+	K^+	Na^+
Triticum vulgare	Root bleeding sap	10.4	6.5	17.6	14.1
	Guttation liquid	Nil	6.3	Nil	6.0
Zea mays	Root bleeding sap	17.1	0.52	19.9	2.6
	Guttation liquid	Nil	0.35	Nil	0.39
Aeluropus litoralis	Root bleeding sap	12.2	7.8	12.6	55.7

[a] Part (A) of table after Schardakov, 1928; Part (B) original data.

According to Chapman (1968), succulents such as *Allenrolfea, Halocnemum,* or *Salicornia* discard portions of their fleshy cortex together with their leaves, thus releasing large quantities of salts from plants (Table 8.5) and enabling their survival.

Table 8.5

SODIUM CONTENT[a] OF YOUNG, MIDDLE AGED, AND SHED LEAVES OF *Atriplex halimus* AND *Nitraria retusa,* AND OF YOUNG AND SHED PORTIONS OF THE FLESHY CORTEX OF *Arthrocnemum glaucum* AND *Salicornia herbacea* PLANTS

Species	Leaves			Cortex	
	Young	Middle aged	Shed	Young	Shed
Atriplex halimus	4.3	7.10	7.86		
Arthrocnemum glaucum				3.5	4.2
Nitraria retusa	0.8	1.47	1.52		
Salicornia herbacea				3.2	5.1

[a] Values in % of dry weight.

Salt Retransportation

Substances accumulated by aerial tissues of plants may eventually be transported back to the roots via the phloem and be lost by them to the surrounding medium. This was well established for the transport of sodium in squash (Cooil *et al.*, 1965), but it is probably also true for many halophytic species as well.

A similar desalinization process was found in *Suaeda monoica* and in *Salicornia europaea*. Diurnal fluctuations in sodium and in chloride content of *Salicornia* were described (von Willert, 1969) for plants growing in two saline habitats, i.e., low-flooded and elevated. In both habitats, the salt content of plants was at its peak twice a day, in the late morning and in the late afternoon. In the early afternoon and night hours, the salt content dropped noticeably. As no external leaching of the plants was observed—which could account for such a phenomenon—it is suggested that Na and Cl ions are transported from the shoot toward the roots, and are periodically lost back to the medium of rooting.

A similar case also seems to hold true for nonsecreting mangroves. *Rhizophora*, for instance, is commonly accepted as a "salt excluding" species (Scholander *et al.*, 1962; Atkinson *et al.*, 1967). Nevertheless, the amount of NaCl transported daily into the leaves is still considerable, and eventually accumulation of NaCl might reach toxic levels. However, as sodium and chloride concentrations in mature leaves do not rise above a level of approximately 600 μEq/ml (Atkinson *et al.*, 1967), a mechanism for salt removal must be sought.

Moreover, salt content of *Aegialitis* leaves decreased as leaves aged. Such changes were selective and probably could be applied only to monovalent ions (Atkinson *et al.*, 1967). However, it is still difficult to say whether such a decrease resulted from an increase in efficiency of the secretion mechanisms or from a faster retransportation of sodium, potassium, and chloride via the phloem back to the roots.

Accumulation of Salts in Salt Hairs

Removal of salts from essential and salt-sensitive metabolic sites in leaf mesophyll may also take place by special epidermal hairs. Though such hairs usually function for only a short period, it was reported by various investigators that their role seems to be important.

Salt-accumulating bladders are well known in various semihalophytic and halophytic species of the Chenopodiaceae e.g., *Atriplex*. In these plants, a salt hair is comprised of two cells; a small stalk cell and a large bladder cell (Fig. 8.13) (Black, 1954; Osmond *et al.*, 1969; Mozafar and Goodin, 1970). The structure of the stalk cell shows great similarities to that of various cells of salt glands. It is composed of thick cytoplasm rich in mitochondria, dense endoplasmic reticulum, and numerous small vesicles. However, unlike salt glands, the salt hairs contain chloroplasts. Stalk cells are connected to adjacent bladder or mesophyll cells by numerous plasmodesmata.

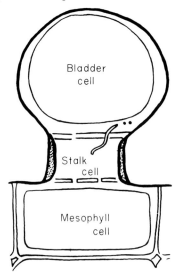

Fig. 8.13. Schematic drawing of a salt hair in *Atriplex*. (After Osmond *et al.*, 1969.)

Salt content and concentration of sodium and chloride in bladders are higher than that of mesophyll cells, and certainly of the external solution (Fig. 8.14). Sodium content usually exceeded that of chloride. The bladders also exhibited a higher electronegativity with respect to the rest of the mesophyll cells and to the external solution (Osmond *et al.*, 1969). Hence the direction of salt movement in such plants is from the medium through the mesophyll into the bladder. Since this process is against a concentration gradient, it probably requires metabolic energy.

Light has an obvious effect on mesophyll cells and on bladders of *Atriplex nummularia* and *Atriplex spongiosa*, i.e., a short-period fast effect on the electropotential difference, and a durable stimulative effect on accumulation and content of chloride. Those effects are of a different nature for various leaf tissues, being more or less linear for mesophyll cells but curvilinear for salt hairs (Osmond *et al.*, 1969). The nature of the light-stimulated transport mechanism of chloride into bladders of *Atriplex spongiosa* was investigated by Lüttge and Osmond (1970). Active uptake of chloride was found to be dependent on noncyclic photosynthetic electron transport. However, as chloroplasts of salt hairs are not very active, chloride uptake by bladders depends for its energy supply on photosynthesis of the leaf mesophyll cells.

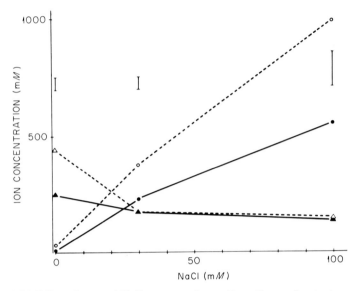

Fig. 8.14. Effect of external NaCl concentration on the sodium and potassium content in the bladders and mesophyll cells of *Atriplex halimus*. Open circles, sodium in bladders; filled circles, sodium in leaf; open triangles, potassium in bladders; filled triangles, potassium in leaf; vertical bars, least significant difference.

Most of the chloride accumulated in salt hairs of *Atriplex* seems to be concentrated in the stalk cells or in the peripheral plasma of bladder cells (Osmond *et al.,* 1969). Vacuoles have a relatively low salt content.

Differences in responses to increasing external NaCl concentrations were observed between annual and perennial species of *Atriplex*. In two annual species (*A. spongiosa* and *A. inflata*), chloride concentrations in leaves were positively correlated with concentrations of the external solutions. Such an accumulation resulted in a conspicuous succulence. The chloride content in the leaves of the perennial species (*A. vesicaria, A. nummularia* and *A. halimus*) reached saturation rapidly and did not increase above a certain level with an increase in concentration of the external solution (Osmond *et al.,* 1969; Rosenblum and Waisel, 1969; Mozafar and Goodin, 1970). Thus, it seems that perennial species have better adaptive mechanisms than annual ones, at least with respect to accumulation of chloride.

Ecological Significance of Salt Removal

Active secretion of toxic ions as well as their passive loss out of live or dead tissues apparently constitute a major contribution for plant adaptation to saline environments. For example, salt secretion is an efficient mechanism which prevents accumulation of large quantities of sodium and chloride inside tissues, while having only little effect on the balance of potassium, calcium, nitrate, and phosphate. It is an adaptive characteristic of nonsucculent halophytes growing in saline habitats. Succulents release accumulated salts by shedding leaves or other fleshy tissues.

Secretion of salts out of glands of various salt-secreting species differs both quantitatively and qualitatively. When growing on nonsaline sand dunes, plants such as *Tamarix aphylla* and *Limonium latifolium* secrete mainly $CaCO_3$, which is an important constituent of such sands. Nevertheless, under saline conditions, those plants were found to secrete large quantities of sodium as well as of chloride. Thus, such mechanisms enable plants to keep a low concentration of those ions inside their tissues.

In other plants, e.g., *Aeluropus,* the secretion mechanism seems to exhibit a higher preference toward NaCl. When NaCl is present in the medium, *Aeluropus* plants secrete droplets containing high concentrations of both sodium and chloride ions. The efficiency of the secretion mechanism was demonstrated in *Limonium* plants. In leaves of this species, chloride content decreased even when leaf disks were placed in diluted salt solutions and despite a concurrent chloride uptake. Values of $1 \text{ ml} \cdot \text{mm}^{-2} \cdot \text{hr}^{-1}$ of liquid secretion and 0.05 mg NaCl $\text{mm}^{-2} \cdot \text{hr}^{-1}$ were reported for *Limonium* (Ruhland, 1915). Even higher values of chloride secretion by mangroves were reported by Atkinson *et al.* (1967). Because of such a high efficiency, chloride concentration of the secreting leaves evidently remains only slightly higher than its concentration in the external medium. The same is true also for sodium. In all examined cases, the amount of sodium secreted from leaves of *Aeluropus litoralis* during 3- or 5-day periods was higher than the sodium content of the very same leaves.

Despite these findings, the secretion mechanisms seem to be limited in their capacity. For *Tamarix aphylla* it was noted that secretion was positively correlated with concentration of external medium. However, when NaCl concentration reached $0.4 M,$ the secretion of both sodium and chloride was reduced. Apparently, in this range of concentrations, flow of xylem sap from the roots to the shoots may be reduced by the high osmotic potential of the root environment and, consequently, NaCl supply to the leaf salt glands could be lowered. It is also possible that under such conditions, the secretion mechanism *per se* is affected. Thus, in spite of the high salt influx under highly saline conditions, only little can be secreted, and such conditions constitute the upper limit where the secretion mechanism is of ecological significance.

In spite of the high capacity for salt removal via the salt glands, most salt-secreting species possess a second safety mechanism in their roots for the screening of toxic ions and for slowing down their penetration inward. A dual selective mechanism, i.e., in the root and in the shoot, was reported for the transport of ions into the guttation fluid, as well as into the secreted fluid of the salt-secreting species *Aeluropus litoralis* (G. Pollak, personal communication).

Accumulation of chloride in bladders seems also to be considerable and high accumulation at high external NaCl concentrations was reported. However, in spite of the ecological significance that one may try to attach to such an accumulation, the total mass of the bladders seems to be too small to remove appreciable amounts of salts from the mesophyll of *A triplex* leaves throughout the plant's life-span.

In spite of the choice of mechanisms found in different plants for regulation of salts in shoots, it seems that each of the mechanisms is limited in its capacity. Evidently, the results are a limitation of plant distribution in saline habitats.

-------------------9

Dispersal and Propagation

Dissemination of Fruits and Seeds

Seed dispersal is one of the factors determining the existence of plants in specific locations, and constitutes a critical phase in the plant life cycle. According to Stebbins (1950), this is one of the phases during which natural selection exerts most of its effects. Several modes of dissemination are found among the various groups of halophytes.

Propagules and seeds of most coastal halophytes are well adapted for dissemination by water, accounting for the circumglobal distribution of certain genera or groups of species. As far back as 1695, it was noted that seeds of foreign plants reached the shores of Scotland and Ireland (Sloane, 1695). There is no doubt that ocean currents aid considerably in such distribution (Uphof, 1941). The high content of vital propagules in sea drifts all over the world serves as supporting evidence (Schimper, 1891).

Certain tissues of most hydrohalophytes contain large intercellular air spaces which enable them to float for considerable periods of time. *Suriana maritima, Gyrocarpus jacquinii,* and *Entada scandens* have air-filled lacunae usually between the cotyledons. *Canavalia obtusifolia* has spongelike intercellular spaces in the cotyledons, whereas *Pancratium maritimum* has an air-containing tissue in the seed coat (Fig. 9.1). In various species of *Salicornia,* seeds are buoyed by air bubbles adhering to the seed coat hairs: propagules of *Thespesia populnea* show hollow parts between the fruit and the seeds; *Scaevola lobellia* has aerenchymatous tissue in the endocarp, and *Cocos nucifera* has such tissue in the

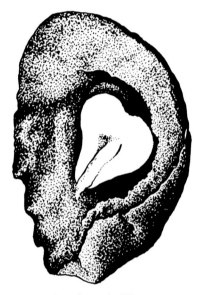

Fig. 9.1. A cross section of a seed of *Pancratium maritimum.*

mesocarp (Warburg, 1898; Schimper, 1891). Fruits of *Halimione portulacoides, Cakile maritima, Crambe maritima,* and of *Suaeda maritima* can float for a long time, because they are buoyed by air-containing tissues. Dispersal of *Salicornia pusilla* occurs while the seeds are immersed in the fleshy cortex of shed stem segments (Fig. 9.2.). These segments are able to float in seawater for periods of up to 3 months, at the end of which the seeds germinate inside the disseminule (Dalby, 1963). Sea currents serve as a disseminating agent of this species, and the distribution of this plant is limited to, or near, the high tidal mark.

Praeger (1913) compiled a detailed list of seed buoyancy capacity of various species. Seeds of approximately 11% of the total 786 species investigated remained buoyant beyond 1 month—most of them marsh or coastal species. Seeds of such halophytes are dispersed in abundance, and their numbers do not seem to constitute a limiting factor in their dispersion. According to Boorman (1968), large numbers of seeds of different species were found along a 15 cm belt transect of the drift-line material in various areas of England (Table 9.1).

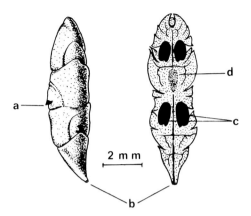

Fig. 9.2. The disseminule of *Salicornia pusilla*. a, Orifice to perianth; b, points of abscission from parental plant; c, pair of seeds in plane of section; d, position of pair of seeds in plane at right angles to section. (After Dalby, 1963.)

Dispersion may be carried out by more than one agent. Long-distance dispersal of water plants depends, in part, on the survival ability and on the maximal period of retention of seeds in the droppings of water birds.

Ridley (1930) showed that the majority of indigenous plants on oceanic islands are dispersed by sea, wind, birds, and bats. Plants dispersed by birds usually have small, fleshy or adhesive fruits and seeds. Although this depends very much on the plant and bird species involved, generally long-distance dissemination can be attributed to internal transport by birds (De Vlaming and Proctor, 1968). Sea dispersion of *Cycas rumphii, Calophyllum inophyllum,* and of various species of *Pandanus* occurs over long distances and along the coast, whereas dissemination by bats takes place over shorter distances and in an inland direction (Van der Pijl, 1969). Dispersal of *Aster tripolium* and of various species of *Statice* (Fig. 9.3) also involves two agents, but in this case, sea and wind.

Fig. 9.3. The disseminule of *Statice pruinosa.*

Table 9.1

THE NUMBER OF SEEDS IN 15 CM BELT TRANSECT OF DRIFT-LINE MATERIAL
IN VARIOUS AREAS[a]

Species	Area[b, c]							Total[d]	Percentage[d]
	1	2	3	4	5	6	7		
Aster tripolium	–	6	12	–	4	26	60	108	15.6
Halimione portulacoides	3	12	8	–	6	7	6	42	6.1
Limonium vulgare	2	6	12	–	8	2	–	30	4.3
Puccinellia maritima	1	–	2	–	16	7	4	30	4.3
Salicornia spp.					4	1		5	0.7
Suaeda maritima					6	2	18	26	3.8
Triglochin maritima	3	–	2	1	102	–	–	108	15.9
Agropyron pungens	–	8	–	15	20	2	–	45	6.4
Atriplex hastata	–	–	–	1	–	–	–	1	0.1
Frankenia laevis	–	–	2	–	–	–	–	2	0.3
Limonium binervosum	–	4	12	–	–	–	–	16	2.3
Plantago coronopus	–	–	–	1	–	–	–	1	0.1
Suaeda fruticosa	2	4	32	9	216	5	–	268	38.8
Tripleurospermum maritimum	–	–	–	–	–	4	–	4	0.6
Geranium pusillum	–	–	–	–	–	–	2	2	0.3
Euphorbia peplus	1	–	–	–	–	–	–	1	0.1
Trifolium hybridum	–	–	–	–	2	–	–	2	0.3
All species	12	40	82	27	384	56	90	691	
Percentage[e]	2	5	11	4	57	8	13		100

[a] After Boorman, 1968.

[b] No. 1, area: Upper Plover, aspect facing: SE; No. 2, area: Hut NE, aspect facing: SW; No. 3, area: Hut SE, aspect facing: SW; No. 4, area: Hut W, aspect facing: NE; No. 5, area: Upper Missel, aspect facing: SW; No. 6, area: Missel Long Hills, aspect facing: SE; No. 7, area: Missel Beach Point, aspect facing: E.

[c] Numbers of seeds of each species in each area.

[d] Total number of seeds of each species for all areas and species contribution expressed as a percentage.

[e] Total number of seeds for each area and area contribution expressed as a percentage.

Muir (1937) reported that viable seeds of *Heritiera* and *Thespesia populnea* were found on the South African coast near Riverdale. The closest source from which the seeds may have been washed is central Mozambique, 1000 miles away. There is no doubt that these seeds were immersed for a long time in seawater before being brought to their final destination. Seeds of *Hibiscus tiliaceus* were still able to float after 120 days in a 3.5% NaCl solution, and those of *Suriana maritima* and *Canavalia* remained floating after 143 and 70 days, respectively (Uphof, 1941). Experiments conducted in the author's laboratory revealed that seeds of *Pancratium maritimum* floated for a longer time on 1 M NaCl solutions than on tap water (100 and 25 days, respectively). It is also known that propagules of *Bruguiera* may spend over 5 weeks in seawater, without losing their viability.

Seeds of various coastal species of *Gossypium* have hard, impermeable seed coats, enabling them to be immersed for a long time and to disperse without damage in seawater.

No generalizations can be made with regard to dispersal of marine halophytic angiosperms. Beside so many plants with buoyant structures on their propagules (Love, 1963), the seeds of other species of marine angiosperms, e.g., *Halophila, Thalassia,* and *Cymodocea* fall in the immediate vicinity of the parent plant, or even remain attached to it (antitelechory of Zohary, 1937). As a result it is unlikely that long-distance dispersal of seeds, seedlings, or vegetative propagules of submerged marine angiosperms can occur (Sculthorpe, 1967).

As in many other groups of plants, the means of dispersal of terrestrial halophytes vary greatly. Some have propagules suitable for long-distance dissemination, and reach distances of hundreds of kilometers, e.g., the propagules of *Tamarix, Phragmites,* and *Senecio.* Plants of these genera have minute seeds and in addition are equipped with an efficient pappus. Other species disperse their seeds alongside the mother plants (*Atriplex, Aeluropus, Suaeda,* etc.) or reproduce vegetatively. For example, the dispersal unit of *Atriplex* consists of a single-seeded fruit with a thin pericarp enclosed between a pair of bracts, which may either be united or remain free (Fig. 9.4). The high mineral content of the bracts makes them hygroscopic and heavy and restricts the dispersal of the disseminule to relatively short distances. On the other hand, it was reported that diaspores of *Atriplex* and *Suaeda* rapidly reached the newly

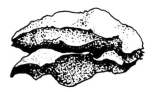

Fig. 9.4. The disseminule of *Atriplex halimus,* side and surface views.

dried Dutch polders, probably by means of zoochory (Van der Pijl, 1969). The spike of *Aeluropus* remains intact, being shed in one piece next to the mother plant.

The effective dispersal distances of most inland halophytic species are obscure and difficult to estimate. From superficial observations it seems that dissemination does not cover considerable distances, even in anemochoric species (e.g., *Salsola, Pluchea,* or *Statice* sp.), zoochoric species (e.g., *Atriplex, Suaeda,* or *Nitraria* sp.), or tumble weeds (e.g., *Atriplex canescens* and *Salsola kali*). Nevertheless, the scattered appearance and the disjunctive distribution areas of most halophytes certainly indicate that the dispersal mechanisms of these plants are effective.

The frequent changes which occur on the borders of saline habitats are lethal for some species, but do not affect others. Deep-rooted species with a good capability for vegetative reproduction (e.g., *Prosopis farcta*) remain behind in places where their seeds are unable to germinate and their seedlings cannot survive the high salt content of the upper soil layers. The increase in salinity does not exterminate such plants and the old specimens remain alive for many years—until the ecological conditions become favorable again. During this period, the entire production of seeds does not have any chance of germination and is consequently lost.

Seeds of some species may be carried for long distances by wide sweeping storms or by sea currents into areas of unsuitable temperature, salinity, or pH conditions and hence are destroyed. Furthermore, long-distance dispersal is also doomed to failure in many other cases because of unsuitable bioecological conditions. Diaspores of *Sonneratia* may germinate far from their native habitats. However, these plants will remain sterile because of the absence of pollinating bats (Van der Pijl, 1957).

Germination

The existence of a plant species in its natural habitat depends, among other things, upon its ability to reproduce under similar ecological conditions. Thus, the effect of salinity as a specific and dominant factor in a saline environment determines, to a great extent, the ability of halophytes to reproduce and perpetuate their existence.

Information regarding the germination behavior of halophytes is still scanty and was mostly accumulated during the last 30 years. For example, in the review of Uphof (1941), only very few papers regarding reproduction of halophytic angiosperms were included (Poma, 1922).

Germination is definitely a critical period in the plant life cycle and inhibition of germination by high salt concentrations may exclude large segments from its potential distribution area. Thus, the control of germination constitutes a major factor in the zonation and inhabitation of saline habitats (Toole *et al.,* 1956; Crocker and Barton, 1953; Koller, 1955; Waisel, 1958).

EFFECTS OF SALT CONCENTRATION, TEMPERATURE, LIGHT, AND WATER REGIME

High salt concentrations do not have much effect upon seeds unless the latter are soaked. At this stage, a salty environment can affect and inhibit germination in two ways: (a) by preventing uptake of water by the embryo due to the high osmotic potential of the medium; and (b) by poisoning the embryo, due to toxic effects of certain ions.

Germination of seeds is usually retarded by high concentrations of salt solutions both in halophytes and in glycophytes. The degree of imbibition (Shive, 1916), delay, or inhibition of germination is proportional to an increase of the external osmotic potential (Schratz, 1934; Ayers and Hayward, 1948; Ayers, 1952).

Toxic effects of certain ions on seed germination and seedling development had been studied by a number of investigators at the beginning of the century. Stewart (1898) and Harris and Pittman (1918, 1919) found that chloride salts were most toxic for germination, sulfate less so, and carbonate the least. However, they pointed out that while toxicity of NaCl and Na_2SO_4 seemed to be dependent on concentration, this did not apply to carbonate. Toxicity of Na_2CO_3 was primarily dependent on the presence of organic matter in the soil. Harris (1915) found the relative toxicity of soluble salts to be in the following descending order: NaCl, $CaCl_2$, KCl, $MgCl_2$, KNO_3, $Mg(NO_3)_2$, Na_2SO_4. He also concluded that salt mixtures were not as toxic in soils as in solution cultures.

Germination of alfalfa seeds was virtually inhibited when NaCl or mannitol concentrations of 12 to 15 atm were used, but reduction and retardation of germination were greater in NaCl than in mannitol solutions (Uhvits, 1946). Such difference in response to the two osmotic media suggested a toxic effect of NaCl, most probably of chloride. Chloride toxicity was also reported for embryos of *Haloxylon* sp. (Strogonov, 1964) and was easily demonstrated in recovery experiments. Seeds transferred from a solution of low NaCl concentration to tap water showed a considerably greater recovery than those transferred from a high NaCl solution.

Sodium chloride exerts both osmotic and ionic effects on germination. Distinction between these effects is usually made in two ways. The first method investigates simultaneously the germination rates in NaCl solutions and in isoosmotic solutions of an inert osmotic medium (Uhvits, 1946). Lower germination rates obtained in NaCl treatments are ascribed to specific ionic toxicities. The second method used to investigate the toxic ionic effect of NaCl is to transfer seeds, which did not germinate in salt solutions, to distilled water, and to determine any additional germination. Uhvits (1946) found that transfer of alfalfa seeds from salt solutions to water did not raise their percentage of germination. The observed inhibition of germination of alfalfa seeds was apparently due to ionic effects. On the other hand, Ungar (1962), using the same approach found that inhibition of germination of a few halophytes in high salt concentrations was due solely to osmotic effects. Also according to Chapman (1968), the effect of salinity on germination of *Aster tripolium, Aster subulatus* and of other species as well, is mostly osmotic. Full germination was obtained in seeds which had been transferred from saline to fresh water.

Seeds of *Limonium vulgare* Mill. and *Limonium humile* Mill. slowly germinated in seawater. However, most of the seeds which did not germinate in such a medium, rapidly germinated after being transferred to fresh water (Boorman, 1968). The same was true also for *Puccinellia nuttalliana* (Macke and Ungar, 1971).

Germination of *Aeluropus* caryopses was not fully recovered even after being transferred to fresh water (Pollak and Waisel, 1972). Thus, such a test implies that toxicity occurred. However, irreversible damage may be caused to seeds even by a pure osmotic stress of the medium. This is supported by the fact that transfer of caryopses of *Aeluropus litoralis* into distilled water, which did not germinate in a mannitol solution, raised germination percentage only slightly, if at all.

According to Uphof (1941), germination of seeds of halophytes and of glycophytes was similar in the presence of various concentrations of seawater, with optimal germination occurring in pure water. Propagules of lower plants, e.g., *Fucus* sp., showed obligatory requirements of salt for germination. These propagules did not germinate in dilute salt solutions below 0.9% NaCl. Although most of the terrestrial halophytes germinated best in fresh water (Chapman, 1960), seeds of some flowering halophytic species (e.g., *Atriplex, Salicornia,* and *Suaeda*) germinated better in 0.5% NaCl than in distilled water (Ungar, 1962; Mayer and Poljakoff-Mayber, 1963).

Conflicting data exist regarding germination of *Salicornia herbacea* in salt solutions. While some investigators reported highest germination percentage in salt-free media (Feekes, 1936), others reported optimal germination in 50% seawater (Walter, 1968). *Aster tripolium* also germinated better in media with a low salt content (Schratz, 1937; van Eijk, 1939). Data presented by Boorman (1968) suggest that *Limonium vulgare* is also one of the few species in which germination is stimulated by salinity. In such cases, germination is spread over long periods.

According to Ayers and Hayward (1948), and Ayers (1952) a correlation between salt tolerance of a plant at germination and its tolerance at other growth phases is not obligatory. This is certainly true for a variety of crop plants. However, different conclusions were reached in investigations of wild halophytes (Waisel, 1958). A positive correlation was observed in a variety of species between salt resistance during germination and salinity of their habitat (Fig. 9.5). Plants growing naturally in sites with a high salt content showed a higher salt resistance in germination than plants from less salty sites. A similar correlation between the ability to germinate under saline conditions and zonation of plants in a salt marsh was also reported by Ungar (1965, 1967b).

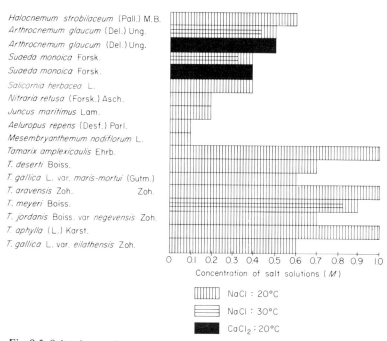

Halocnemum strobilaceum (Pall.) M.B.
Arthrocnemum glaucum (Del.) Ung.
Arthrocnemum glaucum (Del.) Ung.
Suaeda monoica Forsk.
Suaeda monoica Forsk.
Salicornia herbacea L.
Nitraria retusa (Forsk.) Asch.
Juncus maritimus Lam.
Aeluropus repens (Desf.) Parl.
Mesembryanthemum nodiflorum L.
Tamarix amplexicaulis Ehrb.
T. deserti Boiss.
T. gallica L. var. *maris-mortui* (Gutm.)
T. aravensis Zoh.   Zoh.
T. meyeri Boiss.
T. jordanis Boiss. var *negevensis* Zoh.
T. aphylla (L.) Karst.
T. gallica L. var. *eilathensis* Zoh.

Concentration of salt solutions (*M*)

▨ NaCl : 20°C
▤ NaCl : 30°C
■ CaCl$_2$: 20°C

Fig. 9.5. Salt tolerance in germination of some halophytes. (After Waisel, 1958.)

The halophytes *Suaeda depressa* and *Tamarix pentandra* germinated in salt solutions up to 4% NaCl, whereas the glycophytic *Bromus japonicus* tolerated only 0.75% NaCl. Semihalophytic species, e.g., *Haplopappus phyllocephalus* or *Iva annua,* showed intermediate tolerance and could not germinate in solutions containing more than 2% NaCl. Germination percentages of *Aster tripolium* and *Triglochin maritima* seeds are extremely low in seawater. *Plantago maritima* did not germinate in solutions with a concentration above 75% seawater and *Atriplex hastata, Beta maritima,* and *Plantago coronopus* above 50% seawater. In each case, an increased salinity of the medium extended the time required for germination. No germination of *Aster tripolium* was found in a medium with a salt concentration above that of seawater. Most other halophytic species investigated so far were capable of germination in salinities beyond those tolerated by adult plants (Ungar, 1965; Pollak, 1967; Ovadia, 1969; Keren,

1970). Germination in those species did not decrease gradually with increasing salinity, but decreased sharply when the salinity of the medium reached certain concentrations of NaCl (Ungar, 1962). In some species germination may thus occur in their native habitat, even without earlier leaching of the soil. Nevertheless, this does not hold for many other halophytic species which germinate only after reduction in salinity of seed beds (Chapman, 1960; Pollak and Waisel, 1972).

Excess salt is not always a negative factor in plant life. In many halophytic species, control of germination is obtained by the salt content of the disseminule. Beadle (1952) found that regulation of germination in five Australian species of *Atriplex* was osmotic, and controlled by the high chloride content of the bracts enveloping the seeds. Koller (1957) also reported that germination control in *Atriplex dimorphostegia* seeds was osmotic, but claimed that the osmotic effect was not solely due to the presence of chloride. Such control was found only in the halophytic species of this genus. It is interesting to note that in certain nonhalophytic species of *Atriplex,* salt content of the dispersal unit was found to be relatively low (Kadman-Zahavi, 1955). Germination of *Atriplex* seeds is affected by other mechanisms as well. Seeds of *Atriplex dimorphostegia* do not germinate before a water-soluble inhibitor is washed out of the dispersal unit.

Germination of many other xerohalophytic species, e.g., *Zygophyllum dumosum* (Fig. 9.6), was also found to be regulated by the salt content of the dispersal unit (Lerner *et al.,* 1959). Germination of such seeds can be obtained only by thorough rinsing of the seed envelopes, or after rains which are heavy enough to leach the salts out of the close environment of the embryo. Such a mechanism serves, thus, as a rain gauge, and ensures the survival of the young seedlings (Koller, 1957).

Apart from affecting the germination percentage, salinity also slowed down the process, extending germination over a long period of time. This was found to be true for *Aeluropus* as well as for other halophytes, such as *Prosopis farcta* (Eshel, 1966) and *Suaeda monoica* (Ovadia, 1969).

Temperature is a dominant factor in germination, growth, and salt tolerance of plants under saline and alkali conditions (Ahi and Powers, 1938). Germination is affected by extreme temperature conditions, by the duration of such temperature conditions, and by the stage of development at the time of exposure. Interactions between salinity and temperature are well known, and generally, conditions of high temperature reduce the salt tolerance of the seeds (Uhvits, 1946; Hayward and Bernstein, 1958). Ungar (1967b) compared the effects of temperature on seed germination of halophytic and glycophytic species under saline conditions. No group distinction was evident. High temperature conditions stimulated germination of *Salicornia europaea* seeds, a late season species, but inhibited seed germination of the early season plant,

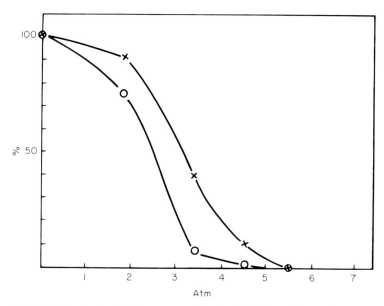

Fig. 9.6. The relationship between germination inhibition and osmotic pressure of crude *Zygophyllum* extracts (○ – ○) and solutions of sodium chloride (x–x). (After Lerner *et al.*, 1959.)

Spergularia. An increase of even $2°-3°C$ in mean temperature reduced germination percentage in various salt treatments (Uhvits, 1946). The upper limit of NaCl concentration which still enabled germination of soybeans was $0.2M$ at $30°C$, but $0.3M$ at $15°C$ (Osaga, 1939). Reduction of germination percentage observed at higher temperatures is beyond the effects of temperature on the osmotic potential of the medium.

Low temperatures also affected salt tolerance of germinating seeds. Seeds of *Triglochin maritima* and of *Plantago maritima* germinated in higher percentages grew faster and tolerated higher salinities after a period of chilling at $3°C$. A minimum of a 30-day chilling period was required in order to obtain such effects (Binet, 1964; Béchet and Binet, 1964).

A continuous 14-day pretreatment of *Limonium humile* and *Limonium vulgare* seeds at $-10°C$ did not change their final germination percentage after being moved to $20°C$. On the other hand, daily cycles of $-10°C/20°C$ (12 hours each) reduced the percentage of germination by over 30% (Boorman, 1968). Different cases were reported as well. According to Chapman (1942), germination of prefrozen seeds of *Spergularia media* reached 80%, while that of the controls was only 6%. Seeds of *Spergularia* germinated in higher salt concentrations, up to 3% NaCl, after a period of exposure to low temperatures $(-10°C)$.

The requirement for light is a common characteristic of hygrophytes. Since a positive requirement for light enables hygrophytes to germinate in shallow water, the significance of this trait for survival of such plants is evident. On the other hand, shallow water or soil surface, comprise in salines sites with highest salinity. Such conditions might thus cause a high seedling mortality instead of preventing it. Light requirement in hydrohalophytes may thus be a relictic characteristic and give indications as to the origin of those plants rather than being of adaptive value.

Light plays a role in regulation of germination in some species of *Atriplex*. Sensitivity to light in *Atriplex dimorphostegia* seeds depends on intact membranes and is lost when the seed coats are punctured (Koller, 1957).

Germination of certain species is not uniform even under constant environmental conditions. Periodic fluctuations in germination capacity were observed in *Atriplex dimorphostegia* (Koller, 1957).

Germination of halophytes is also affected by the water relations of their habitats. Halophytic species of wet marshes, e.g., *Salicornia herbacea* and *Halocnemum strobilaceum* germinate only late in the season, after the water level has dropped off. Nevertheless, during this season salinity is very high. On the other hand, halophytes of drier habitats (e.g., *Nitraria retusa*) germinate during the rainy season when salt concentration is relatively low. Practically nothing is known of the exact requirement of water and the state of hydration that is needed for germination.

Morphology of Seed Germination

In addition to the physiological adaptation of seeds of halophytes, morphological aspects of seeds and seed germination may also contribute to its success.

Seed germination in certain groups of terrestrial halophytic species, e.g., the Chenopodiaceae, differs in a way from normal germination. In many such species, hypocotyls rather than roots emerge first from germinating seeds. As the hypocotyls are covered by heavy cuticular layers, they are far better protected than roots against dehydration and excessive salt uptake. Exposing the hypocotyl first provides a certain protection for the germinating seedling. Only after a certain period of time, which differs among species, and after the lower tip of the hypocotyl adheres to the soil, do root expansion and development begin (Negbi and Evenari, 1961).

Many species of *Atriplex* are known to bear polymorphic dispersal units. *Atriplex hortensis* has four kinds of flowers, each giving rise to fruits of different color and shape (Becker, 1913). Seeds of *Atriplex dimorphostegia* also bear dispersal units of different size which vary in their germination behavior. Moreover, even in similar dispersal units, seeds of *Atriplex* exhibit polymorphism

and vary in germination (Beadle, 1952). Contribution of such polymorphism to
the adaptability of the plants is still obscure.

GERMINATION OF SEEDS OF SELECTED TERRESTRIAL HALOPHYTES

Aeluropus litoralis

Although germination of caryopses of *Aeluropus litoralis* occurs in relatively
concentrated NaCl solutions, it is inhibited even by dilute solutions and highest
germination percentages were obtained in distilled water. Prolonged exposure of
the caryopses to NaCl or mannitol reduced the percentage of seeds able to
germinate (Pollak and Waisel, 1972).

Salt tolerance of *Aeluropus* seeds seems to be high up to an ambient
temperature of 30°C, but becomes narrower at higher temperatures (35°C and
above). Such inhibition is above and beyond the changes induced by
temperature in the osmotic potential of the medium and seems to result from
ionic or osmotic poisoning of the developing embryo (Fig. 9.7).

Since caryopses of *Aeluropus litoralis* tend to germinate only in the light,
natural germination may occur only on the the upper soil surface which
generally contains the highest percentage of salts. Concentration of NaCl in the
soil surface is usually far beyond the concentrations tolerated by the caryopses

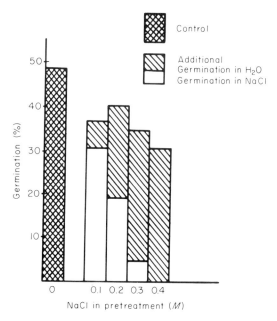

Fig. 9.7. The effect of salt pretreatment on germination percentage of caryopses of
Aeluropus litoralis. (After Pollak and Waisel, 1972.)

of *Aeluropus*. Consequently, without marked elution of salts from the upper surface of the soil, no germination of *Aeluropus* is possible in its natural habitats. In salt marshes of the Dead Sea region, germination therefore seems to be restricted to winter and early spring, when salts are sometimes eluted from the upper soil profile, and day temperatures are near optimum. Nevertheless, it seems that the appropriate combinations of environmental conditions for successful germination do not frequently occur, and natural reproduction of *Aeluropus* is mostly accomplished by vegetative propagation.

Arthrocnemum glaucum

Seeds of *Arthrocnemum glaucum* are highly salt tolerant. It seems that such tolerance is osmotic and does not depend on the composition of the medium (Waisel, 1958). Seeds of *A. glaucum* require relatively high temperatures for full and fast germination. Germination occurs only under very specific ecological conditions (Fig. 9.8).

Atriplex sp.

The genus *Atriplex* is comprised of a group of species with only some of them being halophytic. Thus, it is not surprising to find out that seeds of various species within this genus are sensitive to salinity during germination. Even germination of the halophytic species, such as *Atriplex polycarpa,* was severely reduced already by saline conditions with an osmotic potential as low as approximately −4 bars. At −17 bars, germination of this species was stopped altogether (Chatterton and McKell, 1969). On the other hand, germination of seeds of *Atriplex halimus* seems to be somewhat improved in salt solutions of concentrations below 30 mM (Mayer and Poljakoff-Mayber, 1963). Germination in *Atriplex* is controlled by the hardness and permeability of the seed coat and is also regulated by the salt content of the enveloping bracteoles. Seeds are insensitive to light, and germination occurs in a wide range of environmental pH (Beadle, 1952). Seeds are sensitive to aeration and will hardly germinate when inundated.

Distichlis spicata

Seeds of *Distichlis spicata* undergo a period of dormancy. Germination occurred only after the seeds' requirements for a period of low temperature after ripening were satisfied, or when nitrate was abundantly added to the medium (Amen *et al.,* 1970). Dormancy, and consequently also germination of seeds of *Distichlis,* are controlled by hormones. Dormancy seems to result from a specific endogenous inhibitor that inhibits nitrate reductase activity in the endosperm. Dormancy is overcome either by destruction of the inhibitor during stratification processes or by its leaching (Amen *et al.,* 1970). Seeds of *Distichlis* are not photosensitive and their germination behavior resembles very much that of seeds of *Limonium* (Boorman, 1968), which occupy the same habitats.

Fig. 9.8. Sites of germination of *Arthrocnemum glaucum* in the Dead Sea region, Israel.

Juncus maritimus

Light is an obligatory requirement for the germination of fresh seeds of *Juncus maritimus*. Germination of seeds of this species requires either continuous light or a short night. The need for light is lessened in aged seeds. Sodium chloride solutions in concentration range of 0.2–0.3 M completely inhibited germination (Tadmor *et al.*, 1958; Waisel, 1958). Germination of this species was only slightly affected by temperatures between 15°–30°C.

Phragmites communis

Germination of *Phragmites communis* seeds in salt solutions is inhibited and the degree of inhibition is positively correlated with the increase in salinity. However, seeds collected from halophytic ecotypes of *Phragmites* (from the saline of Sedom, Israel) tolerated higher salinities in germination (up to 0.5M NaCl) than did seeds of glycophytic ecotypes (up to 0.3M NaCl only) (Waisel and Rechav, 1972). Germination was optimal at temperatures between 25°–30°C. At 20°C, germination percentages were about the same as at 25° or 30°C, but germination of seeds at 20°C was delayed. Under higher temperature conditions (35°C), seed germination was accelerated, but total germination percentages were lower. Germination of *Phragmites communis* was better in the light than in the dark, but light was not obligatory (Waisel and Rechav, 1972).

Caryopses of *Phragmites* usually germinated even when inundated by water. Thus, survival ability of seedlings depended on their rates of growth, which enabled the seedlings to reach above the water level (Bittman, 1953; Bakker, 1957). Submerged seedlings remained alive under certain temperature conditions for long periods, although their growth was restricted. Also in this case, seedlings of the saline ecotypes of *Phragmites* survived better and exhibited higher growth rates under saline conditions than did the nonsaline ecotypes.

Salicornia sp.

Germination of seeds of *Salicornia herbacea* seemed to be positively correlated with illumination. Seeds were sensitive to prolonged inundation, and according to Walter (1968), have to be out of water at least 2–3 days in order to germinate satisfactorily. During this period, growth rates of the radicles are high and the radicles become long enough to anchor the seedlings and to prevent their wash-off by the tide. Also germination and seedling development of *Salicornia perennis* occurred in salt marshes of the eastern coast of England, during the neap tides (Chapman, 1960). Seedlings needed 2–3 days free from tidal immersion to become established. According to Walter (1968), seeds of *Salicornia* exhibit an extremely high salt tolerance. *Salicornia* seeds were able to germinate even in 10% NaCl solutions, and their germination period was, thus, not essentially restricted to the rainfall seasons.

Spartina sp.

Germination of *Spartina anglica* seeds is inhibited by light (Hubbard, 1969) and immersion during tides may enhance germination by reducing the light intensity. *Spartina* seeds germinated in full strength seawater. However, under such conditions, seeds germinated only after approximately 1 month of soaking. In freshwater, their germination was much faster. In the salt marshes of eastern United States, best germination of *Spartina* was observed on mud flats with diluted seawater (Teal and Teal, 1969). Seedlings were unable to compete with freshwater plants in freshwater marshes. Viability of *Spartina* caryopses differs among the various species, with that of *S. glabra* being very low (Ridley, 1930).

Tamarix

Ability of seeds to germinate in saline solutions is especially high in various species of the genus *Tamarix*. Fresh seeds of some species of *Tamarix* germinated even in 1 *M* NaCl solutions. However, germination capacity was decreased as seeds aged.

No correlation between salt tolerance in germination and salinity of their respective habitats could be demonstrated for the various species of *Tamarix* investigated. For instance, seeds of *Tamarix aphylla*, a semihalophyte of dunes and wadies, had similar tolerance to salt as did seeds of *Tamarix amplexicaulis* growing in sites of extreme salinity on the shores of the Dead Sea.

Seeds of all species of *Tamarix* are short-lived. They remain viable for long periods only if kept under low temperatures and low humidity conditions (cf., Wilgus and Hamilton, 1962). Under normal temperatures and humidity or under field conditions, the seeds of various species of *Tamarix* maintained viability for a few weeks only (Fig. 9.9). The seeds' life-span was negatively correlated with

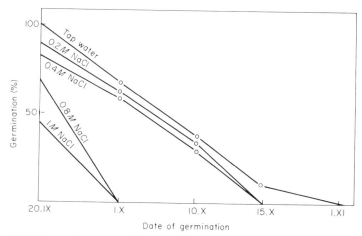

Fig. 9.9. Effects of storage period on germination of *Tamarix aphylla* seeds in tap water and in NaCl solutions of various concentrations. (After Waisel, 1960c.)

the salinity of the habitat, thus indicating that those sections of the populations which were salt tolerant had a shorter life-span than did average seeds.

Seeds of *Tamarix* usually start to germinate within 12 hours after imbibition (Waisel, 1960c; Horton *et al.,* 1960). The hypocotyl elongates and pushes the cotyledons out of the seed coat. Within 24 hours, the radicle begins to grow and a ring of root hairs appears near its tip. Germination was affected by the seed source and differences between the germinability of *Tamarix* seeds produced in the spring and in the autumn were also reported (Hulett and Tmanek, 1961). Usually there is no effect of light on germination and seeds germinated similarly under different photoperiods as well as in complete darkness. The seeds are resistant to wide changes in pH and germinated evenly in the pH range 2–9.

Germination of Seeds of Marine Angiosperms

Mangroves and submerged marine angiosperms comprise another group of halophytes with specific ecological requirements for seed germination. In most mangrove species, seed germination occurs immediately after seeds mature, with no intervening period of dormancy. Many of those plants, e.g., *Rhizophora* and *Bruguiera* are viviparous. The developing embryo, receiving water and nutrients from the mother plant, grows uninterruptedly from fertilization until it is shed (Pannier, 1962). The so-called germination thus starts when the seed is still attached to the mother plant. However, since growth of the embryos is not interrupted by periods of quiescence or dormancy, it is difficult and misleading to speak of "germination" in such a case of viviparous species.

The problem of germination under anaerobic conditions and heavy tides that prevail in coastal mangrove habitats has to be overcome by certain mechanisms. Young developing seeds that would fall into the mud, might sink and could suffocate. Development of the long hypocotyl (20–40 cm), while the seedling is still attached to the mother plant, ensures anchorage of the seedlings in species such as *Rhizophora,* etc. When seedlings attain a certain size and become too heavy, they fall into the mud beneath the tree, with the upper parts of their shoots remaining above water level. Such a mechanism of germination ensures the germinating seedling with a supply of light and oxygen and prevents its wash-off. In addition, seedlings of most species of mangroves can also withstand short periods of inundation.

Seeds which were not anchored immediately, i.e., those dropped during high tide, are carried away by water currents and may remain buoyant for prolonged periods. Initially, the seedlings float horizontally, but as germination proceeds, roots develop and they turn into a vertical position (Walter and Steiner, 1936).

Little information is available on germination control mechanisms or on responses of seeds to various environmental factors for submerged marine angiosperms. Similar to many freshwater submerged hydrophytes, but contrary to mangroves, seeds pass through a period of quiescence and germination of those species is limited to specific conditions. Seeds of *Halophila* or of *Zostera* have a hard and heavily cutinized seed coat. Evidently, as in terrestrial plants, the testa plays a role in preventing immediate seed germination and keeps a constantly low and scattered germination over long periods of time.

GERMINATION AND ECOTYPIC DIFFERENTIATION

The capabilities of halophytic and glycophytic ecotypes of the same taxa to germinate under saline conditions are of special interest. Germination of seeds collected from *Aster tripolium* plants grown on a saline and on a salt-free soil was compared. High-salt seeds germinated in similar percentage as the low salt seeds, but their germination was slower (Schratz, 1934). On the other hand, according to Chaudari (1968), germination of seeds of a saline population of *Lepidium perfoliatum* reached higher percentages under saline conditions than those of the nonsaline population. Seeds collected from halophytic populations of *Limonium pruinosum, Alhagi maurorum, Prosopis farcta,* and *Phragmites communis* germinated better under saline conditions than did seeds from glycophytic ecotypes of the same species (Waisel, 1960d; Eshel and Waisel, 1965; Waisel and Rechav, 1972). However, generalization on this matter is still speculative as different modes of adaptation can be found with various ecotypes, even within one species.

Vegetative Reproduction

Vegetative reproduction substitutes for or at least contributes to the reproductive potential of many plants. This is certainly true for various halophytic species that are restricted to narrow ecological limits either by the production of disseminules or by their germination.

In such species as *Aeluropus litoralis, Prosopis farcta,* and *Tamarix aphylla,* vegetative reproduction seems to have an advantage over reproduction by seeds in assuring establishment of plants in saline habitats. The vegetative reproduction in *Aeluropus* by means of runners is a well-adapted mechanism. Growth of the developing roots has the advantage of rapid penetration through the upper salty soil layers. Furthermore, roots developing at different places along the runners are not dependent on a direct supply of water from the soil (Waisel and Pollak, 1969a). Being well supplied with water by the parent plant, roots can thus penetrate layers of extreme salinity. The vegetative propagation of *Aeluropus litoralis* also was much less affected by salinity and temperature than was germination. Rooting occurred in a wide range of saline conditions, up to $0.6\,M$ NaCl solutions. Low salinity did not enhance root emergence, and in this respect, *Aeluropus* may be classified as a salt-resistant and not salt-requiring species. The absence of apparent temperature effects on rooting of *Aeluropus* also indicated that such reproduction may occur during any season of the year, and unlike germination, is not limited to a narrow range of ecological requirements.

It is important to note that the above mentioned information is related to runners only. These organs contain at least a few formerly differentiated root primordia in the internodes. Thus, their apparent "rooting" ability may mainly reflect interruption of the quiescent period and growth of existing root primordia, but not differentiation of new ones. The fact that in "short-noded" cuttings, i.e., those which lack root primordia, new primordia were not initiated, suggested that a distinction must be made in this species between the effects of salts on root initiation and on root growth (Pollak and Waisel, 1972).

Not only seeds, but also other disseminules exhibit high resistance to frost. Ridley (1930) reported that rhizome fragments of *Spartina glabra* can be embedded in ice for long periods without apparent damage and can be disseminated in this way.

Cuttings of *Tamarix aphylla* are also rather hardy and withstand extreme environmental conditions. Cuttings preserved their rooting ability until 13% of their weight, i.e., one-third of their water content, was lost (Fig. 9.10). Cuttings

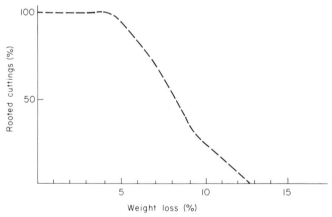

Fig. 9.10. Effect of water loss on the rooting ability of *Tamarix aphylla* cuttings. (After Waisel, 1960c.)

were able to sprout in soils of different textures, but better sprouting was observed in well-aerated light soils. Salinity adversely affected the taking of cuttings of *Tamarix aphylla*. However, cuttings still rooted in NaCl solutions up to a concentration of 0.2 M (Waisel, 1960c). Under 0.3 and 0.4 M NaCl treatments, cuttings sprouted (Fig. 9.11). However, as no roots were produced they died rapidly.

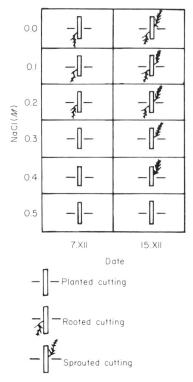

Fig. 9.11. The effect of irrigation with various NaCl solutions on sprouting and rooting of *T. aphylla* cuttings. Cuttings planted on December 1st. (After Waisel, 1960c.)

Vegetative reproduction is the major means of reproduction also for *Limonium vulgare* and *Limonium humile*. Large-scale reproduction from seeds seems to be important only when these species invade a new area (Boorman, 1968).

The presence of salts in the root medium generally depresses rooting ability and root growth in glycophytic and semihalophytic species. On the other hand, several halophytic or semihalophytic species, e.g., *Aster tripolium, Ruppia* sp., *Myriophyllum* sp., and *Suaeda monoica* exhibited optimal rooting and root growth only at a certain range of low salinity (cf. Montfort and Brandrup, 1927).

Many halophytic species, e.g., *Arthrocnemum* sp., *Nitraria retusa,* and *Salicornia* sp. are capable of forming adventitious roots on their twigs. This ability varies between species and during various seasons of the year (Waisel and Pollak, 1969b). However, for some species, this is the only means of reproduction in large portions of their distribution area.

$$\underline{\hspace{5cm}}10$$

Growth

General Remarks

It is commonly accepted that soil solutions with high salt concentrations cause growth retardation in most plants. This is true for glycophytes at NaCl concentrations below 50 mM, as well as for halophytes at relatively high concentrations (Montfort and Brandrup, 1928; Hayward, 1956; Black, 1956, 1960; Bernstein and Hayward, 1958; Gale *et al.*, 1970). Saline conditions affect plant growth in a variety of ways, i.e., inducing poor physical conditions of the soil, decreasing water uptake, causing toxic accumulation of sodium and chloride, and reducing nutrient availability. When salinity increases and water potentials around plant roots decrease, turgor of plant cells declines, and cells cease to divide and elongate. Under water-stress conditions, stomata close, photosynthesis is reduced, protein breakdown is accelerated and, instead of growing, plants lose weight (Kozlowski, 1964; Slatyer, 1967; Kramer, 1969). When exposed to saline conditions, most plants have fewer leaves and show a poorly developed shoot with a smaller number of nodes and shorter internodes. Halophytes must be adapted to all of these environmental effects.

Growth Inhibition and Osmotic Effects

Expansion of cells is primarily correlated with turgor pressure and decreased turgor has long been referred to in explaining reduced plant growth under saline conditions (Magistad, 1945; Wadleigh and Gauch, 1948). Growth of common crop plants is affected by salinity when the electrical conductivity of the soil paste is less than 2 mmhos/cm. Salt-tolerant crop plants may still grow, when soil conditions permit, even when conductivity increases to 8 mmhos/cm. Some exceptionally salt-resistant plants may grow satisfactorily, even when conductivity of the soil solution is 10 mmhos/cm. Only halophytes can survive when conductivity of the saturated soil extract exceeds these values (Richards, 1954).

Various components of the soil solution may inhibit growth. However, it is the total soil water potential (Ψ) which actually affects plant growth. This hypothesis was supported by experimental evidence showing that a similar inhibition of growth was caused by a specific level of soil water potential whether Ψ soil was composed of soil matric or osmotic potentials (Wadleigh and Ayers, 1945; Wadleigh *et al.*, 1946).

Data presented by Wadleigh and Ayers (1945) (Fig. 10.1) show that growth of bean plants was reduced similarly by a wide combination of matric and osmotic soil water potentials. However, reduced water availability, *per se,* cannot explain reduction in growth, because as mentioned previously, plants undergo osmotic adaptation.

According to Greenway (1968), the rate of plant growth under saline conditions is affected by the concentrations of salt inside the cells. Even though the rate of salt uptake in halophytes is high, the concentration inside their cells remains low, due to dilution. Consequently, growth rates of those plants are not reduced by salinity.

Growth in this respect must be expressed in terms of dry, rather than fresh, weight, because the increasing succulence of plants caused by NaCl may result in false interpretation of the effects of salinity. Greenway (1962a) compared, under saline conditions, the growth and yield of a relatively salt-tolerant variety of barley with that of a salt-sensitive variety. Not only was total dry weight increment lowered by salinity treatments, and differently in the two varieties, but also various components of grain yield were lowered differently (Fig. 10.2).

Dry matter production under conditions of low water potentials is expected to decrease because of reduction in photosynthesis combined with increase in respiration, i.e., suppressing the net assimilation rates.

Ion retention and osmotic adjustment by plant tissues are, of course, of considerable significance for plant growth in saline habitats. Adaptation of glycophytes to increased osmotic stress of the medium has been well documented (Bernstein, 1961, 1963; Slatyer, 1961, 1967). Wherever osmotic

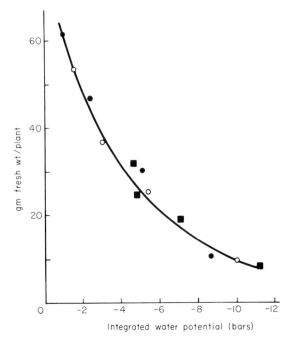

Fig. 10.1. Growth of bean plants in a combination of matric and osmotic water potentials. Filled circles, low tension series; open circles, medium tension series; squares, high tension series. (After Wadleigh and Ayers, 1945.)

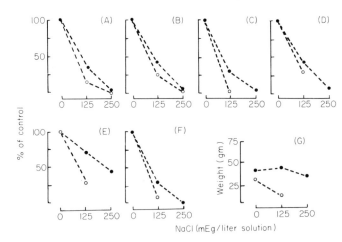

Fig. 10.2. Effects of salinity on growth and yield components of a relatively salt-tolerant (●, var. *Bolivia*) and a relatively salt-sensitive (○, var. *Chevron*) variety of barley. (After Greenway, 1962a.) A, Whole plant; B, shoot excluding grain; C, grain; D, ears/plant; E, grain/ear; F, grain/plant; G, weight of 1000 grains.

stress was caused by high solute concentrations in the growth medium, the OP of plant tissues increased concomitantly by an amount equal to the osmotic potential of the solute added to the medium. Hence reduction in plant growth in a medium of low water potential cannot be solely due to inadequate internal osmotic adjustments or difficulties in water uptake. Under a high external osmotic stress, the rate of ion accumulation might not be adequate for fast osmotic adjustment, resulting in temporary inhibition of growth (Greenway and Thomas, 1965). In both cases, little is known about actual sites where adjustment occurs. It was suggested that even under what appeared to be osmotic equilibration, a high proportion of ions in the tissues is present in the cell wall, and protoplasts of at least some cells within the tissues are actually under water stress (Oertli, 1966; Y. Waisel, A. Eshel, and A. Ramati, unpublished data). The reasons for growth inhibition under such conditions are thus obvious.

When a plant is subjected to external osmotic stress it passes through the following sequential events (Greenway and Thomas, 1965): (1) osmotic stress; (2) reduction in growth; (3) increase in internal solute concentration; (4) partial recovery of growth; and (5) stabilization of the internal level of osmotically active substances. Consequently, at the last stage, the water content of leaves and shoots should be sufficient for normal functioning, even though the rate of ion uptake from the medium might be inadequate to sustain rapid growth. It should also be emphasized that a high concentration of solutes in plant cells may be brought about not only by uptake of solutes from the external medium, but also by cellular synthesis of osmotically active compounds. Synthesis of oxalate in *Atriplex* leaves can serve as an example (Williams, 1960; Osmond, 1967). For sugar beet, optimal growth and highest sugar content were obtained either under dry or saline conditions (Hunter and Jungen, 1952; Waisel and Bernstein, 1959; El Sheikh *et al.*, 1967). Similar results were reported for spinach by Lehr and Wybenga (1958). High growth rates of halophytes in saline culture solutions may be partly due to their high rates of ion absorption and fast osmotic adjustment.

Under conditions of high evaporation, a decrease in water potential of the soil solution, together with reduced permeability of roots, eventually results in a water stress in plant tissues. Under such conditions, growth inhibition can result not only directly from an environmental water stress and from decreased turgor, but also indirectly because of reduced translocation of growth regulators from roots to shoots (O'Leary, 1969).

Ionic Effects

Growth is checked under saline conditions, by total ion activity in the soil solution, as well as by the relative contribution of some ions. A few ppm of boron are toxic and affect plants more than a 1000-fold higher content of other ions.

However, while under saline conditions the growth of most crop plants is inhibited, and growth and development of wild halophytic plants are either unaffected or even stimulated (Montfort and Brandrup, 1928; Van Eijk, 1934, 1939; Chapman, 1960; Pollak, 1967).

Very little is known about mineral metabolism of these plants. However, in view of the fact that such plants are unaffected by the high NaCl content of the environment, certain mechanisms appear to have developed in those plants that enable their adaptation to saline conditions. Adaptation of halophytes to salty conditions includes high tolerance for the negative effects of salinity, as well as a positive reaction toward it.

In some plants sodium affected growth, regardless of the counter anion (Dorph-Peterson and Steenbjerg, 1950). On the other hand, some investigators claimed that responses of *Suaeda monoica* plants to NaCl were much higher than to Na_2SO_4 (Ovadia, 1969). Similar results were obtained for *Atriplex halimus* by Gale and Poljakoff-Mayber (1970). When sodium was applied as $NaNO_3$ the positive effects of sodium and of nitrate in accelerating the growth of *Suaeda* plants were cumulative.

There is no doubt that yield of a semihalophyte, such as sugar beet, is substantially increased by addition of NaCl to the soil (Tinker, 1965). Water culture experiments have unequivocally demonstrated this (Ulrich and Ohki, 1956; El Sheikh *et al.*, 1967). With potassium in the range of 0.5–12 mEq/liter addition of as little as 0.5 mM sodium chloride could increase both fresh and dry weight of sugar beet plants. Stimulation of growth can be ascribed in part to chloride ions and in part to sodium ions (Ulrich and Ohki, 1956). Sucrose content of sugar beets increased over a wide range of sodium chloride concentrations (Ulrich and Ohki, 1956; Waisel and Bernstein, 1959), thus contributing to the osmotic adjustment of the plants.

Sodium chloride is probably the oldest chemical used for crop fertilization, and was used in Europe for hundreds of years. Addition of fertilizers containing chloride caused a significant increase in yield of tomatoes and cotton (Eaton, 1942). Similar results were obtained with garden beets grown in water culture (Raleigh, 1948). Necessity of chloride for plant growth was emphasized by the fact that in chloride-free growth media, chloride deficiency symptoms and injuries appeared early in the life of plants (Broyer *et al.*, 1954). Symptoms of chloride deficiency included wilting of tips of leaflets, progressive chlorosis, change in color (bronzing), and necrosis of leaves. In severe cases of chloride deficiency, failure to produce fruit was reported. When low concentrations of chloride were added to the culture solution of deficient plants, the symptoms of deficiency disappeared and growth increased proportionally to the concentration of chloride.

Growth of halophytes can be negatively affected, unaffected, or stimulated by NaCl. Only salt-requiring species, whether obligatory or preferential, are included in the last group (cf., Chapman, 1960; Adams, 1963; Webb, 1966).

In certain halophytes (e.g., *Artemisia maritima, Plantago maritima*) growth rates, plant heights, and leaf sizes were higher in nonsaline habitats than in saline ones. In other species, e.g., *Limonium carolinianum,* growth was unaffected by low salt concentrations.

Plants such as *Salicornia stricta, Suaeda monoica, Nitraria retusa,* and *Atriplex halimus* exhibited restricted growth, abnormal leaf development, and high sensitivity to various infections in sodium-deficient media. Addition of sodium ions stimulated their growth, but the concentrations required were of magnitudes of 50–150 mM NaCl, i.e., far above those of micronutrients.

The stimulative effect of salinity on growth of halophytes was also observed under field conditions. According to Schratz (1936), production of a stand of *Salicornia* increased proportionally with soil salinity up to a salt content of 6% of the soil solution (Table 10.1).

Table 10.1

FRESH WEIGHT OF *Salicornia* PLANTS (gm/m^2) GROWING IN HABITATS WITH DIFFERENT SALT CONTENT[a]

% salt in soil solution	Weight
1.6–2.0	58
2.1–2.5	85
2.6–3.0	223
3.1–3.5	232
3.6–4.0	242
4.1–5.0	250
5.1–6.0	250
>6	293

[a] After Walter, 1968.

Optimal salt concentrations for growth vary among species, and are different for various stages of ontogenetic development. Growth of *Suaeda monoica* was optimal at the seedling stage when 50 mM NaCl were added to the medium, and at 150 mM for adult plants. The anion content of the medium affected growth only slightly, though better growth was obtained with NaCl or NaNO$_3$ than with other salts.

While *Salicornia herbacea* plants require a 2% NaCl solution for optimal growth, the requirement of *Frankenia pulverulenta* for NaCl was 1% only (Stocker, 1933). *Atriplex inflata* and *A. nummularia* showed an increase in dry weight when sodium chloride, up to a concentration of $0.6\,M$, had been added to the culture solution (Ashby and Beadle, 1957). *Atriplex inflata* showed maximum growth in a solution containing $0.05\,M$ sodium chloride. The dry weight of plants under such treatments was twice as high, and weight of plants grown in the solution containing $0.6\,M$ sodium chloride was some 1.7 times higher than weight of plants grown in the NaCl-free solution. Data for *A. nummularia* were similar, with maximum dry weight produced in the solution containing $0.2\,M$ sodium chloride. *Atriplex vesicaria* (Black, 1960) and *Halogeton glomeratus* (Williams, 1960) showed similar growth responses to addition of sodium chloride to their growth medium.

Growth of *Salicornia olivieri* and *S. ramosissima* was greatest when 2–3% NaCl was present in the medium (Halket, 1915). *Glaux maritima* also grew better whenever NaCl was added to the growth solution. Sodium chloride did not enhance growth of *Suaeda maritima* to the same degree as other plants in the same habitat were affected, although this species is highly salt-tolerant. Similar responses were found for *Spergularia media* and *Plantago maritima,* which grew best without salt in the soil (Terras, J. A. cited by Halket, 1915). Thus, salt resistance, growth response, and the ability of plants to exist in saline habitats are not essentially intercorrelated.

Increased growth represents increased synthesis of dry matter. *Atriplex hastata* plants grown on Hoagland's solution, with the addition of sodium chloride, exhibited faster rates of leaf thickening and extended the period in which this occurred, than plants which were grown on a salt-free medium. The increased thickness observed was found to be caused by an increased size of the palisade cells. Sodium chloride also affected leaf area and the surface area of epidermal cells. Maximum leaf area occurred in $0.1\,M$ NaCl and minimum area in $0.6\,M$ (Black, 1958). Epidermal cells had a maximum surface area in $0.2\,M$ NaCl—double the area in treatments devoid of sodium chloride, and in those containing $0.6\,M$ NaCl. The number of epidermal cells per leaf decreased progressively with increasing concentrations of sodium chloride.

Growth of *Atriplex* sp. and *Halogeton glomeratus* was affected similarly by Na_2SO_4 and by NaCl, although the dry weight values of sulfate-treated plants were somewhat lower. Thus, halophytic species show specific growth responses to high concentrations of sodium ions.

The exact physiological role of salts in metabolism of halophytes is still obscure. Jennings (1968a) assumed that sodium excesses cross the plasmalemma without any need for energy expenditure and that the overall process of sodium transport into the vacuole results in ATP production. It was suggested that such ATP is later used for additional growth, thus explaining growth stimulation caused by NaCl in halophytes.

Another way in which sodium may improve the growth of halophytes in saline media is by improving their ecological rather than their physiological conditions.

Leaves of *Nitraria retusa* plants grown in a salt-deficient medium showed typical infection. Since leaves of plants grown in NaCl solutions remained healthy, it seems that a high content of salt in the leaves prevented such leaf infections. Similar data were also reported by Ashby and Beadle (1957) for *Atriplex nummularia* and by Ovadia (1969) for *Suaeda monoica.*

According to Strogonov (1964), *Suaeda glauca* responded more vigorously to sulfate than to chloride salinity. Plants grown in sulfate salinity were at least twice the size of those grown in chloride treatment. However, since salts in those experiments were applied on a percentage—rather than on an equivalent basis—such a response is still questionable.

In some halophytes, the stimulatory effect of solutions with high ionic strength was not restricted to sodium, and acceleration of growth could be induced by other monovalent ions. However, when two cations with a stimulatory effect were added simultaneously to the growth medium, e.g., sodium and potassium, the presence of one cation depressed the growth stimulation induced by the other one.

Addition of potassium chloride to the basal nutrient solution resulted in increase in dry weight of *Atriplex inflata* and *A. nummularia* (Ashby and Beadle, 1957), or *Atriplex halimus* (Fig. 10.3) (Rosenblum and Waisel, 1969; Mozafar *et*

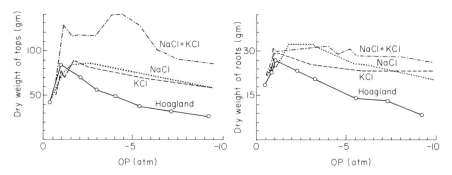

Fig. 10.3. Effects of salinity (NaCl; KCl; NaCl + KCl) on growth of *Atriplex halimus.* (After Mozafar *et al.,* 1970a.) Reproduced by permission of the American Society of Agronomy, from the *Agronomy Journal.*

al., 1970a). *Lepidium crassifolium* required a high Na : K ratio in its leaves for normal growth (Weissenbock, 1969). Growth of *Aster tripolium* and of *Salicornia herbacea* was also greatly improved by addition of potassium to the saline medium (Baumeister and Schmidt, 1962).

Halophytes and glycophytes differ in the shape of their growth curves (Montfort and Brandrup, 1928). Growth of corn (Fig. 10.4) dropped more or less linearly with increase in salinity of the medium. On the other hand, *Aster tripolium* or *Salicornia stricta* plants gave an optimum type of growth curve, with the peak for *Salicornia* being wider. The exact location of the peak on the

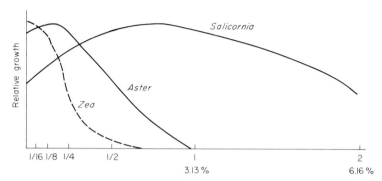

Fig. 10.4. Relative growth rates of *Zea mays, Aster tripolium,* and *Salicornia stricta* in solutions of varying concentrations of Brenner's solution substituting seawater. (After Montfort and Brandrup, 1928.)

concentration curve depends on salt tolerance of the species and on its growth phase. This is presented schematically in Fig. 10.5.

Most hydrohalophytes are obligatory halophytes. However, differences among species are still encountered. Optimal salinity for growth of *Ruppia rostellata* was found to be 2% NaCl. Nevertheless, the plants grew well even in nonsaline water (Montfort and Brandrup, 1928). Growth of other species, e.g., *Lepidium crassifolium,* was optimal only in soils containing low salt concentrations.

Plants of different ages and growth phases also responded differently to salinity. While a saline medium affected seed germination of halophytes and of glycophytes similarly, developing seedlings showed marked segregation in their growth response (Uphof, 1941).

Growth of germinating *Rhizophora mangle* plants in diluted seawater was found to be three to five times higher than in rainwater or in undiluted seawater (Pannier, 1959). This is in full accord with the optimal range of salinity in the natural habitats of these plants, i.e., estuaries and other habitats with partly diluted brackish water.

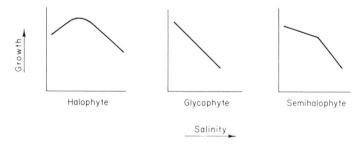

Fig. 10.5. Schematic representation of the effects of salinity on the relative growth of halophytes and glycophytes.

Atriplex polycarpa seedlings exhibited higher salt tolerance at advanced growth phases, than during the seed germination stage (Chatterton and McKell, 1969). Although growth was somewhat inhibited by increasing salinity, complete cessation of growth was not recorded even in solutions with an osmotic potential of −25 bars. Osmotic potentials of this magnitude also limited growth of *Tamarix aphylla* (Waisel, 1961). However, as postulated by Gale *et al.* (1970), such an effect is restricted to conditions of dry atmosphere and high moisture deficits, where salts are needed for osmotic adjustment of plants. In a humid atmosphere, sodium chloride inhibited growth of the halophyte *Atriplex halimus*, even at low salt concentrations, i.e., an OP of only −3 bars (Fig. 10.6).

Fig. 10.6. Effects of low (1) and high relative humidity (2) on growth of *Atriplex halimus* plants at 0 (control), −3, −5, −10, and −20 atm osmotic potentials. The various osmotic potentials were obtained by adding differing amounts of NaCl to Knop nutrient solutions. (After Gale *et al.,* 1970.)

Effects of Salinity on Growth of Various Organs

Salt requirements of various organs of growing halophytes generally differ. However, available data are inconsistent. Optimal shoot growth of *Aster tripolium* occurred when plants were in a salt solution, whereas roots grew best in fresh water (Montfort and Brandrup, 1927, 1928) (Fig. 10.7). On the other hand, Stocker (1960) and Troughton (1960) claimed that an increase in root

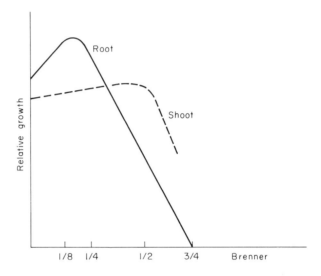

Fig. 10.7. Effects of increasing salinity (Brenner solution substituting seawater) on growth of roots and shoots of *Aster tripolium*. (Adapted from Montfort and Brandrup, 1927.)

growth and decrease in shoot growth were obtained under high water stresses induced by NaCl, whereas total growth of the plants was reduced. Troughton (1967) found that sodium chloride inhibited both growth initiation and rate of growth of roots and shoots of *Lolium perenne*. Sodium chloride also seems to affect growth of various plant organs in a different way than soil matric or soil osmotic water stress.

Shoot growth of *Aster tripolium* was best in 2% NaCl, but roots grew best at only 1% (Bickenbach, 1932). In *Salicornia herbacea,* both organs responded similarly to salinity. However, in *Atriplex polycarpa,* salinity inhibited shoot growth more than the root growth, thus increasing the shoot/root ratio. Root growth was already inhibited by media with water potentials lower than −25 bars (Chatterton and McKell, 1969).

Shoot and root growth of *Limonium vulgare* seedlings was equally inhibited by seawater, with shoot length and root length reduced by about 60% (Boorman, 1968).

GROWTH OF ROOTS AND SUBTERRANEAN SHOOTS

The interrelationships between growth and ionic or osmotic environments are especially marked in roots. It is well known that roots are able to absorb only a minor fraction of the available soil minerals in the root vicinity. Such a low efficiency in nutrient exploitation is related to the randomness of root distribution and sparse growth. Thus, reduction in root growth causes a decrease in uptake of nutrients, which in its turn reduces growth of all other plant organs. Moreover, root growth affects the growth and development of the entire plant. However, such relations are somewhat different in the case of halophytes. The shoot/root ratio in various halophytic species has, according to Repp (1939), the values shown in the following tabulation:

Suaeda maritima	64.3
Salicornia herbacea	35.5
Camphorosma ovata	12.1
Aster pannonicus	7.1
Triglochin maritimum	3.0

Evidently, the ratio in halophytes is high in comparison to that of glycophytes, although "roots" in Repp's investigation included other underground plant parts as well.

Very little information is available on growth rates of individual roots of halophytes and the time factor of their activity.

While growth of corn and pea roots decreased linearly with increase in salinity of the growth medium (Fig. 10.8), root growth of various halophytes was only slightly affected at low salt concentrations (*Aster tripolium*) or even stimulated (*Salicornia herbacea*) by dilute salt solutions (Montfort and Brandrup, 1928). Also experiments conducted by Wilcox and Waisel (unpublished data) show that NaCl had an accelerating affect on growth of individually treated roots. When roots of *Suaeda monoica* plants, grown on a nonsaline medium, were exposed for a few hours to NaCl solutions as concentrated as 1 M, their growth rates doubled.

Since both cell division and cell elongation occur close to the root tip, sodium chloride must affect that site. In fact, a study of the pattern of sodium distribution along *Suaeda* roots revealed that sodium tended to accumulate at root tips (see also Chapter 7).

One of the most common means of adaptation of plants to saline habitats involves capacity of plants to grow roots into nonsaline niches of a generally saline habitat. Plants with a deep root system may thrive under epipercolative water regimes because their roots remain below the saline layers. On the other hand, such plants cannot survive under a subpercolative regime with a salty horizon deep in the soil profile. The reverse is true for shallow rooted plants.

Salicornia sp. and *Allenrolfea occidentalis* are examples of plants which grow on soils with a highly saline top layer, but with only moderately saline lower

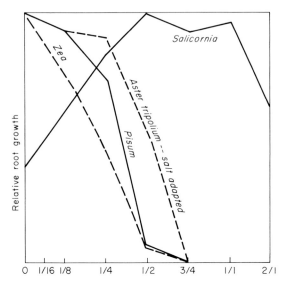

Fig. 10.8. Effects of salinity (Brenner's solution substituting seawater) on root growth in two glycophytes (corn and peas) and two halophytes (*Aster tripolium* and *Salicornia stricta*). (Modified after Montfort and Brandrup, 1928.)

horizons. *Kochia vestita* serves as an indicator for soils that are nonsaline in the upper 30 cm, but contain saline layers at deeper levels (Weaver and Clements, 1938).

In the Neusiedler See region, the roots of *Lepidium crassifolium* reached a depth of 1 m, and extended below the saline horizons. Under the same conditions, roots of *Camphorosma ovata, Suaeda maritima, Salicornia herbacea,* and *Puccinellia peisonis* were concentrated in the saline A_2 horizon, i.e., not deeper than 20 cm (Repp, 1939). Evidently, the latter species are salt-requiring plants. Roots of *Triglochin maritimum* and *Plantago maritima* reach a depth of 30–50 cm. In the coastal salt marshes of the eastern coast of the United States, most species have a very shallow root system (Steiner, 1935). Only *Juncus* roots penetrated the soil as deep as 100 cm (Fig. 10.9).

The root systems of halosucculents generally are poorly developed, although they vary from plant to plant and from site to site. Roots of *Arthrocnemum glaucum* descended vertically to a depth of 60–70 cm in the Dead Sea region, Israel; branching began at a depth of 15–25 cm, and reached maximum branching at 30–50 cm (Stocker, 1928; Shmueli, 1948). *Suaeda monoica* also has a deep root system with roots reaching a depth of 3 m. In contrast *Suaeda palaestina* has a superficial root system. *Prosopis farcta* and *Alhagi maurorum* have very well developed rhizomes and their root system reaches some 15 m or more (Shmueli, 1948). *Nitraria retusa* has a strong and highly branched root system penetrating to a depth of over 3 m.

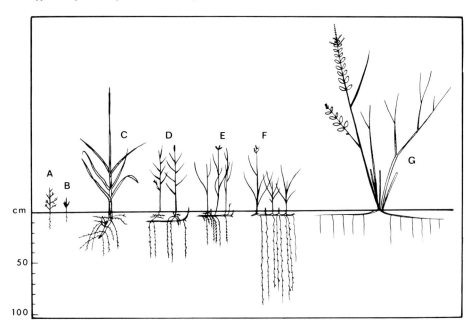

Fig. 10.9. Typical root systems of various salt marsh plants. (After Steiner, 1935.) A, *Salicornia europaea;* B, *S. mucronata;* C, *Spartina glabra;* D, *Distichlis spicata;* E, *Spartina patens;* F, *Juncus gerardi;* G, *Iva ovaria.*

Rhizomes of *Distichlis stricta* were found inside the profile of a saline-alkali soil in Nevada, not deeper than 65 cm (Robertson, 1955). Fine roots of this species as well as of *Agropyron elongatum* were found at even lower depths, down to 3 m below the soil surface. The ability of halophytes to avoid salinity by extending their roots into regions of low salinity is thus a common phenomenon.

Root growth in halophytes is usually a seasonal phenomenon. According to Stocker (1928), Hungarian "szik" soils do not contain active roots whenever their osmotic potential is above 28 atm. Under such conditions, only adult suberized roots were found across those soils. However, when sufficient rains fall to leach out some of the salt, new roots develop in the leached horizons.

This is also true of the rooting capability of various plants. Best rooting occurred in the halophytic plants of Sedom, Israel, during the winter. Very poor rooting if any, was found during the mid-summer months, even when the medium in both cases was moist and nonsaline (Pollak and Waisel, 1972).

Rapid growth of roots is of ecological significance for halophytes. Root development of seedlings of halophytic ecotypes of *Prosopis farcta* exceeded root growth of glycophytic ecotypes under similar conditions of salinity (Waisel, 1960d). Roots of the halophytic ecotypes were only slightly affected by salinity above 0.1 M NaCl, whereas growth of the glycophytic ecotypes was adversely affected by any addition of NaCl to the root medium. Thus, rapid root growth appears to be an important adaptive characteristic under saline conditions.

GROWTH OF LEAVES

According to Strogonov (1964), salinity, especially chloride salinity, inhibits leaf initiation, differentiation, and growth in glycophytes, but stimulates it in halophytes. *Atriplex halimus* plants had a higher leaf area, when grown in media containing NaCl than in those containing Na_2SO_4. According to Gale and Poljakoff-Mayber (1970), such responses result from a more intensive sprouting of lateral leaf buds. *Suaeda monoica* seems to differ from *Atriplex* in this respect. Greater leafiness of *Suaeda* plants grown under saline conditions, results from better growth of individual leaves rather than from an increase in their number.

CAMBIAL ACTIVITY

The cambium is very sensitive to salinity. Cambial activity and xylem production in stems of tomato (Hayward and Long, 1941) or *Populus euphratica* (Liphschitz and Waisel, 1970a, b), were low under saline situations. Under such conditions, a drop in total production was observed, as well as a shift from the differentiation of vessel members to the formation of parenchyma cells. The effect on the quantity of conducting tissues must eventually affect the water relations and growth of plants.

FLOWERING

Flowering is a very sensitive phase in a plant's ontogenetic development. Most plants require a readily available water supply during periods of flower initiation, anthesis, and early fruit production. On the other hand, at least one investigator (Alvim, 1960) claimed that a period of exposure to low water potentials was required for *Coffea* plants in order to induce full flowering.

Very little is known about factors controlling flowering of halophytes. According to Halket (1915), *Salicornia olivieri* plants do not flower unless salt is present in their growth medium. The seasonal behavior of various halophytes in the field does not differ considerably from that of a similar population of glycophytes (Fig. 10.10), and suggests that photoperiod, thermoperiod, light intensity, and plant age may play a role in the control of flowering in both groups.

Spartina townsendii and *S. anglica* are long-day plants and do not flower unless the day length exceeds 10 hours (cf. Hubbard, 1969). However, the few

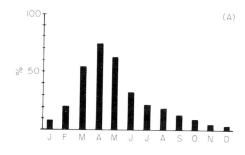

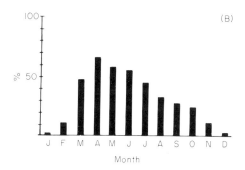

Fig. 10.10. Flowering spectrum of the glycophytic (A) and halophytic (B) flora of Israel.

species of terrestrial halophytes which were investigated in Israel under controlled conditions (*Nitraria retusa, Suaeda monoica,* and *Aeluropus litoralis*) showed no photoperiodic response. No flowering was found in young plants under constant temperature conditions, either under long- (16 hours) or short-day (8 hours) conditions. Neither did *Nitraria* plants flower when kept for 2 months under long-day conditions, and then moved to short days or vice versa.

The flowering responses of various species which are exposed to water stresses seem to vary. *Nitraria* plants usually flower late in spring, when the soil still contains available water. During the summer, when high water deficits prevail, no flowers are initiated in *Nitraria* plants. On the other hand, *Suaeda monoica* flowers in the autumn following a long period of high water stress. *Aeluropus litoralis* flowers in the field throughout the year, provided sufficient water is available. Coastal halophytes seem to behave more uniformly and most species are late flowering. Other species are not restricted to specific seasons in their flowering. Thus, generalizations at this stage are still remote and more research is needed.

Aftereffects of Salinity

Aftereffects of salinity on plant growth and metabolism are well known. Stimulation of growth of shoots and roots is very common in plants which were transferred from saline to nonsaline conditions. Thus, it is not only a direct adaptation of plants to stress, but also their release from temporary stresses which affect growth.

Productivity

Because of the adverse conditions for growth in saline habitats, the natural productivity of the native plants is usually low. Productivity of the takyr salines of central Asia (Rodin, 1954) was only $\frac{1}{100}$ of the productivity of similar adjacent nonsaline sites (Table 10.2). Marshes are the most productive of saline habitats. A *Spartina* marsh utilized approximately 6% of the incident radiation energy (Teal and Teal, 1969). Production of organic matter by a *Spartina* salt marsh may reach values as high as 10 tons/acre/year. Among the various types of salt marshes, the highest yields were obtained on estuaries, where water was somewhat diluted and salinity was reduced. Very little is known of the primary

Table 10.2

PRODUCTIVITY OF THE *Takyr* SALINES AND THE ASH AND NITROGEN
CONTENT OF THEIR PLANT MATERIAL[a]

Zone	Annual production of plant cover (dry weight) (kg/ha)	Nitrogen content of the plant cover (kg/ha)	Ash content of the plant cover (kg/ha)
I Low and saline spots, very few annuals	100	1.5	6–7
II Average habitats of the takyr, few annual plants only	300	9–10	20–22
III a. Spots rich in annuals transition to nonsaline conditions	1200–1600	20–30	90–140
b. Semi-desert vegetation with annuals on seroziem soils	10,000–12,000	90	475

[a] Adapted from Rodin, 1954.

productivity and energy relationships in other halophytic ecosystems. Productivity of *Salicornia* stands increases with salinity (Table 10.1). Productivity of submerged marine angiosperms is also high and dry weight production of 5 ton/ha has been reported (Bacalbasa, 1951). Also scattered data on productivity of mangroves have been reported (Troll and Dragendorff, 1931; Chapman, 1954; Scholander *et al.,* 1955; Golley *et al.,* 1962). However, productivity of most saline habitats is still unexplored.

Metabolism

‒ | |

High concentrations of soluble salts in the root medium influence plant metabolism in a variety of ways. For most glycophytic species, high salt concentrations are toxic. For others, the presence of at least some salts is essential for normal growth. Several questions thus arise about differences in metabolism which eventually lead to different responses of the two groups of plants. What are the specific effects of various ions which are toxic to glycophytes, but which do not affect facultative halophytes? What is the nature of the various metabolic mechanisms in obligatory halophytes which require salts for normal operation, etc?

Salinity and Respiration

The effects of low water potentials on respiration of different species of plants were found to be diverse. Subjecting pine seedlings to water potentials ranging from 4 to 44 atm resulted in fluctuations in their respiration. A rise in water deficit up to 12 atm caused a decrease in respiration, then an increase, and consequently above an osmotic potential of 28 atm, another decrease in respiration was found (Brix, 1962).

The correlation between respiration and salinity of the growth medium varied among different groups of plants. According to Takaoki (1957), different relationships are typical of each group (Fig. 11.1). In halophytes, an increase in salinity of the medium caused an increase in the osmotic potential of plant tissues, as well as in their respiration rates. Similar treatments caused only a limited increase in the osmotic potential of glycophytes, but a decrease in their rates of respiration. Nevertheless, in both cases, information is available for very few species and generalizations can be misleading. According to Kreeb (1964), no changes in respiration and photosynthesis were caused in *Salicornia* or *Suaeda* plants by addition of NaCl to the growth media, in spite of induced stimulation of growth.

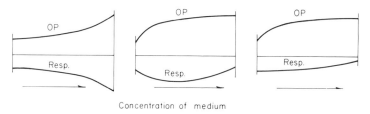

Concentration of medium

Fig. 11.1. Relationships between osmotic potential and respiration in a few groups of plants; left, halophytes; center, semihalophytes; right, glycophytes. (After Takaoki, 1957.)

Differences in response to salinity were also observed in glycophytes, not only between species, but also between organs. Salinization usually affected leaves more than the stems and roots. Sodium chloride in the growth medium increased respiration of pea leaf segments by over 30%, and that of stems and roots by only 10–15%. Stimulation of respiration by salts was found in normally grown plants, as well as in plants treated with low concentrations of DNP (Livne and Levin, 1967).

Halophytic species show very similar trends with respect to the responses of growth and respiration to salinity (Poljakoff-Mayber and Meiri, 1969). Growth of *Tamarix tetragyna* saplings was affected by saline media with an osmotic potential as low as −3 bars, although the water content of their shoots was affected only by salinity with an OP higher than −10 bars. Respiration of *Tamarix tetragyna* roots differed when plants were grown under various conditions of salinity. Under salinity treatments with osmotic potentials of −3 bars, rates of endogenous respiration in such roots were lowered to about 50% of the rates in plants grown in nonsaline solutions. A consequent increase in salinity, up to a potential of −12.4 bars, caused no further change in respiration. As expected, respiration rates usually were higher when measurements were made in slightly saline media, than in distilled water. However, respiration rates of plants which had been previously grown under nonsaline conditions, did not change even when measurements were conducted in saline solutions.

Similar relationships were also reported for other halophytes. Van Eijk (1939) found higher rates of respiration in halophytes treated with various salt solutions than in control plants. Respiration rates of *Plantago coronopus* and *Phragmites communis* tissues also increased when plants were subjected to NaCl solutions (Takaoki, 1957).

Respiration of plant tissues can be divided into two components: ground respiration and anion respiration (Lundegårdh and Burström, 1933, 1935; Lundegårdh, 1945). The intensity of the latter depends on amounts of ions taken up (A) and on a constant induction rate (K) typical of each anion. Such a constant represents the amounts of CO_2 released from roots for each anion taken up. Such relationships can be expressed by the equation:

$$R_t = R_g + KA$$

where R_t = root total respiration and R_g = root ground respiration.

The K values for *Aster tripolium* roots decreased with increasing external concentration of the medium, and were much lower than K values obtained for the glycophytic wheat (Table 11.1).

Table 11.1
K VALUES FOR ROOT RESPIRATION[a]

Plant	Anion	K value
Wheat	NO_3^-	1.5–3.2
	Cl^-	2.4–4.8
	SO_4^{2-}	12
Aster tripolium	Cl^-	0.076–0.67
	SO_4^{2-}	0.32–1.08

[a] Compiled by Adriani, 1958 for data of Lundegårdh and Van Eijk.

Respiration of 2-week-old seedlings of the halophytes *Salicornia europaea, Aster tripolium,* and *Plantago maritima,* was compared with that of three glycophytic species of *Plantago* and two of *Aster* (von Willert, 1970). The rates of oxygen uptake increased only slightly in the halophytes when NaCl was added in concentrations of up to 100 mM. At greater concentrations, up to about 600 mM, respiration of the halophytic species increased rapidly. Glycophytes differed in this respect, with sodium chloride inflicting damage on their respiratory mechanisms after only short exposure (von Willert, 1970).

Changes in respiration, induced by salinity, may also be qualitative. A decrease in respiratory quotient (RQ) was frequently observed under such conditions. Such a decrease in RQ reflects higher production of organic acids by the treated tissues. Among the organic acids, malate is probably the most important in glycophytes, whereas in halophytes it is oxalate.

RQ values of respiring *Tamarix* roots are high. They increased from 1 to 2 upon a change from a nonsaline medium to one of -15 bars. An increase in RQ suggests conditions which are similar to those which occur in plant tissues under anaerobiosis. Such an assumption is also supported by accumulation of lactic acid in roots grown under saline conditions (Poljakoff-Mayber and Meiri, 1969).

The various components of respiration also differ in different types of plants. Mangrove seedlings normally have an appreciable anaerobic component in their respiration (Chapman, 1962). Nevertheless, submerged organs of *Avicennia* seedlings had lower rates of respiration than did emerged ones. In both cases, respiration increased with age (Bharucha and Shirke, 1947).

Salinity and Carbohydrate Metabolism

Metabolism of carbohydrates in plants is affected by a general increase in salinity, as well as by the types of ions present.

Effects of chloride and sulfate on the respiratory mechanism of pea roots were compared by Porath and Poljakoff-Mayber (1968). Differences in the metabolic paths which convert glucose to CO_2 were expressed by the C-6/C-1 ratio. Variations in this ratio suggest changes of respiratory paths between the glycolytic and the TCA cycles, on the one hand, and the pentose cycle, on the other hand (Katz and Wood, 1963). The ratio in such types of experiments is computed from the yield of labeled CO_2 and calculated as percentage of the radioactive glucose absorbed by plant tissue. The C-6/C-1 ratio in pea tissues showed an obvious pattern of changes either with increasing salinity of the medium or with changes in its type (Porath and Poljakoff-Mayber, 1968). A general decrease in the C-6/C-1 ratio was found with increasing salinity. Two types of salinity, e.g., sodium sulfate, and sodium chloride, depressed rates of absorption of external glucose, as well as respiration rates. Chloride salinity depressed CO_2 evolution from C-6 labeled glucose, but did not affect CO_2 evolution from C-1 labeled glucose. Sulfate salinity depressed both.

Although changes in the C-6/C-1 ratio are not a direct function of the entire metabolism of glucose in tissues, such a ratio certainly suggests changes induced by salinity in the activity of various enzymatic systems. The relative percentage of the pentose–phosphate pathway increased in plants exposed to NaCl salinity, but was practically unaffected in plants exposed to Na_2SO_4 salinity.

Activity of many respiratory enzymes extracted from roots or pea plants grown in saline media was depressed. The activity of glucosephosphate isomerase was unaffected by sulfate salinity. Moreover, activity of this enzyme was even accelerated more than tenfold by chloride salinity (Tables 11.2 and 11.3). As shown in these tables, not all enzymes investigated were affected similarly. Activity of phosphogluconate dehydrogenase remained practically unchanged under different salinity treatments. Sulfate, as well as chloride salinity, increased the NADP-linked activity of glucose-6-phosphate dehydrogenase in the soluble fraction. The mitochondrial fraction of the same enzyme was left unaffected.

Table 11.2

EFFECT OF SALINIZATION OF GROWTH MEDIUM WITH NaCl ON SPECIFIC ACTIVITY OF SOME GLYCOLYTIC ENZYMES IN ROOT TIPS OF PEA[a,b]

Enzyme	Salinity (atm)			
	0.0	1.0	3.0	5.0
Glucosephosphate isomerase	0.04	0.10	0.30	0.60
Fructokinase	0.18	0.11	0.11	0.12
Triosephosphate isomerase	0.20	0.16	0.15	0.12
Pyruvate kinase	0.27	0.13	0.12	0.12

[a] After Porath and Poljakoff-Mayber, 1968.

[b] Results are given as units OD/mg protein/min. The results are means of at least three different experiments, each of them in duplicate.

Table 11.3

EFFECT OF SALINIZATION OF GROWTH MEDIUM WITH Na_2SO_4 ON SPECIFIC ACTIVITY OF SOME GLYCOLYTIC ENZYMES IN PEA ROOT TIPS[a,b]

Enzyme	Salinity (atm)			
	0.0	1.0	3.0	5.0
Glucosephosphate isomerase	0.04	0.03	0.03	0.04
Fructokinase	0.18	0.13	0.14	0.14
Triosephosphate isomerase	0.20	0.13	0.11	0.07
Pyruvate kinase	0.27	0.21	0.09	0.06

[a] After Porath and Poljakoff-Mayber, 1968.

[b] Results are given as units OD/mg protein/min, salinity is expressed in atmospheres. The results are means of at least three different experiments, each of them in duplicate.

Respiration of exogenous substrates by *Tamarix* roots was also affected by salinity (Poljakoff-Mayber and Meiri, 1969). Comparison between the behavior of a glycophyte pea and a halophyte (*Tamarix*) showed that up to a salinity of about $0.1 M$ NaCl, both species behaved similarly. However, at higher salt concentrations, in which peas did not survive, the halophyte survived despite a drastic metabolic change (Poljakoff-Mayber and Meiri, 1969).

Starch metabolism in plants may be altered variously by chloride. Montfort (1931) compared the starch metabolism in a glycophyte (*Syringa*) and in a halophyte (*Aster tripolium*). Only minor effects of salinity on metabolism of starch were observed in *Aster,* whereas in *Syringa,* starch content dropped to half, by a saline treatment below $^1/_8$ the concentration of North Sea water (Fig. 11.2).

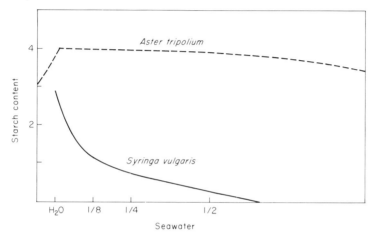

Fig. 11.2. Hydrolysis of starch by sliced leaves of a glycophyte (*Syringa vulgaris*) and a halophyte (*Aster tripolium*). Leaves were placed for 5 hours in the sun with their petioles in water of different salt contents. (After Montfort, 1927.)

In addition to salinity, inundation also has an important role in metabolism of *Salicornia stricta* plants. According to Langlois (1971), *Salicornia* plants that were inundated twice a day had a low content of sugars, whereas plants that were not inundated accumulated large quantities of sugars in their shoots.

Effects of Salinity on Activity of Some Enzymes

Since some enzymes are more sensitive to water stresses than others, shifts in metabolism are observed under saline conditions.

A high salt content of plant tissues is known to affect enzymes which are involved in starch metabolism. A high salt content in glycophytes inhibits starch synthesis and promotes its breakdown. However, such processes in halophytic species were only slightly affected (Montfort, 1927). Effects of salinity on starch metabolism depend primarily on monovalent ions, and are inhibited by divalent ions (Mg^{2+} or Ca^{2+}) (Stocker, 1933). The minimal, optimal, and maximal concentrations of salts which affect activity of certain enzymes are far higher in halophytes than in glycophytes. Starch decomposition in guard cells of the halophyte, *Glaux maritima,* began at $0.4\,M$ NaCl, whereas in cells of the glycophyte, *Rumex acetosa,* it started in 0.05 M NaCl (Iljin, 1932 cited by Stocker, 1933).

Activity of a number of isozymes of malate dehydrogenase was investigated in ten species of plants under different salinity conditions (Weimberg, 1968). Among the species investigated, only radish (*Raphanus sativus*) can be regarded as a semihalophytic one. It is thus interesting to indicate that specific activity of malate dehydrogenase in this species was far lower than in any of the other species investigated. No isozyme of malate dehydrogenase was common to all tissues or to all species investigated.

Production of isozymes also differed with the type of salinity. Two isozymes of malic dehydrogenase were electrophoretically separated from pea plants grown on nonsaline and Na_2SO_4-containing media. A third isozyme was obtained from plants treated with NaCl (Porath and Poljakoff-Mayber, 1968). Activity of all isozymes of malic dehydrogenase was stimulated by addition of NaCl up to a concentration of 0.02 M. However, above this concentration, NaCl had an inhibitory effect (Weimberg, 1967).

Although growth of pea plants was retarded under saline conditions, levels of specific activity of eighteen different enzymes remained constant in plants grown on nonsaline media, or on media salinized by NaCl, KCl, Na_2SO_4, or K_2SO_4 (Weimberg, 1970).

Sodium chloride had a stimulatory effect on the activity of amylase when given in concentrations of 5–40 mM (Dixon and Webb, 1957). According to Strogonov (1964), this enzyme was activated similarly by NaCl and Na_2SO_4. However, sodium and chloride are not always the most effective ions. Starch synthetase obtained from various plants had an apparent requirement for potassium. The efficiency of sodium in activating the enzyme was only 21% of that of potassium (Nitsos and Evans, 1969).

Enzymes isolated from salt-tolerant bacteria remained active in media containing higher salt concentrations than enzymes taken from salt-sensitive strains. However, this is not a general occurrence. Comparisons of enzymes taken from different species of *Atriplex* yielded similar levels of salt tolerance (Greenway and Osmond, 1970).

Salts may change rates of activity of some enzymes, but not of others. Usually, ions at low concentrations have stimulatory effects on cytochrome oxidase activity, with monovalent ions being more active than di- and polyvalent ones. At high salt concentrations, inhibitory effects were observed, with monovalent ions being less inhibitory than polyvalent ones (Peterson, 1961). Several ions (Cl^-, SO_4^{2-}, Br^-, NO_3^-, and I^-) exhibited the same effects on enzyme activity when applied at similar activities.

Activity of acid phosphatase was inhibited in *Suaeda monoica* plants that were exposed to saline media, whereas activity of PEP-phosphatase was stimulated (Ahmad and Hewitt, 1971). Addition of Cu ions accelerated, in that species, the activity of both enzymes.

SALINITY AND ATPASE ACTIVITY

As mentioned, salinity influences enzymes which are involved in energy transport.

Various ATPases with different characteristics were extracted from plant cells. Dodds and Ellis (1966) isolated a cation-activated ATPase from the particulate fraction of pea and carrot root homogenates. Activity of such a system was stimulated by potassium chloride, potassium nitrate, or by sodium, rubidium, lithium, and ammonium ions. Matile *et al.* (1967) found an ATPase attached to fragments of the plasmalemma of yeast cells. There is some evidence that a sodium–potassium-activated ATPase with similar properties is also located on the plasmalemma of cells of *Nitella translucens* (MacRobbie, 1964, 1965, 1966; Spanswick and Williams, 1964). An enzyme with similar properties was extracted from bean roots by Gruener and Neumann (1966). However, in that case, most of the activity was in the supernatant, and not in the particulate fraction of the preparation.

Activity of ATPase depends upon many things, including ionic environment. According to El Sheikh *et al.* (1967), ATPases exhibit higher rates of activity when at least one of their active sites is bound to sodium, than when both sites are bound with potassium ions. Even at that point, the available information is not clear. A nonsoluble ATPase preparation (pellet of sugar beet homogenate centrifuged at $20,000\,g$ for 60 minutes) responded to magnesium ions. An additional stimulation was obtained by the addition of $Na^+ + K^+$ at a constant ionic strength. Optimal synergism was obtained at an Na : K ratio of 1 : 1 and minimal stimulation at a ratio of 1 : 8 (Hansson and Kylin, 1969).

Higher rates of ATPase activity must eventually reduce the content of ATP. In fact, data reported by Hasson-Porath and Poljakoff-Mayber (1971) showed that the level of ATP in pea roots decreased while that of ADP increased with increasing salinity. The level of AMP remained unaffected.

A $Na^+ + K^+$-activated ATPase was also found in leaves of *Avicennia nitida* (Kylin and Gee, 1970). Activity of the enzyme depended on both ionic strength of the medium, as well as on the ratio $Na^+ : K^+$ (Fig. 11.3).

Lipids comprise an essential part of ATPases and enzymes become inactivated when stripped of their lipids. Moreover, nonspecific ATPase of certain fractions of bean preparations became a more specific sodium- or potassium-activated enzyme when reconjugated with phosphatidyl choline (P. J. C. Kuiper and A. Kylin, personal communication).

Only little information is available on general influences of saline media on enzymatic activity in animals (Augenfeld, 1969; Thuet *et al.,* 1969), and plants (Zhukovskaya and Matukhin, 1964; Porath and Poljakoff-Mayber, 1968). Some data on ATPase activity concerns its role in ion transport (Skou, 1964; Donting and Caravaggio, 1966; Atkinson and Polya, 1967; Fischer and Hodges, 1969; Williamson and Wyn Jones, 1971). Although it was suggested that such enzymes

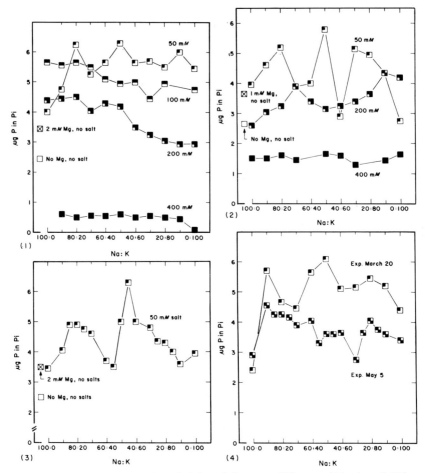

Fig. 11.3. (1) ATPase activity of *Avicennia* leaves at different salt levels and different Na : K proportions. Fraction obtained between 7000 *g* and 20,000 *g*. After several months with nutrient solution made up in 50% seawater, the plants had been grown a week in nutrient salts alone. Preparation made January 17, test run January 23. (2) ATPase activity of microsomal fraction of *Avicennia* leaves at different salt levels and different Na : K proportions. After several weeks in nutrient solution made up with 50% seawater, the plants were grown 10 days in nutrient solution made up with 25% seawater. Preparation from March 17, test performed March 20. In the interest of legibility the curve for 100 m*M* salts has not been drawn. It was intermediate between the curve for 50 and the curve for 200 m*M* salts. (3) Effects of freeze storage at −20°C on the ATPase activity in *Avicennia* leaves. Same microsomal preparation as in (2), but test performed May 5. Total monovalent salts 50 m*M*. (4) Effect of freeze storage on the ATPase activity of the microsomal fraction of *Avicennia* leaves. These plants had been grown 10 days in nutrients made up in deionized water. Previously they had been cultured several weeks in nutrients made up with 50% seawater. Preparations from March 17, tests from March 20 and May 5. Conditions as in (2) and (3) (MgCl$_2$, no salt) controls at 3.1 and 3.4 μg of P, respectively. (After Kylin and Gee, 1970.)

are important for plant adaptation to saline conditions (Jennings, 1968a), and contribute to salt resistance (Kylin and Gee, 1970), very little is known of the presence of various ATPase systems in plants of different ecological groups (Horovitz and Waisel, 1970a,b). With regard to the response of ATPase to sodium, it was suggested that halophytes differ from mesophytes only quantitatively, but not qualitatively (Jennings, 1968b). This view is based on observations that increased light, aridity, and high sodium concentrations had similar effects (cf. Biebl and Weissenbock, 1968), namely that they increased production of ATP and the succulence in plants of various ecological characteristics. In order to gather at least some information on this topic, ATPase activity of two glycophytic species (*Phaseolus vulgaris* c.v. Brittle wax and *Zea mays* c.v. White horse tooth) was compared with that of two halophytic species (*Suaeda monoica* and *Atriplex halimus*). Comparisons were made in growth media and reaction mixtures of different salinities (Horovitz and Waisel, 1970a,b). Addition of NaCl to the growth medium stimulated activity of the soluble ATPase of glycophytic bean and corn roots, but inhibited such activity in the roots of halophytic *Suaeda* and *Atriplex* plants (Table 11.4).

Table 11.4

THE EFFECT OF NaCl IN THE GROWTH MEDIUM AND IN THE REACTION MIXTURE
ON THE ATPase ACTIVITY OF VARIOUS PLANTS[a,b]

Plant species	Reaction mixture		Growth media			
			−NaCl (A)		+NaCl (B)	
			% from a		% from a	% from (A)
Phaseolus vulgaris	−NaCl	a	4.78		7.00		146
	+NaCl	b	5.06	106	8.04	115	159
Zea mays	−NaCl	a	2.67		4.85		182
	+NaCl	b	5.42	203	11.30	233	208
Suaeda monoica	−NaCl	a	4.04		1.87		46
	+NaCl	b	2.53	63	1.20	64	47
Atriplex halimus	−NaCl	a	7.02		4.33		62
	+NaCl	b	4.76	68	2.60	60	55

[a] After Horovitz and Waisel, 1970b.

[b] Data in nmoles P_i/30 min/mg protein.

Addition of 100 nmoles of sodium chloride to the enzymatic reaction mixture stimulated activity of ATPase in the supernatant of bean and corn roots (Na^+-activated), but inhibited that of *Suaeda* and *Atriplex* roots.

Evidently, results of such experiments suggested that a basic qualitative difference in response to sodium exists between the ATPase systems of the two glycophytic and two halophytic species. The question whether such a difference is of a general nature is still open.

It is not easy to say what role ATP might have in increasing succulence of plants. ATP might be involved in synthesis of new wall material or in increasing cell wall extensibility. Such an effect would also explain the role of phosphate in producing increased succulence of plant shoot tissues (Jennings, 1968b).

There is little difficulty in explaining increased production of ATP due to an increased rate of photophosphorylation. Effects of low water potentials seem to be more indirect. According to Jennings (1968b), such effects could also result from high uptake rates of sodium under arid conditions.

Understanding the role of sodium in nutrition and in physiology of halophytes requires additional study. The mechanism by which sodium ions induce ATP production is suggested as reversal of activity in a membrane-bound ATPase; an enzyme which otherwise acts as a sodium pump. If sodium-activated ATPase is present in the plasmalemma of higher plants, one would anticipate synthesis of ATP under conditions of high external concentrations of NaCl, but low internal ones (Jennings, 1968b). Such an assumption is based on the work of Garrahan and Glynn (1967) who succeeded in reversing the sodium pump in red blood cells, so that ATP was synthesized. No such data are available for plants, but under certain conditions, such reversal was thought to be feasible (Jennings, 1968b). However, for the enzyme to act in this manner, sodium ions within the cytoplasm must be maintained at low concentrations. Jennings (1967, 1968b) suggested that this could be achieved by a second pump at the tonoplast, which transports sodium ions into the vacuole. It should be realized that such a system of two pumps need not lead to a net synthesis of ATP, because ATP synthesis at the plasmalemma could well be cancelled out by ATP breakdown by a sodium pump at the tonoplast.

No evidence for the presence of a pump in the tonoplast has been reported. On the contrary, data obtained by X-ray microanalysis (Waisel and Eshel, 1971) indicate that the cytoplasm of *Suaeda* leaf cells is relatively rich in sodium. Thus, although appealing, the theory suggested by Jennings (1968b) for succulence still needs supporting experimental evidence.

Organic Acid Metabolism

Absorption of cations by plants is balanced partly by concurrent absorption of inorganic anions, and partly by endogenous synthesis of organic anions. A need for balance by organic anions arises during growth processes, when nitrate and sulfate are assimilated, or when cations are absorbed in excess of anions. An organic acid balance of excess cations, principally of oxalate, was found in many halophytic species. In certain species, e.g., *Atriplex* sp., it is known that organic acids may even be produced in excess. Plants with such a type of metabolism may contain appreciable deposits of calcium oxalate crystals in their leaf tissues. Moreover, chemical analyses of leaf samples collected from plants in the field showed that oxalate levels reached the order of 50–400 mEq/100 gm dry weight, and were in excess of amounts accountable as calcium oxalate (Osmond, 1967). Such data were found for leaves of fifteen different species of *Atriplex, Bassia, Kochia, Rhagodia,* and *Halogeton.*

The organic acid content of *Atriplex halimus* leaves is composed mostly of oxalate. The amount in the plant remains more or less constant through a wide range of high NaCl or KCl concentrations in the growth medium. Only in salt solutions of relatively low ionic strength was a low acid content found (Mozafar *et al.,* 1970a). A low content of organic acids was also noted in other plants, when the water potential of their tissues was also low (Zholkevitch and Koretskaya, 1959).

The expressed sap of *Atriplex nummularia* and *A. vesicaria* contained approximately 150 mEq. soluble oxalates per 100 gm dry weight, and the pH of the cell sap was 6.12 and 6.25 respectively. According to Osmond (1963), oxalate is present in the leaf cell sap as a counter-ion of sodium and potassium. As a matter of fact, most organic acids of plants, e.g., formate, acetate, and succinate are present in a mobile form. According to Jackson *et al.* (1970), such acids in the ionic form can even be taken up in large quantities by tissues such as barley roots. Oxalic acid in vesiculated leaf hairs of *Atriplex halimus* reached concentrations as high as $1.36 M$ (Goodin and Mozafar, 1970). The mean oxalate content in an individual hair was calculated as $4.1 \pm 0.5 \ \mu$Eq.

As may be seen in Table 11.5, organic anions, principally oxalate, also make

Table 11.5
IONIC BALANCE IN *Atriplex* spp.[a,b]

Plant material	Anions					Cations			
	Oxalate	Total acids	Cl	PO$_4$	Total anions	Na	K	Ca+Mg	Total cations
A. vesicaria[c] (young leaves)	74	147	116	31	294	196	57	98	351
A. vesicaria[c] (old leaves)	151	260	38	29	327	137	30	144	311
A. inflata[d]	312	392	58	6	456	189	182	149	520
A. spongiosa[d]	395	455	45	7	507	182	185	156	524

[a] After Osmond, 1963.

[b] All values expressed as mEq/100 gm dry weight leaf tissue.

[c] Grown in solution culture sodium chloride 5 mEq/liter+½ strength Hoagland's solution.

[d] Grown in sand irrigated with sodium chloride 5 mEq/liter+full strength Hoagland's solution.

an important contribution to the ionic balance of other species of *Atriplex*. Such a contribution is relatively small in young leaves, but increases with leaf age. An inverse relationship seems to exist between chloride tolerance and organic acid production in different species. Osmond (1963) showed that organic acid content in mature leaves of chloride-tolerant perennials, e.g., *A. vesicaria,* was lower than in chloride-sensitive annuals, such as *A. spongiosa.* Salt-tolerant and salt-sensitive ecotypes of *A. vesicaria* showed a similar pattern.

Compartmentation regarding different systems of ionic balance can be found in different cells within one tissue (Osmond, 1963). Organic anions appear to balance cations in mesophyll cells of *Atriplex* leaves, whereas inorganic anions mainly balance cations in the vesicles of the same leaves.

Oxalate synthesis may be geared to the same processes involved in accumulation of inorganic ions. Formation of calcium oxalate crystals in some mesophyll cells may be regarded only as a visible manifestation of accumulation processes common to vacuolated cells, and may shed light on the chemical properties of certain constituents of the vacuolar sap. It is unlikely that such crystalline deposits function either as a store of calcium, or as a means of disposal of "toxic" excess quantities of calcium or oxalic acid (Osmond, 1963; von Lötsch and Kinzel, 1971).

Oxalate production in four species of *Atriplex* was markedly reduced at low light intensities (Osmond, personal communication). This could possibly imply that a certain photosynthetic product comprises the immediate substrate for oxalate synthesis. Evidence from other Chenopodiaceae indicates that glycolate, an early photosynthetic product, is a precursor of oxalic acid. The exact biochemistry of oxalate metabolism in *Atriplex* is still obscure. However, in view of the high oxidation level of oxalic acid, and its accumulation to high levels in leaves, it seems unlikely that oxalate is involved in energy yielding metabolic processes. Still, some of it may be involved in generation of active formyl groups.

Dark CO_2 Fixation

It is well established, that certain glycophytic plants assimilate ^{14}C when exposed to [14]CO_2 in the dark. Dark assimilation results in net accumulation of Krebs cycle acids in leaves of Crassulaceae (cf. Ranson and Thomas, 1960) or is associated with excess cation uptake by grass roots (Sutcliffe, 1962). In all reported instances, malic acid was found to be the principal labeled compound, although other organic and amino acids were also accumulated. It is generally accepted that CO_2 fixation in the dark occurs after CO_2 reacts with phosphoenol pyruvate (PEP) to form oxaloacetate and, thereafter, malate and citrate. According to Webb and Burley (1965), a pool of aconitic acid builds up in NaCl-treated plants. Acids of such a pool only slowly acquire ^{14}C from newly supplied $^{14}CO_2$. Such response thus supports the idea that several compartments exist even within one cell.

Acid metabolism occurs in various members of the Crassulaceae, Asclepiadaceae, Euphorbiaceae, Cactaceae, Begoniaceae, and Compositae. It is found in plants which grow in different climates (Milburn *et al.*, 1968). In all such plants, *RQ* is less than 1, meaning a high consumption of oxygen, but a low release of CO_2. The acid content of such plants is higher in the morning than in the evening. High acid content of terrestrial plants can be attributed primarily to accumulation of malate. However, in some species, citrate also accumulates. During the day, plants undergo photodeacidification and their acid content decreases.

Dark CO_2 fixation under saline conditions has been investigated in various algae (Craigie, 1963), spinach (Joshi *et al.*, 1962), *Bryophyllum* (Karmarkar and Joshi, 1969), *Batis maritima,* and *Salicornia europaea* (Webb and Burley, 1965). Under saline conditions, a shift from the normal pattern of acid accumulation was observed and most of the oxaloacetate turned into amino acids.

Labeled malic acid was either absent or appeared only in traces in marine plants which were exposed to $^{14}CO_2$ in the dark. In such plants, the principal labeled compounds were aspartic, glutamic, and amino acids. In *Zostera marina,* the main products of dark fixation were also amino acids. However, in this species, malic acid had also accumulated in significant quantities.

Dark fixation of CO_2 in *Batis maritima* and *Salicornia europaea* was reported by Webb and Burley (1965) and in *Nitraria retusa* by Keren (1970). Dark fixation of labeled CO_2 in four salt marsh succulent halophytes indicated a disparity of the label in favor of amino acids (Table 11.6). The remainder of the radioactivity was found in neutral compounds.

Rates of dark fixation of CO_2 by *Nitraria* plants were four times higher than those of tomato, even though both species were grown under similar conditions. Moreover, when *Nitraria* plants were treated with a 0.1 *M* NaCl solution, their rates of dark CO_2 fixation at least doubled the rates exhibited by plants grown in NaCl-deficient media (Keren, 1970).

Table 11.6

PERCENTAGE DISTRIBUTION OF RECOVERED ^{14}C AFTER DARK FIXATION OF $^{14}CO_2$ BY SALT MARSH HALOPHYTES[a]

Species	Amino acid fraction	Organic acid fraction
Salicornia virginica	63.5	12.5
Salicornia europaea	52.5	25.0
Batis maritima	73.7	9.6
Borrichia frutescens	46.8	22.1

[a] After Webb and Burley, 1965.

The final product into which CO_2 is incorporated in dark fixation differs among species and ecological conditions (Tables 11.6 and 11.7). The mean rate of incorporation of supplied ^{14}C into the ethanol soluble fraction of *Spartina alterniflora* was 3.6 times higher for salt-deficient than for salt-containing plants (Webb and Burley, 1965).

Table 11.7

PERCENTAGE DISTRIBUTION OF RECOVERED ^{14}C AFTER DARK FIXATION
OF [14]CO_2 BY *Spartina alterniflora*[a]

	Amino acids		Organic acids	
Time in hours	−NaCl	+NaCl	−NaCl	+NaCl
0.5	59.8	66.8	40.2	30.4
1	41	59.4	59	36.7
1.5	37.4	56	62	41
3	11.4	46.5	88.6	50.5
4.5	−	37	−	60

[a]After Webb and Burley, 1965.

Since most species of salt marsh succulents, in which dark CO_2 fixation was investigated so far, responded similarly, such phenomena actually seem to be associated with salinity of the environment. Effects of NaCl on dark fixation may be exerted either indirectly or directly on the CO_2 carboxylating reactions and closely associated metabolic processes. Dark fixation may enable some plants to tolerate, or even require, high internal salt levels.

The ratio of ^{14}C amino acids to ^{14}C organic acids calculated from published data of dark fixation products was generally greater than 1 for halophytes (Rho, 1959; Joshi, *et al.*, 1962; Craigie, 1963; Webb and Burley, 1965), and less than 1 for glycophytes (Saltman *et al.*, 1957; Kunitake *et al.*, 1959). *Spartina alterniflora* showed a combination of patterns of glycophytic and halophytic plants (Table 11.7), depending on the duration of the experiments. The amino acid : organic acid ratio was typically "halophytic" at short experimental periods changing toward a "glycophytic" ratio with increasing time of exposure. Such behavior was observed in saline, as well as in salt-free media.

According to Karmarkar and Joshi (1969), the dominant enzyme in such processes under nonsaline conditions was malic dehydrogenase. When NaCl was added, the activity of malic dehydrogenase was inhibited and the equilibrium was shifted by increasing the activity of transaminases. Joshi *et al.* (1962) also suggested that NaCl, *per se*, did not affect the primary pathway of dark CO_2 fixation, but either activated the oxaloacetate-aspartate and other transaminase enzymes, or inhibited malic dehydrogenase. Such inhibition could be caused by altering the affinity of the coenzyme for the enzyme. Craigie (1963) also agreed that NaCl did not affect the carboxylation reaction directly, but appeared to induce a strong tendency for transamination reactions.

Dark fixation of CO_2 in halophytes seems to be highly significant, in view of the fact that it is one of the few metabolic differences which was established between halophytes and glycophytes. Furthermore, under high water stresses which prevail in salines, and when stomata are closed, dark CO_2 fixation remains the only means for carbon supply for such plants. Thus, it is obviously of considerable importance for plants growing under extremely dry or saline conditions (Nishida, 1963; Klinge and Fischer, 1967).

Effects on Assimilation

It is well established that halophytic species can survive in concentrated salt solutions, not because they avoid salinity, but mostly despite it. Thus, stomata of *Aster tripolium* remained functional and rates of photosynthesis and of starch production of these plants remained unaffected even in a 4.2% NaCl solution (Montfort, 1926). The same was true for other halophytes.

Low growth rates, or net assimilation rates of glycophytes exposed to saline conditions may result either from accelerated respiration and faster consumption of metabolites, from reduced photosynthesis, or from the combined effects of both.

Low water potentials in plant tissues may affect photosynthesis in several ways: (1) they may affect diffusion of CO_2 into binding sites, (2) alter the structure and function of organelles, which are responsible for the light reaction, (3) change chemical processes of the dark reaction, and (4) inhibit removal of assimilated products. Most frequently, all four processes are affected by low water potentials simultaneously, although not to the same extent. Consequently, rate-limiting processes differ occasionally.

Boyer (1965) reported that photosynthesis of cotton plants was reduced to about 70%, when plants were grown under saline conditions (-8.5 bars). Such reduction occurred despite the fact that stomata remained open and CO_2 movement into the leaves was unaffected. Thus, salinity must have either affected permeability of other barriers to diffusion of CO_2, or reduced any of the other constituents of photosynthetic activity. Reduction in net photosynthesis also was recorded in sugar beet plants grown under high water stresses.

The stage at which photosynthesis is inhibited by low water potentials is obscure. In certain cases, it was suggested that CO_2 diffusion through the stomata could be limiting. However, it is questionable to what degree even closed stomata actually inhibit CO_2 movement into leaves (Gaastra, 1959). In other cases, diffusion across plasma membranes of the mesophyll cells was affected, but in still other cases, rates of various biochemical processes involved in CO_2 assimilation were reduced.

According to Talling (1961), succulent halophytes do not show distinct peculiarities regarding photosynthesis. Nevertheless, rates of photosynthesis expressed on a fresh weight basis are naturally low, and the ratios of rates of respiration to photosynthesis, calculated on a unit area basis, are unusually high (Neuwohner, 1938; Beiler, 1939).

Salts, in many cases, accelerate rather than inhibit assimilation in halophytes. Assimilation in submerged sea plants (e.g., *Zostera marina*) was optimal in solutions containing salt concentrations equivalent to those of seawater. Photosynthesis of other submerged brackish-water or saltwater plants (e.g., *Ruppia rostellata*) was only slightly affected, even by changes of salinity from traces to about 1.8 M NaCl.

Effects of salinity on photosynthesis of various species of terrestrial halophytes (*Frankenia, Tamarix, Phragmites,* etc.) were investigated by Stocker (1954). The plants exhibited high rates of photosynthetic activity, even in saline media, provided that their water content was not limiting. Thus, it seems that in such plants salinity did not have toxic effects. On the other hand, small alleged decreases in photosynthetic activity were reported for various other species of halophytes (e.g., *Salicornia* sp. and *Arthrocnemum* sp.) grown under highly saline conditions (Arnold, 1955; Adriani, 1958).

Salinity affects assimilation in halophytes much less than it does in glycophytes. Experiments conducted by Montfort (1927) showed that starch accumulation in leaves of *Aster tripolium* was only slightly affected when plants were subjected to various concentrations of Brenner's solution (substituted seawater). By comparison, under the same conditions, starch accumulation by leaves of the glycophytes, *Syringa* and *Helianthus,* was markedly affected (Fig. 11.4).

Variations in metabolic paths of photosynthesis are also found between halophytes and glycophytes, as well as among closely related halophytic species. Osmond *et al.* (1969), Björkman *et al.* (1969a,b), and Osmond (1970), reported that a few species of *Atriplex* which were investigated differed in their photosynthetic CO_2 fixation systems. While in *Atriplex hastata,* the initial products of photosynthesis were 3-carbon compounds (3-PGA), malate and aspartate of the 4-carbon system were found in *Atriplex spongiosa* and in *A. semibaccata* (cf., Hatch and Slack, 1966, 1970; Osmond, 1967; Johnson and Hatch, 1969). According to Laetsch (1968), such CO_2 fixation is correlated with a specific structure of the leaf, the so-called "kranz" type of assimilating tissue (cf. Downton and Tregunna, 1968) which is probably most efficient under conditions of high water stresses. Such plants have higher photosynthetic maxima and lower compensation points than plants with the common C-3

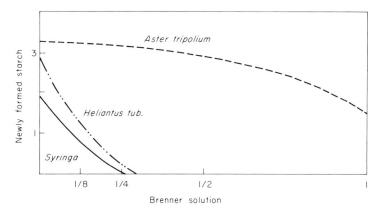

Fig. 11.4. Assimilation by leaves of glycophytes (*Syringa* and *Helianthus*) and a halophyte (*Aster tripolium*). Starch-free leaves were pretreated for 3 hours in water of different salt contents. (After Montfort, 1927.)

pathway of CO_2 fixation. The C-4 type plants are commonly found in habitats with high light and high temperature conditions. It is interesting to note that in plants with a C-4 carbon pathway, the uptake of chloride was stimulated by light. Such stimulation was not frequently found in C-3 plants (Lüttge *et al.*, 1971).

Plants with the C-4 type of carbon assimilation, e.g., *Atriplex spongiosa*, seem to lack photorespiration. Since photorespiration increases eightfold in C-3 type plants, e.g., in *Atriplex hastata*, between 20°-35°C, the absence of this type of respiration must be of ecological significance (Hofstra and Hesketh, 1969; Osmond *et al.*, 1969).

Rates of CO_2 fixation were three times faster in *A. spongiosa* than in *A. hastata*, due to lower intercellular resistance and lack of photorespiration in the former (Osmond *et al.*, 1969).

Osmond (1970) reported that *Atriplex spongiosa*, a species with the C-4 type of CO_2 assimilation, also had a low carboxydismutase and low glycolate oxidase activity, as well as a high PEP carboxylase activity.

A low compensation point is often linked to low rates of photorespiration.

As light saturation of mangroves is relatively low and occurs at 3000 ft.c, it is believed that many mangrove species have low rates of photorespiration. According to Laetsch (1968), the C-3 type of CO_2 fixation, where CO_2 forms oxaloacetate and malate first, was observed in many halophytes, and probably is typical of plants subjected to high water stresses. As this system is efficient, it enables plants to assimilate rapidly during the short periods when their stomata are open.

Photosynthetic activity differs even among ecotypes of a single species. Gabrielsen and Larsen (1935) showed that photosynthetic responses varied between different ecological forms of the halophyte *Aster tripolium*. Similar data were reported by Beljakoff (1930), Ferchau (1959), Milner *et al.* (1959), and Björkman *et al.* (1969a).

Carbon dioxide assimilation is influenced not only by constant salinity, but also by fluctuations in salinity, causing an increase in assimilation when plants are transferred from concentrated to diluted solutions. Nevertheless, after several periods of such fluctuations, an ultimate decrease in assimilation was observed (Stocker, 1954).

Leaves of *Salicornia* or *Arthrocnemum* plants turn light green when plants are grown in high salt concentrations. Such changes in color are due to a low chlorophyll content (cf. Table 11.8). Both salinity of the medium and the degree of root inundation affect the chlorophyll content of *Salicornia herbacea* plants and subsequently their assimilation.

Table 11.8

CHLOROPHYLL CONTENT ($a + b$) OF *Salicornia herbacea* PLANTS GROWN UNDER
DIFFERENT WATER LEVELS AND SALINITY (mg/cm^3)

Growth media	Water level		
	All roots drained	Only upper roots drained	Inundated
Hoagland's solution	121.2	188.0	79.6
Hoagland's solution + 200 mM NaCl	58.8	116.5	57.6

The exact role of chloride in plant metabolism is still unknown. However, it has been claimed that chloride is required for photosynthetic reactions, which involve oxygen evolution (Warburg, 1948; Bove *et al.*, 1963).

Effects of Salinity on Metabolism of Proteins and Growth Regulators

Plants which grow in saline environments absorb ions from the growth medium, but in different proportions. Such a change in ion content and ion composition of plant cells induced changes in activity of certain metabolic systems. One of the first effects of a high ion content in plant cells is likely to induce changes in protein hydration, because ions affect the nature of the hydration shell which surrounds the protein molecules (Klotz, 1958). This may affect activity of some active groups such as sulfhydryls, influence viscosity of protoplasm, etc. Various ions have different effects on hydration of enzyme proteins and may alter their activity (Laties, 1954; Hackett, 1961).

It was suggested that sodium ions disturb the organized structure of the hydration water of proteins, whereas potassium does not (Kavanau, 1966). Such changes may have serious consequences for membrane proteins. Under extreme conditions of salinity, proteins are precipitated. In certain cases, alterations in protein structure were shown to be at least partially osmotic, because similar effects of low water potentials on enzymatic activity were induced by both sucrose and NaCl (Kessler *et al.*, 1964).

Responses of enzymes to saline conditions vary in accord with their individual characteristics. Activity of some is impaired, and of others is accelerated. For example, Kessler *et al.* (1964) reported changes in enzyme activities at low concentrations of sucrose or NaCl. They also found that salinity strongly suppressed RNA and DNA content, particularly when salinity treatments were first imposed. A decrease in RNA was attributed to intensified activity of cytoplasmic RNase, whereas DNA content was probably reduced as a result of impaired synthesis. Reduction in size of leaves of bean plants treated with NaCl was assumed to result from inhibition in the RNA-protein metabolic path (Nieman, 1962). DNA content, under such conditions, was also negatively affected, but, in this case, it was due to a disturbed synthesis. Similarly to the activity of RNase, activity of various respiratory enzymes was also accelerated under saline conditions (Nieman, 1962).

Protein content of various plant tissues declined under drought or saline conditions, because of increased proteolysis and decreased protein synthesis (cf., Vaadia and Waisel, 1967). Such conditions also inhibited transamination processes (Zholkevitch and Koretskaya, 1959), and caused accumulation of ammonia (Strogonov, 1964).

Effects of salinity on incorporation of amino acids into pea root tip proteins were investigated by Kahane and Poljakoff-Mayber (1968). Apparently, such incorporation was of dual nature: in the presence of salts, uptake of leucine was depressed and normal metabolic pathways were disturbed (Fig. 11.5). If roots were grown in high salt solutions, uptake and incorporation of amino acids were affected, even if measurements were consequently carried out in the absence of salt. In such cases, NaCl or Na_2SO_4 affected uptake, incorporation, and metabolism of ^{14}C-labeled leucine in different ways; Na_2SO_4 caused a higher increase of leucine in the fraction of amino acids which was used as substrate for respiration. Pea roots grown in various types of salinity presumably also synthesized different proteins. Kinetin inhibited incorporation of amino acids into nonstressed and Na_2SO_4-stressed roots, but promoted uptake and incorporation of amino acids into proteins in NaCl-stressed tissue (Kahane and Poljakoff-Mayber, 1968). This was in contrast with the findings reported for tobacco leaves (Benzioni *et al.*, 1967).

Reduction in protein content under low water potentials is mostly due to increased protein destruction. According to Gates and Bonner (1959), Kessler *et*

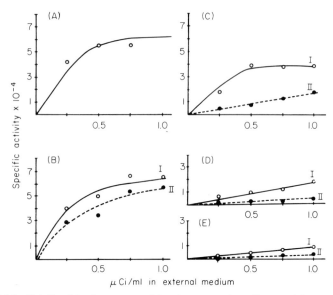

Fig. 11.5. Relationship between activity in external medium and incorporation of leucine-1-^{14}C. Incorporation experiments were carried out in the medium of the same composition as the growth medium. Specific activity of leucine was 6.6 mCi/mmole. I, Chloride salinity; II, sulfate salinity. A, Control; B, medium salinized to 1 atm; C, medium salinized to 3 atm; D, medium salinized to 5 atm; E, medium salinized to 7 atm. The results are expressed as specific activity of the TCA insoluble precipitate, i.e., cpm/mg protein. (After Kahane and Poljakoff-Mayber, 1968.)

al. (1964), and Shah and Loomis (1965), high water deficits in plant tissues enhanced RNase activity, although such deficits also caused decrease in RNA synthesis. This seems to be true for glycophytes, such as tomatoes and peas, and for halophytes, such as beets. A faster decrease in nucleic acid content was caused in plants treated in Na_2SO_4 rather than in K_2SO_4 solutions (Rauser and Hanson, 1966). They reported that such treatments caused leakage of divalent cations (calcium and magnesium) which normally acted as stabilizers of ribosomes and nucleic acids against endogenous nucleases. As a result of such a leak, lower contents of mRNA, rRNA, and DNA were found.

According to Strogonov (1964), the binding between chlorophyll and chloroplast proteins depends upon the ion content of cells. Under high salinity, such links are loosened and as a result more chlorophyll can be destroyed. The degree of such bonding was determined by tissue extraction in 96% and 60% ethanol (Osipova, 1947). While 96% ethanol extracted all the chlorophyll, only the weakly bound one was extracted by 60% ethanol. Thus, the ratio 96% extracted/60% extracted served as a measure of the tightness by which chlorophyll was held by the chloroplast proteins. More chlorophyll could be

Table 11.9

EXTRACTABILITY OF CHLOROPHYLL FROM NORMAL COTTON LEAVES AND
FROM LEAVES SUFFERING FROM SALT BURNS[a]

Soil salinity	Leaf	Chlorophyll in ethanol extracts (%)	Chloride content (%)
Weak	Normal	18.07	1.21
	Salt burnt	28.30	2.29
Strong	Normal	11.36	3.36
	Salt burnt	25.13	4.42

[a]After Strogonov, 1964.

extracted from damaged leaves than from healthy ones (Table 11.9), and the calculated ratio expresses, to a certain degree, the salt resistance of the tissue.

It is interesting to note that losses of chlorophyll were reversible. Cotton leaves treated in a 0.2% $NaHCO_3$ solution had a lower chlorophyll content than did normal leaves. However, when the same plants were moved from pH 9 to 6.4, they regained a dark-green color. Such a reversible phenomenon was termed *necrobiosis* by Strogonov (1964).

Addition of cytokinins to salt-affected plants caused retention of the levels of protein synthesis in the shoots, which otherwise drop with aging. As cytokinins are normally supplied to the shoots by roots, their concentration in the xylem sap may eventually regulate protein synthesis and shoot growth (Kende, 1964). Chromatographs of the bleeding sap of sunflower roots yielded two fractions with a capability of chlorophyll retention. Only one of the two bands was capable of stimulating cell division in soybean callus (Kende, 1965). When roots of bean and sunflower plants were exposed to saline media, or to any other environment with low water potential, the amount of bleeding sap which could be collected decreased and no cytokinin activity was found in such sap. When the growth medium was later changed to a salt-free one, recovery of cytokinin activity in the bleeding sap was observed (Weiss and Vaadia, 1965; Itai, 1967).

As cytokinins had an antagonistic effect on senescence and proteolysis in plant tissues and were stimulatory for protein synthesis, it was suggested that at least some toxic effects of salinity might be attributed to inhibition of cytokinin transport from roots to shoots (Itai, 1967). On the other hand, cytokinin secretion should remain unaffected in roots of halophytes which normally are exposed to saline soils. In fact, a comparative study of two glycophytes (*Phaseolus vulgaris* and *Zea mays*) and a few halophytes (*Atriplex halimus*, *Aeluropus litoralis*, *Suaeda monoica*, and *Tamarix aphylla*) indicated that, while a saline pretreatment decreased the cytokinin content in the bleeding sap of the glycophytes (cf. also Itai, 1967), only negligible changes occurred in sap of halophytes (Table 11.10) (Waisel and Shapira, unpublished).

Table 11.10

SOYBEAN CALLUS GROWTH (GM/30 DAYS) IN MEDIA CONTAINING KNOWN
QUANTITIES OF KINETIN OR BLEEDING SAP CONCENTRATES OF *Aeluropus
litoralis* AND *Tamarix aphylla*[a]

| | *Aeluropus litoralis* | | *Tamarix aphylla* | |
R_f	+NaCl	−NaCl	+NaCl	−NaCl
1	0.063 ± 0.044	0.045 ± 0.020	0.079 ± 0.025	0.103 ± 0.014
2	0.057 ± 0.030	0.053 ± 0.001	0.114 ± 0.018	0.117 ± 0.025
3	0.060 ± 0.022	0.064 ± 0.008	0.111 ± 0.010	0.108 ± 0.017
4	0.067 ± 0.025	0.053 ± 0.018	0.125 ± 0.002	0.095 ± 0.041
5	0.068 ± 0.026	0.087 ± 0.032	0.085 ± 0.031	0.122 ± 0.033
6	0.084 ± 0.026	0.143 ± 0.028	0.112 ± 0.031	0.070 ± 0.050
7	0.244 ± 0.104	0.053 ± 0.023	0.105 ± 0.031	0.100 ± 0.044
8	0.155 ± 0.015	0.098 ± 0.032	0.103 ± 0.044	0.091 ± 0.025
9	0.136 ± 0.036	0.022 ± 0.006	0.079 ± 0.010	0.121 ± 0.054
10	0.146 ± 0.038	0.032 ± 0.015	0.068 ± 0.020	0.069 ± 0.035

[a] Controls:

Kinetin content (ppm)	Callus weight
0	0.076 ± 0.014
0.001	0.130 ± 0.020
0.01	0.166 ± 0.020
0.1	0.377 ± 0.085

Preliminary data also suggest that besides the quantitative differences between the two groups of plants, qualitative differences, as reflected by different R_f values of the active fractions, also exist between the two groups.

Other growth substances probably are also affected by salinity. Sprouting of lateral buds of *Atriplex halimus* plants grown under NaCl salinity was implied as being involved in the removal of apical dominance. Salinity may thus induce auxin decomposition, and consequently change the entire hormonal balance of salt-affected plants (cf. Itai *et al.,* 1968).

Not very much is known of gibberellin metabolism in halophytes. However, it is interesting to note that one of those plant regulators (GA_{25}) was first isolated from leaves of the halophyte *Sonneratia apetala* Ham. (Ganguly *et al.,* 1970).

Miscellaneous Effects

A high content of chloride in plant tissues causes changes in the nitrogen metabolism of plants. Such disturbances resulted in accumulation of free amino acids in leaves of corn, barley, sunflower, and broad beans grown under saline conditions (Strogonov, 1964).

Accumulation of nonprotein nitrogen in leaves, occurs in many salt-affected plants. However, halophytes seem to differ in this respect and the nonprotein fraction of nitrogen in various species of *Atriplex* is low. Moreover, such fraction remained low also in *Atriplex polycarpa* plants which were grown in saline media (Chatterton *et al.*, 1971).

The reddish appearance of plants of highly saline habitats is due to accumulation of β-cyanin caused by disturbed nitrogen metabolism. Under conditions of high salinity the uptake of nitrate is inhibited. When *Salicornia* plants were sprayed with 1% urea, the plants remained green (Hoffmann and Sachert, 1967).

Putrescine $[NH_2(CH_2)_4NH_2]$ is known to accumulate in potassium-deficient plants and cause necrotic spots on the leaves. A high content of putrescine was found in *Vicia faba* plants grown under saline conditions, but with no potassium deficiency. Accumulation of putrescine may contribute to the explanation of salt toxicity symptoms in various cases (Strogonov, 1964).

Unlike productivity of agricultural crops which is greatly reduced by salinity, production of rubber by guayule is an exception. Higher production of rubber seems to fit the general trend of accumulation of alkaloids, anthocyanins, essential oils, and fats in plants, which are subjected to low water potentials (cf. Evenari, 1962; Vaadia and Waisel, 1967). Such end products are assumed to be synthesized from decomposition products arising from increased hydrolysis of proteins and carbohydrates.

Salts in the growth medium thus have various divergent effects on plant metabolism. Nevertheless, because of differences in their effects on various enzymes, cells, organs, and on various species of plants, it is difficult to ascribe effects of salinity to one key process alone. It is also difficult to distinguish between the osmotic and specific ionic effects of salinity on various metabolic processes. An increased knowledge of where ions are located and the enzymes with which they come in contact would contribute to progress in this field.

$$\text{——————————————12}$$

Salt Resistance

Definitions

Definitions of salt resistance are not easy to assess and sometimes seem to be somewhat vague.

Plants cope with salinity problems in various ways. Most of them avoid salinity, some evade salinity or resist it, and a few others tolerate it. Salt avoidance is usually accomplished by limiting germination, growth, and reproduction to specific seasons during the year, by growing roots into nonsaline layers, or by limiting salt uptake. Salt evasion can be achieved by accumulating salts in specific cells or by secretion of excess salts.

Salt tolerance is attained only in plants in which the protoplasm functions normally and endures a high salt content without apparent damage (Iljin, 1932a). Salt tolerance differs among various organs of the same plant, among tissues, and among various stages of plant development. It also varies with salt composition and concentration.

When responses of plants to salinity are discussed, not only maximal concentrations of salts tolerated have to be considered, but also minimal requirements and the long time averages of salinity in the root zone. These three cardinal points provide information on the amplitude of plant resistance to salinity. Accordingly, plants can be divided on an environmental basis into *euryhaline,* i.e., plants which resist a wide range of salt concentrations, and *stenohaline,* i.e., plants with a narrow range of resistance. Such characteristics control plant distribution and zonation in salines.

Salt resistance of plants can be expressed in other forms, and three criteria were used by Hayward and Wadleigh (1949) and Richards (1954):

1. Ability of plants to survive and to complete their life cycle under saline conditions. This is an ecological category which measures only survival ability and disregards size, appearance, growth, and yield of plants. It is an important criterion for wild plants, because many species are able to endure long periods of high salinity with no additional net growth.

2. Relative ability of a species to grow on saline, in comparison to its growth on nonsaline soils. This is a physiological–ecological category of salt resistance. Such a criterion determines adaptability of plants, more than their ultimate resistance.

3. Relative ability of a species to grow and yield on saline soils in comparison to other species. This is an agricultural category. Although such a criterion may also be of importance from an ecological point of view and gives clues for explanation of plant zonation, it is largely unsatisfactory. For example, plants may have a relatively high yield on saline soils when cultivated, but fail to exist under natural conditions.

Salt Resistance, Growth, and Differentiation

Two maximal concentrations of NaCl limit either growth or existence of plant species and their effects have to be considered separately. Although such limiting concentrations usually differ, they may overlap in many cases (Binet, 1960). In this respect, differences between species are frequently encountered. The upper concentration of salts which limits the existence of *Suaeda maritima* is around 20,000 ppm. However, plants already stop growing at 10,000 ppm NaCl. The same is true also for *Cochlearia anglica,* although plants of this species stop growing at an external concentration of 5000 ppm (Binet, 1960). Growth intensity of *Tamarix aphylla* plants markedly decreased with increasing NaCl concentration of the irrigation solution. Although growth ceased at a concentration of $0.3\,M$ NaCl, the plants remained alive, even when the concentration of salts in the medium was increased to 0.5 or $0.7\,M$ (Waisel, 1961).

Most halophytes can endure long periods in high salt concentrations, with little or no additional growth (Hayward and Wadleigh, 1949). Moreover, sometimes such exposure to salinity is even followed by dying of older tissues.

While ability of plants to grow and reproduce under saline conditions is essential for perpetuation of a species in such habitats, the ability to survive, even with no net growth is of no lesser importance. Such a situation can be looked upon as a period of dormancy which plants undergo, while keeping their growth potential for periods of lower salinity. Only plants that can grow satisfactorily, compete with other species in the same habitat, and complete their life cycle, can be regarded as halophytic.

Exposure of plants to salinity may be brief, as in passage of a growing root through a saline layer of soil. It can be temporary, as in occasional inundation by saline water, or it can be permanent and affect plants throughout their life cycle (cf. Repp, 1961).

Salt tolerance of plants varies greatly during different phases of growth and development (Hayward, 1956). Salt tolerance of germinating seeds of halophytes is usually lower than during later stages of their development. Optimal germination of *Salicornia herbacea, Suaeda maritima,* and *Aster tripolium* is found in nonsaline media, while their growth is optimal only in presence of NaCl. Sugar beet, a species with a relatively high salt tolerance during vegetative growth, is more sensitive to salinity during germination than corn, which is salt-sensitive during growth. Variations are encountered also during other phases of growth. Salt tolerance of barley is twice as low during grain production as in earlier growth phases (Hayward and Bernstein, 1958). Different varieties of rice exhibited satisfactory vegetative growth even in media containing about 6000 mg/liter salt. However, under such conditions, plants did not produce grains.

Some observations on the influence of salinity on sex expression of plants were made by Shakhov (1952). He reported that staminate specimens of aspen (*Populus* sp.), willow (*Salix* sp.) and some other dioecious woody plants were more frequently encountered on soils with a higher salinity than pistillate ones. Similarly, in monoecious plants, such as birch (*Betula* sp.), salts promoted an increase in production of staminate flowers and a decrease of pistillate ones.

Salt resistance varies also between cells and tissues. Epidermal cells of halophytes exhibited a higher salt tolerance than did green mesophyll cells of the same leaves. Stomatal guard cells seem to have a very high salt tolerance and can survive high salt treatment far longer than any other leaf cell (Montfort, 1926).

Ion Composition and Salt Resistance

Toxic effects of salinity may be direct, i.e., poisoning of certain metabolic mechanisms, or indirect, i.e., due to disturbances in uptake and metabolism of various essential elements. Such effects may occur outside the plant tissues or inside them.

According to Bowen (1966), ions may cause toxicity in various ways: (1) acting as antimetabolites; (2) binding or precipitating various metabolites; (3) catalyzing rapid decomposition of essential elements; (4) combining with cell membranes and affecting their permeability; and (5) replacing essential elements but failing to fulfill their functions. Classification of sodium or chloride according to such a scheme is difficult and such ions usually cannot be classified as toxic substances in halophytes.

Salt resistance appears to involve a high degree of ion selectivity and can be related to regulation of ion uptake (van den Berg, 1952). Several studies have indicated that salt resistance in glycophytes was associated with relatively low rates of absorption of toxic chloride and sodium ions (Hayward and Wadleigh, 1949; Ehlig, 1960, Greenway, 1962a,b). The barriers which control permeation of sodium and chloride ions are located in roots alone, in roots and stems, or in roots, stems, and leaves. A luxurious supply of potassium may affect the rates of sodium uptake, and thus contribute to better growth in saline media (Bernstein and Ayers, 1953). Sodium chloride can be tolerated by various plants as long as their potassium requirements are met (cf. Hewitt, 1963). Also Rains and Epstein (1967) suggested that salt resistance in leaves of *Avicennia* can, at least partly, be attributed to their unhampered metabolism of potassium.

Contradictory results were reported as well. Bower and Wadleigh (1948) investigated the effect of different levels of sodium on crop plants supplied with adequate calcium and magnesium. The species most tolerant to high sodium levels, e.g., Rhodes grass (*Chloris gayana*) and garden beets (*Beta vulgaris*), were those which absorbed considerable amounts of that element. Moreover, selective accumulation of sodium is sometimes considered a typical characteristic of halophytes. In this respect, the capacity of sodium to increase succulence of plants is significant (cf. Gauch and Wadleigh, 1951).

Salt resistance of plants varies with different ions. Some species are resistant to sodium and others to potassium or calcium. Many plants which are resistant to sodium do not tolerate high concentrations of potassium or calcium. Cells of glycophytes are easily killed by a high content of sodium ions, whereas they withstand a considerable concentration of calcium. On the other hand, halophytes die in a weak calcium salt solution if sodium is absent (Iljin, 1932b).

Only very few investigators distinguished between the effects on plants of total content of salt in the soil and of the sodium adsorption ratio (SAR). Soils with a low salt content and high SAR values can sometimes be more toxic to plants than those with a low SAR and a high total salt content (Lagerwerff and Holland, 1960).

Calcium was shown to enable a higher salt resistance in various glycophytic and halophytic species (Hyder and Greenway, 1965; MacNae, 1966; LaHaye and Epstein, 1969). According to Elzam and Epstein (1969a,b), the calcium metabolism of roots was the key factor in control of salt resistance of two species of *Agropyron*. In addition to having other effects, calcium antagonized effects of sodium by lowering its rate of uptake (Epstein, 1961; Waisel, 1962a).

Regulation of ion concentration is relevant to salt resistance of many plants, and accumulation of toxic levels of salts in the shoots can be avoided in the following ways: (1) limitation of uptake and transport of ions into the shoot by some mechanism located in the roots; (2) export of ions from shoots and leaves; and (3) increased succulence and dilution of high concentration of salts.

Halophytes do not behave uniformly in this respect. The protoplasm of some species exhibits low permeability. Cells of some plants may even remain plasmolyzed in hypertonic solutions for days, and die before enough salt will penetrate the cells to ensure osmotic adaptation (Repp, 1961). Even many salt-requiring plants have relatively low permeability of salts. Otherwise the transpiration stream would have carried in enough ions to kill even the most resistant plant. However, other species into which salt penetrates abundantly are also known.

Once sodium and chloride ions entered leaves of *Phaseolus vulgaris. Hordeum vulgare, Atriplex hastata,* and *A. nummularia* they were exportable only to a limited extent (Greenway and Thomas, 1965; Greenway *et al.,* 1966). Mature leaves which contained high concentrations of such ions showed further accumulation from external sources, while retranslocation of chloride and sodium, which were previously absorbed, was small. Similar data were also found by Ovadia (1969) for *Suaeda monoica,* but only under certain conditions of transpiration and salt concentration.

Salt resistance is thought to be, at least partly, dependent on the capacity of the plants to mobilize energy for sodium extrusion and accumulation of potassium (cf. Norkrans and Kylin, 1969). Sometimes, the total salt content in such plants is below the expected amount for osmotic adjustment, and other substances must be produced by plants for osmoregulation and salt tolerance (Norkrans and Kylin, 1969). For most halophytes, it was suggested that ion concentrations might be regulated by metabolic export either from the whole shoot (Hylmö, 1953), or from mature leaves (Black, 1956). Only *Juncus gerardii* seems to be an exception, absorbing salts continuously up to the point where salt content is lethal (Walter, 1968). This does not apply to all ions. Studies with *Hordeum vulgare* have shown that substantial retranslocation of potassium (Greenway and Pitman, 1965) and phosphate (Greenway *et al.,* 1966) occurs. Most of such retranslocation took place in the phloem.

Succulence is believed to be one of the major factors in salt tolerance. Increased succulence has a diluting effect on the toxic ion content of cells, thus enabling plants to withstand effects of larger quantities of salts. However, such a system also has many disadvantages. It must be accepted that an increase in the cells' osmotic volume dilutes not only sodium and chloride, but also all other nutritional ions. Thus, lowering the concentration of calcium and potassium in salt-affected succulent tissues may eventually prove more toxic than in nondiluted nonsucculent tissues, even though the absolute salt content remains unchanged.

Salt precipitation in shoots serves as another mechanism of salt evasion. Ahmad (1968) found that various granular substances were accumulated in certain cells of leaves or pith of the shoot of two halophytic species (*Suaeda fruticosa* and *Haloxylon recurvum* of West Pakistan. Spectrographical analysis revealed that the precipitated crystals consisted of sodium, potassium, calcium, magnesium, and quartz. As in the vesiculated hairs of *Atriplex,* also in those plants, salts were accumulated and inactivated only in certain loci within the plant organs, thus leaving the rest of the tissue with a low salt content and with an unhampered metabolism.

Salt Damage and Salt Tolerance

Injuries which occur to salt-affected plants, are manifested in several ways. High concentrations of salts which accumulate in the sap of cells can affect water relations and metabolism of plants. Salt accumulation seems to reduce activity of certain enzymes, and thus leads to a decreased metabolic rate. Accumulation of chloride ions also decreases water absorption and transpiration and gives rise to succulence. Such effects are not observed when sulfate and bicarbonate ions accumulate. Direct injuries to cells may also be noticed when salts affect the structure and function of the cytoplasm. In many cases, salinity causes chloroplast disintegration, depresses the elasticity of protoplasm, and enhances its viscosity (Henkel and Strogonov. 1961).

Salt-tolerant species are able to withstand the metabolic disturbances which occur as a result of high salt content. Their protoplasm is not readily injured by high ionic concentrations and disintegration of their organelles is slow. Halophytes can tolerate chloride toxicity, i.e., leaf edge burns, necrotic spots on the leaves, and chlorophyll bleach. The links between chlorophyll to protein molecules of the chloroplasts are much stronger in halophytes than in glycophytes.

Repp (1939, 1961) and Biebl (1953) investigated protoplasmic tolerance in various halophytes. Tolerance was defined as survival ability of plant cells after being submerged for 24 hours in a salt solution. Survival was determined by plasmolysis and deplasmolysis or by the tetrazolium test. Comparative data on salt tolerance of glycophytes and halophytes are presented in Table 12.1.

Higher tolerance to salinity was shown by coastal plants and other halophytes than by glycophytes. The epidermis of such plants survived immersion in a $0.6–1.4\,M$ NaCl, whereas glycophytes died in solutions above $0.2–0.3\,M$ (cf. also Biebl, 1953).

Various ecotypes exhibit differences in salt resistance, with coastal ecotypes showing high resistance. Spraying plants with seawater or submerging plant tissues in such waters caused less damage to tissues of coastal halophytes than to those of inland ones (see also Chapter 16).

Table 12.1

PROTOPLASMIC SALT TOLERANCE OF VARIOUS PLANT SPECIES[a, b, c]

Species	Salt solution (M)											
	0.2	0.3	0.4	0.5	0.6	0.7	0.8	1.0	1.2	1.4	1.6	1.8
Halophytes												
Halogeton glomeratus	L	L	L	L	L	L	L	L	L	−L	+L	+
Salsola kali	L	L	L	L	L	L	L	L	−L	−L	+	+
Beta vulgaris	L	L	L	L	−L	+L	+	+	+	+		
Glycophytes												
Lotus tenuis	L	L	L	+L	+	+	+	+	+	+		
Ervum ervillia	L	L	+L	+	+	+	+	+	+			
Vicia cracca	L	L	+L	+	+	+	+	+	+			
Medicago sativa (France)	L	−L	+L	+	+	+	+	+				
Medicago sativa (U.S.)	L	+L	+	+	+	+	+					
Lens lenticula	−L	+L	+	+	+	+	+					
Vicia faba	+L	+	+	+	+	+	+					

[a] Adapted from Repp, 1961. Reproduced with permission of UNESCO from *Arid Zone Res.* xiv, 1961, p. 158.

[b] Stem cross sections submerged for 24 hours in a series of NaCl solutions. Viability determined by plasmolysis.

[c] L, Plasma living, plasmolysis normal; −L, plasma living, some structural changes, plasmolysis still smooth; +L, over 50% of the cells are dead; +, nearly all cells are dead.

Salt resistance varies also with the organ affected. Some plants resist salts in their root zone but not in their shoots, whereas, for other plants, the reverse is true (Benecke, 1930b). *Glaucium flavum* is a coastal dune halophyte (Fig. 12.1) which is highly resistant to salt spray, but cannot withstand NaCl concentrations exceeding 5000 mg/liter in the root zone.

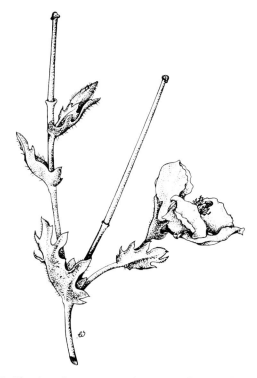

Fig. 12.1. *Glaucium flavum*. A branch bearing a flower and young fruits.

The physiological amplitude of salt tolerance of plants is always wider than their ecological amplitude, because in natural habitats factors other than salinity also affect plant survival. Exclusion of plants from saline habitats does not always indicate that such species have a low salt resistance (cf. Lagerwerff, 1969). Comparisons between various populations of *Festuca rubra* and *Agrostis stolonfera* also suggested that exclusion of *Agrostis* from various saline habitats was not the result of low salt resistance (Hannon and Bradshaw, 1968).

The frequent changes in salinity and water availability that occur at the edges of salines are lethal for some species (stenohaline), but they only slightly alter the behavior of other plants. Deep-rooted species with high capacity for vegetative reproduction (e.g., *Prosopis farcta*) can be found in places where their seeds cannot germinate and seedlings cannot survive the high salt content of the soil.

Raising the Resistance of Plants to Salts

Many investigators have concluded that repeated growing of plants on saline soils has raised their resistance to salinity (Shakhov, 1952). Consequently, higher yields were obtained on saline soils when the seeds used were obtained from plants which had previously been raised under similar conditions. Furthermore, Henkel (1954) proposed a method of increasing salt resistance of cotton and several other plants by pretreating the seeds for a short time with a solution of NaCl. This simple technique was reported to be effective for various crop plants grown under saline field conditions. However, experiments conducted by other investigators did not show any effects of pretreatments on increasing salt tolerance, as long as the seeds were genetically pure. Thus, it seems that the increase in salt resistance reported by Henkel (1954) reflects selection of salt-resistant specimens rather than induction of a high physiological salt tolerance.

The upper limits of salinity which halophytes can withstand vary among species, and are affected by rates of salinization. For fast changes in salinity, the upper limit was about 0.7 M NaCl for *Suaeda monoica, Aeluropus litoralis,* and *Nitraria retusa.* However, limits were far higher when salinity had been gradually increased. Under such conditions, the upper limits of salinity reached 1.1 M NaCl for *Aeluropus* (Pollak, 1967) and 1.8 M for *Nitraria* (Keren, 1970). A slow increase in salinity enabled plants to become osmotically adapted, as well as to develop high protoplasmic tolerance to salinity (Monk and Wiebe, 1961). Such adaptation involves many biochemical changes which harden the cytoplasm and keep the mitochondria, chloroplasts, and other subcellular organelles functional. According to Strogonov (1964), adaptation to salinity involves accumulation of albumins and other specific proteins in the cytoplasm. Such hydrophilic proteins are believed to protect protoplasm from coagulation under high concentrations of electrolytes (Strogonov, 1964). Such mechanisms seem to be autocatalytic, and occur only in plants which were preexposed to salinity. However, the experimental procedures suggested by Strogonov (1964) need further validation.

Several chemicals may also affect resistance of plants to salinity. Miyamoto (1962) reported that the resistance of wheat to salinity was increased by trimethylammonium chloride. Wyn-Jones (1971) also reported that similar compounds (e.g., choline chloride and Betaine hydrochloride) increased the resistance to NaCl in corn and other agricultural plants. Nevertheless, it is still difficult to conclude whether such effects are of a general nature or if they are limited to glycophytes.

Resistance to Salt, Drought, and Alkalinity

Protoplasmic tolerance of salts is not always correlated with other tolerances such as frost or drought tolerance. Usually, xerophytes have a lower protoplasmic salt tolerance than halophytes. Salt-burned leaves are commonly found on xerophytes, which were exposed to salinity (Stadelmann, 1957). On the other hand, halophytes seem to have at least some resistance to drought, because of their high osmotic potentials and hydrophilic proteins. Both dehydration and salt tolerance are high in growing apices of shoots and roots, and exceed the tolerance of most other cells of such plants (Wilson, 1948; Waisel, 1962b, c; Vaadia and Waisel, 1967).

Salinity and alkalinity very commonly occur together in salt-affected soils. Nevertheless, the physiological basis of resistance to alkalinity has not been adequately studied, and information on plant behavior under high pH conditions is scarce (cf. Place *et al.,* 1971). For halophytes, it is well known that there is no general correlation between resistance to salt and alkalinity (Hayward, 1956). *Allenrolfea occidentalis* and *Salicornia subterminalis* are highly resistant to salt but sensitive to alkalinity. A change from saline to sodic soils may kill such plants. Also the highly salt-resistant *Suaeda monoica* suffers badly as pH increases above 9.5 (Ovadia, 1969). On the other hand, other halophytic species such as *Sarcobatus vermiculatus* and *Sporobolus airoides* thrive on both saline and sodic soils.

13

Formative Effects of Salinity

Classification of Succulents

The appearance and structure which characterize a certain group of plants sum up to a great extent their ecological and physiological means of adaptation. Halophytes are no exception in this respect because of specific and typical structural characteristics which make them distinguishable from other groups of plants. Many halophytes are leafless or articulated and usually succulent. Exposure to salinity induces several structural changes in halophytes including thicker leaves, larger cells, especially those of the spongy parenchyma, smaller intercellular spaces, higher elasticity of the cell walls, fully developed water storing tissues, smaller relative surface area (surface/volume ratio), low chlorophyll content, and smaller and fewer stomata per surface area (Stocker, 1933). Coastal halophytes are covered with a heavy layer of trichomes, whereas inland halophytes often have glabrous shoots with a bluish-green color. Most nonsucculent halophytes have salt glands.

Changes in plant structure brought about by salts had been recorded in the second part of the 18th century. However, most investigations on effects of salt were performed on glycophytic crop plants. Relatively little information on natural halophytes has been accumulated.

One of the most striking features of halophytes is that they show a close relationship between uptake of alkali ions and succulence. Such correlation, which is most pronounced in the presence of sodium chloride or other sodium salts, often has been referred to as the morphogenetic effect of sodium and chloride.

High salinity also causes severe internal water deficits, and plants which are easily affected by such stresses may sometimes develop xeromorphic structures under such conditions. In fact, because of the occurrence of such characteristics as succulence, reduced leaf surface, hair cover, etc., in both desert plants and halophytes, Schimper (1903) evolved his theory of "physiological drought." He reasoned that reduced water absorption brought about those modifications in plant anatomy.

Salt-induced succulence can be expressed in terms of increase in cell size, decrease in extension growth, reduction in surface area per volume of tissue, and high water content per unit of surface area (cf. Delf, 1911; Bickenbach, 1932; Frey-Wyssling, 1935). Whereas glycophytes show some succulence, halophytes accumulate large amounts of salts throughout their shoots and show the highest degree of succulence. Salt-secreting species which maintain a relatively low salt content in the shoot generally are nonsucculent.

Differences in structure are well known between plants of dry saline habitats and of wet ones. Consequently, two types of succulent halophytes are classified as succulent mesomorphs and succulent xeromorphs on the basis of their water relationships (Weissenbock, 1969). In the first type—typical of hydro-halophytes—all leaf cells, including epidermal cells, are succulent and the number of stomata per unit area is low (Biebl and Kinzel, 1965; Weissenbock, 1969). The xeromorphs which usually grow in habitats with high water stresses are characterized by succulent and large mesophyll cells, an enlarged cortex, small epidermal cells, and abundant stomata. Succulent xeromorphs are additionally characterized by a highly developed and a highly lignified vascular system (Rumyantseva, 1959). The cell sap of succulent xeromorphs usually contains more organic acids than the sap of succulent mesomorphs, which contain more inorganic ions.

The types of salinity to which plants are exposed, also markedly influence their structure. Formative effects of various salts were investigated in *Salicornia herbacea* by van Eijk (1939) who compared the effects of $NaCl$, $CaCl_2$, $MgCl_2$, $NaNO_3$ and $NaSO_4$, or of various salt mixtures, on succulence. The effects of various ions on succulence seem to parallel the lyotropic series. Sodium chloride was found to be the most efficient salt in promoting succulence. According to van Eijk (1939), chloride rather than sodium played the major role in its induction. Similar conclusions were reached by Keller (1925), Walter and Steiner

(1936), and Williams (1960). Decrease in succulence, together with increase in xeromorphic characteristics, were observed in plants exposed to Na_2SO_4. However, responses to sulfate salinity vary among different species of plants. Halophytes usually respond to NaCl as well as to Na_2SO_4, whereas glycophytes react more to NaCl.

Unlike sodium the presence of high potassium in the cell sap does not affect succulence (Weissenbock, 1969). Concentrations of sodium in leaf sap of *Aster, Artemisia,* and *Plantago* plants grown on saline soils were similar to concentrations of potassium in plants grown on a nonsaline garden soil. Whereas plants of saline soils which contained NaCl showed succulence, those grown in nonsaline soils and dominated by K content in the sap remained nonsucculent.

The question as to whether succulence is a direct response to NaCl or is secondarily induced by a shift in the balance of ions or of organic acids, is still open. According to Watson (1942) and Jennings (1968b), there is no doubt that the effect of sodium on succulence is likely to be a general phenomenon in glycophytes, xerophytes, and halophytes, and may suggest that increased succulence produced by high light intensity, aridity, and sodium ions is caused by substantially the same mechanism. According to Jennings (1968b), succulence is caused by the shift in ATP metabolism induced by sodium transport.

Arnold (1955) suggested that succulence depends on the ratio of absorbed to free ions in plant cells, rather than on the absolute amounts of sodium, chloride, or sulfate present. Succulence is induced only after accumulation of free ions in an organ increases above a critical level. Appearance of succulence is a consequence of metabolic paths and can also be induced by some other factors such as radiation, lack of nitrogen, and high light intensity. An increase in concentration of inorganic ions in older leaves is also correlated with high succulence (Repp, 1939; Biebl and Kinzel, 1965).

Salinity may affect plants at different sites of the plant body. According to Boyce (1951), succulence in *Iva imbricata* results mainly by increased size of nonassimilating mesophyll cells. According to Stocker (1928), chloride causes swelling of the protoplasm of halophytes, thus contributing to their succulence. Tullin (1954) suggested that chloride primarily affected the cell walls, causing destruction of certain binding complexes, enabling expansion of the wall, and thus leading to increasing size of cells.

According to Pokrovskaya (1954, 1957) salinity inhibited cell division and stimulated cell elongation in *Statice gmelini* and *Atriplex tatarica.* Such effects cause the familiar decrease in cell numbers and increase in their size, so typical of succulents. According to Repp *et al.* (1959), succulence is positively correlated with salt tolerance of plants, and the degree of their development can serve as an indicator of the ability of a plant to survive in highly saline habitats.

Succulent plants may be divided into two types: those with succulent leaves and those with succulent stems. In some plants, distinction between the two organs is difficult. There has not been general agreement on the origin of fleshy tissues of articulated plants. Some investigators (De Fraine, 1912; Halket, 1928; James and Kyhos, 1961) regarded fleshy tissues as being foliar in origin, whereas others (Fahn and Arzee, 1959; Fahn, 1963) considered them to be derived from the stem cortex. In any case, these tissues are highly responsive to salt and, under saline conditions, may become highly succulent. Sizes of cells of these tissues in *Salicornia* are positively correlated with salinity of the medium, up to approximately 0.3 *M* NaCl.

Modification of Leaves

Succulent plants have thick leaves, largely because of increased size of mesophyll cells (Fig. 13.1). Such leaves have smaller intercellular spaces and their mesophyll cells have fewer chloroplasts than are found in leaves of nonsucculents. Not all mesophyll tissues respond similarly to salt. Of the two layers of assimilating tissue in *Suaeda monoica* leaves, only the inner one expanded under saline treatments, whereas the subepidermal one remained more or less the same size. Increasing succulence is associated with a decrease in the outer leaf surface per unit volume.

Treatment of plants with NaCl solutions does not always simulate saline conditions. Chloroplasts of *Atriplex* plants grown under laboratory conditions in

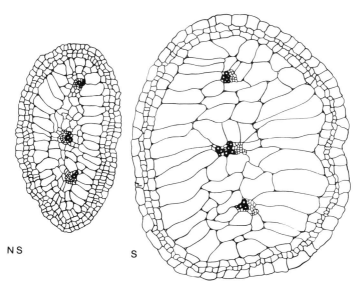

Fig. 13.1. Cross sections of *Suaeda monoica* leaves taken from plants grown under saline (S) and nonsaline (NS) conditions.

saline media showed aberrations, but chloroplasts taken from leaves which were collected in the field were normal (Blumenthal-Goldschmidt and Poljakoff-Mayber, 1968).

The numbers and sizes of mitochondria in leaf cells are also affected by salt. According to Siew and Klein (1968), there were about six times as many mitochondria in developing bean plants treated with NaCl (0.2 M) as in control plants. Mitochondria of salt-affected plants also were larger. Plants grown on saline soils were found to have mucilaginous cell contents, a characteristic of many halophytes (Walter, 1936a; Shmueli, 1948).

The major leaf veins in many species of *Atriplex* are surrounded by a sheath of cubical, thick-walled cells which contain chloroplasts, larger in size than the chloroplasts of other mesophyll cells of *Atriplex* (Laetsch, 1968; West, 1970). Such assimilating tissue was named the "kranz" type (Moser, 1934) and is found in many tropical grasses. Recently it was found that many other halophytes also contain the "kranz" type tissue.

Stomata of succulent leaves frequently are sunken into the epidermis. Their numbers are usually low. For example, the number of stomata reported for various halophytic species of the Neusiedler See region, Austria (Repp, 1939) and of the Dead Sea region, Israel (Shmueli, 1948), is presented in Table 13.1. Evidently, only in one case have stomatal counts exceeded 200/mm^2.

Succulence varies with type of salinity. As mentioned, plants grown in a medium containing NaCl have leaves with small surface area but with large cells (Fig. 13.2). In contrast, plants affected by Na$_2$SO$_4$ have small leaf cells but large

Table 13.1

NUMBER OF STOMATA IN LEAVES OF VARIOUS HALOPHYTIC SPECIES[a]

Species	Number of stomata/mm^2		
	Upper epidermis	Lower epidermis	Average
Plantago maritima	117	212	
Lepidium crassifolium	150	165	
Camphorosma ovata	120	120	
Triglochin maritima	77	103	
Aster pannonicus	47	67	
Suaeda maritima	38	50	
Alhagi maurorum	107	136	
Prosopis farcta	58	70	
Nitraria retusa	–	–	64
Arthrocnemum glaucum	–	–	57
Suaeda monoica	–	–	48–75
Suaeda fruticosa	–	–	11–23

[a] Compiled after Repp, 1939 and Shmueli, 1948.

SALINE   NONSALINE

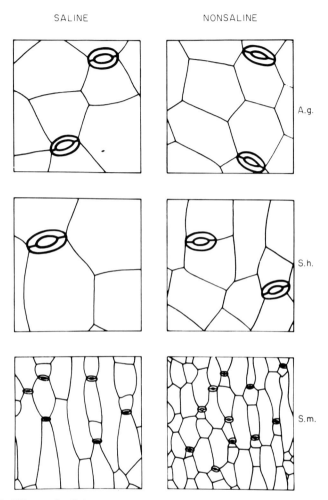

Fig. 13.2. Effects of salinity on the development of epidermal cells of *Arthrocnemum glaucum* (A.g.), *Salicornia herbacea* (S.h.), and *Suaeda monoica* (S.m.).

leaves (Fig. 13.3). Sulfate influences cell division more than cell expansion (Strogonov, 1964), whereas chloride affects expansion more than division.

Some halophytic species, e.g., *Salicornia* sp., contain large sclereids which extend from terminal leaf veins to the epidermis (De Fraine, 1912; Fahn, 1967). These sclereids seem to attain larger sizes in plants grown under saline conditions than in those grown on a medium free of sodium chloride (Fig. 13.4).

Leaves of coastal halophytes (aerohalophytes) show an additional mode of adaptation to their habitats. Their leaves and stems are abundantly covered with

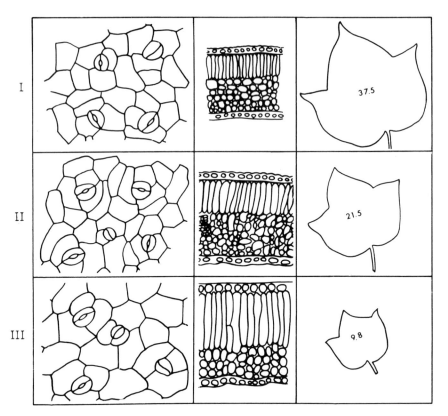

Fig. 13.3. Effects of Na$_2$SO$_4$ and NaCl on leaf growth and cell development in cotton. (After Strogonov, 1964.) I, Control; II, sulfate; III, chloride. Left, Upper epidermis; center, leaf cross section; right, leaf area (cm^2).

trichomes, giving the plants a grayish appearance. Such trichomes, which consist of simple and branched hairs of various types, exhibit marked variations in form (Fig. 13.5). The epidermis of xerosucculents and coastal halophytes is characterized also by a thick cuticle, and a cover of waxy layers (Uphof, 1941).

Trichomes may exert a protective function in plants by:

1. Affecting their water economy—reducing air movement near stomata and consequently reducing transpiration.

2. Affecting the temperature of their leaves—increasing back radiation and providing a layer of insulation.

3. Contributing to their capability to avoid salt spray—preventing seawater droplets from reaching live tissues of the leaves.

It is generally accepted that trichomes have only small effects on rates of stomatal transpiration. Their influence in reducing cuticular transpiration is real

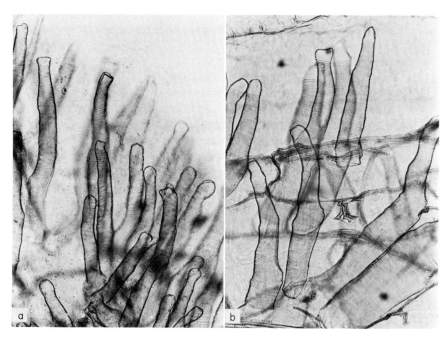

Fig. 13.4. Sclereids of cleared portions of cortex of *Salicornia herbacea* plants. (a) Plants grown on a salt-free medium. (b) Plants grown on a 100 m*M* NaCl solution.

(Weaver and Clements, 1938; Mangum-Shields, 1950), but as cuticular transpiration comprises only a minor fraction of total transpirational loss, the effectiveness of trichomes in reducing loss of water is small. In any case, only dead trichomes were found to have a protective value.

Leaves covered with trichomes were shown to have lower temperatures during exposure to high solar radiation, than bare ones, due to their higher reflectability (Wuenscher, 1970). Again, only dead and air-filled trichomes were really effective.

Protection against wind-borne salt spray seems to be of primary importance for plants growing on exposed coastal habitats. Dead trichomes serve in this respect, both as a wind break and as a salt filter. Leaves with a thick cover of trichomes seem to be well protected against salt spray and only a small fraction of the salts deposited on outer surfaces of such leaves comes in contact with live tissues.

Submerged marine angiosperms are among the very few species of halophytes that do not become succulent. Leaves of these plants are thin and have a poorly developed vascular system and frequently a green epidermis. Leaves of submerged halophytes, like those of most submerged hydrophytes, are also adapted directly to absorb water and nutrients from the medium. It seems that

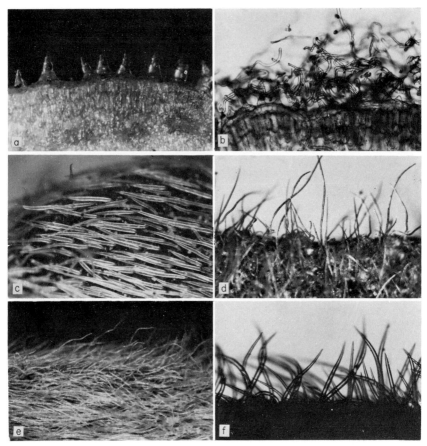

Fig. 13.5. Leaf trichomes of some coastal halophytes: (a) *Silene succulenta*, (b) *Diotis maritima*, (c) *Lotus creticus*, (d) *Convolvulus secundus*, (e) *Medicago marina*, (f) *Oenothera drummondii.*

under submerged marine conditions, structural adaptation of plants to water was superimposed on their adaptation to salinity.

Leaf epidermis in many mangrove species is usually thick, having straight lateral walls. The cuticle is well developed on both upper and lower sides of the leaves. Upper epidermal cells may contain oxalate crystals. Stomata are confined to the lower epidermis and their numbers range in various species between 100–150 stomata/mm^2. These stomata are typically xerophytic. The palisade cells may constitute several layers of narrow cells with intercalated tannin and oil cells. Leaves of mangrove species are either dorsiventral or isolateral.

Usually shading is not a problem in saline habitats, because of sparse distribution of plants. Nevertheless, when a plant was partially shaded by another plant, differences in leaf development were noted (Shmueli, 1948). Leaves of *Alhagi maurorum* plants growing under *Nitraria* bushes had larger surface areas, and their thorns were shorter, than in unshaded plants.

Leaves of many species are dotted with local formations of cork—"cork warts." Leaves of *Sonneratia* and *Aegiceras* and *Nitraria* contain a well developed aqueous tissue (Mullan, 1932).

Modification of Stems

Many halophytic members of the Chenopodiaceae have succulent stems but their leaves are reduced in size and development. Young branches of these plants are made up of succulent assimilating internodes which give the plants an articulated appearance (Fahn and Arzee, 1959). In hydrohalophytic Chenopodiaceae, succulent tissues of articulated branches are comprised of large epidermal and palisade cells as well as water storage and vascular tissues (Halket, 1928; Metcalfe and Chalk, 1950; Fahn and Arzee, 1959; James and Kyhos, 1961). According to Bickenbach (1932), salinity also causes thickening of cortical parenchyma and extensive lignification in the stele. In many halophytic desert shrubs (e.g., *Nitraria retusa*) the xylem also contains living fibers. It is believed that the existence of living fibers in the vascular tissues of shrubs contributes to a better adaptation of such plants to extremely arid conditions (Fahn, 1964). Living fibers may give similar advantages also to halophytes under saline conditions.

Lateral meristems of shoots are also affected by salinity. For example cotton plants (*Gossypium,* sp.) subjected to chloride salinity, had a thicker cortex and less xylem in stems than did plants grown on a salt-free soil (Strogonov, 1964). Such responses to salinity also occur in some obligatory hydrohalophytes. The xylem of *Salicornia herbacea* growing under wet saline conditions had fewer and narrower vessels than when grown under nonsaline conditions.

Effects of salinity on cambial activity of trees were also noted. *Populus euphratica* growing in saline habitats in the Negev of Israel produced a ring porous type of wood, whereas wood of plants in nonsaline habitats was of the normal diffuse porous type (Liphschitz and Waisel, 1970a,b) (Fig. 13.6). Similar changes in wood type were also reported for cotton in which the pattern of xylem production was modified by salt treatments attaining a ringlike structure (Strogonov, 1964). Formation of a semi-ring type of xylem reflects growth periodicity and supposedly results from periods of high water stresses, alternating with periods of low ones (Strogonov, 1964).

In most Chenopodiaceae, and probably also in other species with a thick cortex, the phellogen develops deep inside the axis (Fahn, 1963). However, very little specific information is available on phellogen initiation and activity in

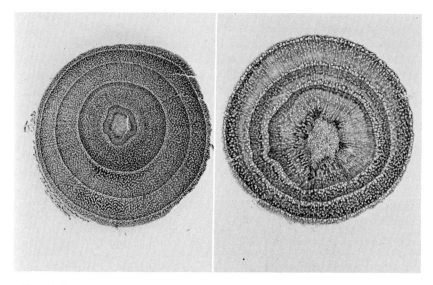

Fig. 13.6. Type and annual amount of xylem produced in *Populus euphratica* trees. Samples taken from plants growing naturally on saline (right) and nonsaline (left) habitats in Avdath canyon in the Negev of Israel. (After Liphschitz and Waisel, 1970a.)

halophytes. In *Suaeda monoica* the phellogen develops next to the phloem from pericyclic or phloem parenchyma cells (Fig. 13.7). Thus, it is formed only two to four cell layers deep in the roots, but ten to twenty cells deep inside the fleshy cortex of the shoot. A tier of initiating phellogen cells may be observed along the hypocotyl (Waisel and Liphschitz, unpublished data).

The anatomical features of various mangrove species were thoroughly studied by Mullan (1932). Most of the characteristics are similar in the different species. In young stems, the cuticle is thickly developed. The epidermal cells contain large quantities of tannins and oil droplets. Oxalate and lime crystals are found in the cortex and such crystal-containing cells are arranged in rows. The cortex of older stems becomes lacunar and serves as an aerating tissue. Such tissue becomes highly developed on the lower parts of the trunk. Internal H-shaped hairs and various sclereids in *Rhizophora* support the lacunar cortex.

Rhizomes of *Spartina* and other salt marsh species also have large air-filled lacunae. In some species, such lacunae are absent in the above-ground stems, or are present only in a reduced form.

Modifications of Roots

Another developmental feature that characterizes halophytic species is the marked reduction in development of the cortex of primary roots. According to Ginzburg (1964), the cortex of *Suaeda monoica* primary roots was only about

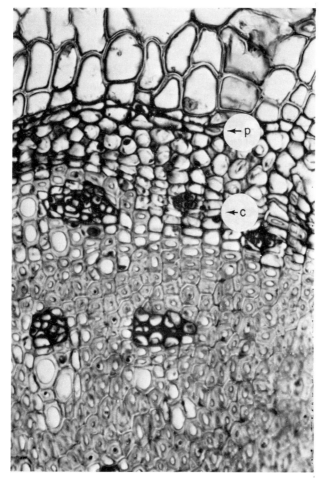

Fig. 13.7. Cross section in the stem of *Suaeda monoica*. Phellogen is initiated next to the phloem. p, phellogen; c, cambium.

two to five cells thick, whereas in various glycophytes, e.g., tomato or beans, it was twenty to thirty cells thick (Fig. 13.8).

Casparian strips of both hydrohalophytes (e.g., *Suaeda monoica*) and xerohalophytes (e.g., *Zygophyllum dumosum, Reaumuria palaestina*) are distinctive and highly developed (Fahn, 1964; Ginzburg, 1964). In certain species, Casparian strips cover almost the entire radial walls of the endodermis as compared to less than one-third cover of these walls in glycophytic dicotyledons (Fig. 13.9).

Like most water plants, hydrohalophytes which are subjected to anaerobic conditions, develop specific adaptive characteristics. For example, an extensive

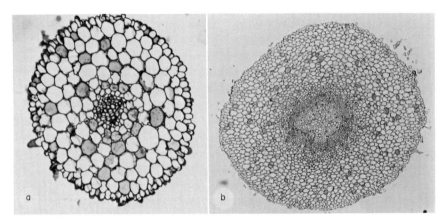

Fig. 13.8. Cross sections of the primary roots of *Suaeda monoica* (a) and *Vicia faba* (b) showing thicker cortex of the latter.

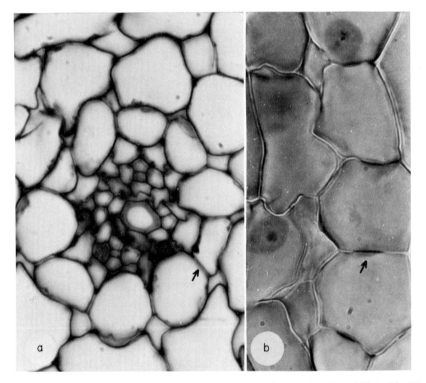

Fig. 13.9. Cross sections of the primary roots of *Suaeda monoica* (a) and *Vicia faba* (b). Differences of development of Casparian strips in the endodermis can be seen. Arrows point to the Casparian strips.

aerenchyma tissue is formed. Aerenchymatous roots help in anchoring plants in the mud. This is an efficient system, since it increases the root surface without an increase in the quantity of respiring material (Williams and Barber, 1961). In some plants, such as mangroves, aerating tissues are highly developed (Fig. 13.10).

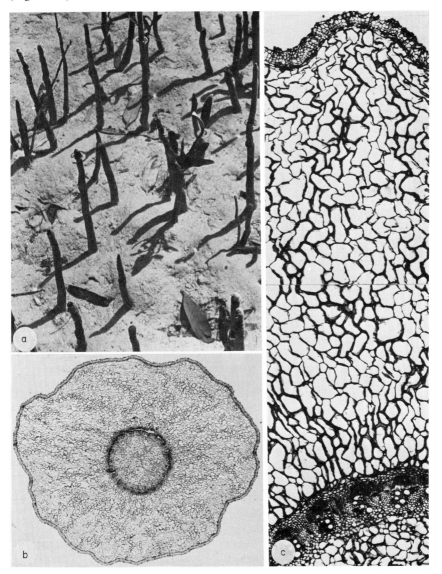

Fig. 13.10. Pneumatophores (a) and cross sections of a pneumatophore of *Avicennia officinalis* showing the spongy cortex (a, x$\frac{1}{3}$; b, x10; c, x52).

The anatomical characteristics of various aerial roots of mangrove species are generally similar. Nevertheless, differences in structure of aerial and subterranean portions are known. The stele of terrestrial roots is reduced, but together with the lignified cells of the outer pith, forms a strong cylinder of mechanical tissues. The cortex forms an aerating tissue with wide lacunae. Depending on species, this tissue is schizogenous or lysigenous in origin. Many strands of lignified sclerenchyma appear in the roots. The endodermis in those plants is distinct and contains starch. Aerial roots transport air into submerged organs exposed to anaerobic conditions. This is discussed in more detail in Chapter 14.

Submerged marine angiosperms (*Cymodocea, Halophila,* and *Zostera*) also have additional adaptive special characteristics. In sites which withstand pounding waves and fluctuations in water level, plants possess creeping rhizomes which, together with clusters of roots growing out of each node, anchor the plants (Dawson, 1966). Roots of such plants develop collenchyma or sclerenchyma tissues.

The question of whether succulence or other anatomical modification of halophytes is a mere reaction to salinity, or whether such characteristics are responses of adaptive value, is still obscure. A long chain of events separates the cause, salt entry, and final structural changes. Ions must penetrate cells and induce major metabolic changes (cf. Jennings, 1968b) before apparent anatomical or morphological modifications can be observed. However, as formative changes seem to be an integral part of halophyte development under saline conditions, and as survival and succulence seem to be linked in many salt-tolerant species, it is tempting to assume that such modifications are of adaptive value.

14

Marine Halophytes

Seacoasts are among the habitats which are destroyed most by the modern human society. They are widely used for recreation for a continuously increasing number of people, are exploited for certain building materials, and polluted by oil and other chemicals that are disposed of in the oceans. In addition, rates of reproduction and growth of coastal plants are low and the sites spoiled by human interference are hardly reestablished. Thus, the study of coastal plants seems to be more a source of inventory information for future generations.

Coastal angiospermous halophytes can usually be subdivided into the following groups:

1. Submerged marine halophytes, *hydrohalophytes.*
2. Low coast plants, *hygrohalophytes.*
   a. Swamp halophytes, mangroves
   b. Marsh halophytes
3. High coast plants, *aerohalophytes.*

Each of these groups has specific ecological requirements and is restricted in its distribution to certain ecological niches. In this chapter, we shall discuss the specific characteristics of those three groups of plants and of their respective habitats.

Submerged Marine Angiosperms

Submerged marine angiosperms constitute a distinct and specific group of halophytes with limited distribution in certain coastal water habitats. Like other marine angiosperms, these plants have evidently evolved from freshwater ancestors. Since their return to marine life is secondary, marine angiosperms should be regarded as the most advanced form among the angiospermous hydrophytes. The submerged angiosperms are taxonomically very uniform and were long thought to be restricted to two families of monocotyledons: Potamogetonaceae (eight genera) and Hydrocharitaceae (three marine genera). However, recently taxonomists have distinguished six families within this group (Willis, 1966):

1. Cymodoceaceae
   Amphibolis C. Agardh
      Three species distributed over the Red Sea, Indian Ocean, tropical Africa, Australia, and Tasmania.
   Cymodocea Koenig
      Ten species of the tropic and subtropic regions (including *Thalassodendron* (Forssk.) Den Hartog).
   Halodule Endl. (= Diplanthera Thou.).
      Seven species.
   Syringodium Kütz.
      Two species of the Caribbean, Indian Ocean, and West Pacific.
   Thalassia Soland ex Koenig
      Two species of the Caribbean and Pacific Ocean.
2. Hydrocharitaceae
   Halophila Thou.
      Various species distributed in the Red Sea, Indian Ocean, western Pacific, Australia, and Tasmania.
   Enhalus Rich.
      One species of saline water of the Indo-Malaysian region.
3. Posidoniaceae
   Posidonia Koenig
      Two or three species of the North Mediterranean and Australian coasts.
4. Ruppiaceae
   Ruppia L.
      Two species of the temperate and subtropical seas.
5. Zannichelliaceae
   Althenia Petit.
      One species of the western Mediterranean and South Africa.

6. Zosteraceae

Phyllospadix Hook.

Two species of Japan and eastern Asia, on one side, and the Pacific coasts of North America, on the other.

Zostera L.

Ten species spread over the tropical, subtropical, and temperate regions of the northern hemispere, on both the Pacific and Atlantic coasts.

Den Hartog (1970) classified the sea grasses into two families with several subfamilies and new genera.

Most species of this group of marine plants form the upper sublittoral belt, and although they grow in shallow water, the plants are seldom exposed to the atmosphere by the low tides. Only *Phyllospadix* plants grow profusely in deep water. Two of the genera, i.e., *Posidonia* sp. and *Phyllospadix* sp., are usually limited to rocky habitats, whereas others are confined to softer sandy or muddy substrates. Out of the twelve genera, *Ruppia* (Fig. 14.1) and *Althenia* are not restricted specifically to marine habitats, but are also frequently found in inland saline marshes.

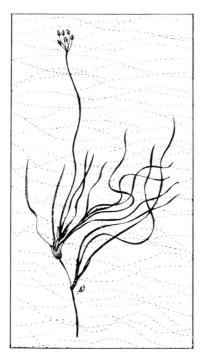

Fig. 14.1. *Ruppia spiralis* L.

The number of species of all six families is not much over forty-five; most of them are of the tropical chorotype. The tropical and subtropical genera (*Cymodocea, Thalassia, Halophila, Halodule,* etc.) are concentrated either in the Indo-Malaysian region or along the Caribbean coasts (Sculthorpe, 1967). Other tropical coasts, especially those subjected to effects of cold sea currents in West Africa and along the western coast of South America, are devoid of them. Most species of marine angiosperms tolerate only minor changes in water temperatures. According to Setchell (1929), those species are restricted to habitats with temperature fluctuations of the upper water layer below a 5°C mean monthly maximum. Only various species of the genera *Zostera* and *Phyllospadix* are more confined to temperate regions and therefore subjected to higher fluctuations in temperature.

According to Love (1963), most marine angiosperms are adapted to dispersal by seawater only. However, in contrast to long-distance dispersal mechanisms of coastal halophytes and mangroves, the submerged plants are dispersed for short distances only. Dispersal of these plants is a slow process, due to the scarcity of seed production, the practical lack of buoyancy, the short-distance drift of the disseminules, and the poor survival of seedlings. Thus, enlargement of their distribution area occurs gradually in steps—a short pace at a time. Nevertheless, distribution of most species is disjunctive. Even the widely distributed species, e.g., *Halodule tridentata* and *Thalassia hemprichii,* do not have a continuous distribution, and their distribution area is interrupted by wide gaps. Some species exhibit clear disjunction between the Far East and the Caribbean.

Overland dispersal of submerged marine plants is extremely limited and even narrow land bridges form an efficient barrier to their distribution. Opening of land barriers to the flow of seawater also opens for those species new chances for distribution and new sites to inhabit. It is well known that *Halophila stipulacea* (Forssk.) Aschers. (Fig. 14.2), *Cymodocea serulata* (R.Br.) Aschers. et Magnus, and *Cymodocea rotundata* Aschers. et Schweinf., are tropical marine angiosperms, whose northern distribution limit was in the Red Sea. During the last hundred years, following the opening of the Suez Canal, those species invaded the Mediterranean.

Three of the genera of submerged sea plants have a limited distribution and are restricted to the Indian Ocean and to the western part of the Pacific Ocean. The genus *Phyllospadix* is confined in its distribution to the coasts of the northern Pacific Ocean. According to Den Hartog (1970), *Enhalus* is probably the youngest among them, a genus which still did not have an opportunity to establish stands in the Caribbean. On the contrary, distribution of *Thalassodendron* is relictic and reduced as compared to what it was, probably due to changes into unfavorable ecological conditions.

Sculthorpe (1967) believed that various marine grasses such as *Halodule, Syringodium,* and *Thalassia* have evolved either from brackish water, or from freshwater progenitors.

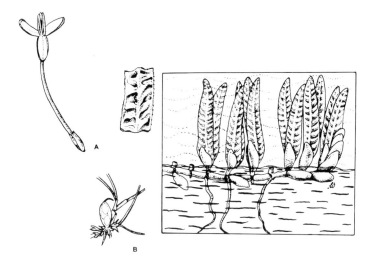

Fig. 14.2. *Halophila stipulacea.* (A) A staminate flower; (B) a young fruit. (After Dawson, 1966.)

The genera *Halodule, Halophila, Syringodium,* and *Thalassia* originated in the Indo-Malaysian region. At the present time, those genera are also well represented in the Caribbean. However, due to the difficulties encountered in long-distance transport, it was suggested that ancestors of these plants migrated to the Caribbean region across the Pacific Ocean before the Miocene, i.e., before the Panamanian Isthmus was formed (Den Hartog, 1964). This idea seems too farfetched because of the fact that sporadic transport of disseminules might have occurred even in historical time, long before the first records of plants were made. This assumption is further supported by the present existence of at least two very closely related taxa in the species, *Halophila decipiens* and *Halodule wrightii,* both in the Caribbean, on the one hand, and in Africa and the Indo-Malaysian regions, on the other hand. It does not seem feasible that such closely related—if not identical—taxa have developed independently a thousand miles apart. Thus, though slowly, distribution of marine angiosperms might occur over great distances.

Marine angiosperms are salt-requiring euhalophytes. Though the question whether they are obligatory or preferential halophytes is still undecided, it seems that most of them are obligatory. Nevertheless, this characteristic varies among species. Even though various marine angiosperms are able to grow also in brackish water, most species almost never appear in inland saline water ponds. It seems that biotic factors, rather than physical ones, govern the distribution of those species. Most of these plants exhibit optimal growth in diluted seawater. In certain cases, a few species may survive even in higher concentrations of inland saline lakes. Distribution patterns of five species of marine angiosperms

(*Thalassia testudinum, Halophila engelmani, Diplanthera wrightii, Ruppia maritima,* and *Syringodium filiforme*) are at least partly correlated with their tolerance to salinity. Tolerance of *Diplanthera* was highest among the five species investigated and that of *Syringodium* the least (McMillan and Moseley, 1967).

Submerged marine angiosperms differ in their mineral nutrition requirements from most other halophytes. The former are characterized by a high preference of potassium over sodium (cf. Steward and Martin 1937), although this selective ability is not so high as in algae. The ratio of K : Na is 0.04 in seawater, 1.15 in *Zostera marina,* and 10.45 in *Ulva latuca* plants. It is unknown whether this ratio is kept high due to an extremely slow penetration of sodium or due to an active metabolic pump secreting sodium outward. Evidence supporting the latter idea was reported during the last few years (Scott and Hayward, 1954, 1955; MacRobbie, 1964, 1965, 1966).

Stands of marine angiosperms can sometimes be very dense and highly productive. According to Walter (1968), the productivity of *Zostera marina* may reach 5 tons dry weight/ha/year. However, such high rates are usually rare and are encountered only under limited conditions of light, substrate, and temperature.

Various sea grasses, e.g., *Zostera marina, Zostera nana,* and *Phyllospadix scouleri,* have some economical uses in eastern Europe and in the Far East. In Rumania and in Russia, such plants are used for mattress fillings, as insulation material, as forage for cattle, and as a fertilizer substitute. Extracts of *Zostera* can also replace agar agar (Bacalbasa, 1951). Utilization is restricted because of difficulties in rinsing the salt out of plants and then drying them.

Owing to the difficulties in reaching their native habitats, marine angiosperms are probably one of the least explored groups of plants (Dawson, 1966). Nevertheless, at least some information has been accumulated on some of the following species.

CYMODOCEA KOENIG

This is a group of herbaceous leafy plants with horizontal rhizomes bearing a few leaves on each internode. Leaves are linear, some 10 cm long, attaining larger sizes in deep water, and are compressed at the base. Flowers are very much reduced. The staminate flowers appear on short stalks having two anthers, while the pistillate flowers are sessile with two ovaries. The fruits are elliptical with a hard pericarp.

The genus consists of a few species inhabiting the tropical and subtropical seas of Africa and Asia. The species segregate into different distribution areas. *Cymodocea rotundata* and *C. serrulata* are practically confined to the Indian and the Pacific Oceans. *Cymodocea nodusa* covers habitats with colder water in the

Mediterranean and the eastern Atlantic (west coast of Africa and the Canary Islands). *Cymodocea nodosa* is one of the pioneer species among the submerged marine angiosperms. Its competitive ability is low and on sandy bottoms, or on substrates which are somewhat enriched in organic matter, it loses its sites to *Posidonia oceanica* (Den Hartog, 1970). Productivity of *Cymodocea nodosa* reached values of 300–400 gm/m^2 in the northern Mediterranean.

Cymodocea rotundata is another species of the Indian Ocean which penetrates the Red Sea up to approximately the 28° latitude.

THALASSODENDRON CILIATUM (FORSSK.) DEN HARTOG

These are plants with erect, unbranched, or slightly branched stems and a thick horizontal rhizome. Wide leaves are concentrated at the stem tips giving plants an appearance of a Date palm (Fig. 14.3). The plants are dioecious. Many stands of *T. ciliatum* are constituted either of staminate or of pistillate plants. The species is common in the Red Sea, and the eastern coast of Africa, as well as in the West Pacific (Den Hartog, 1970).

Thalassodendron ciliatum grows in shallow water below the low tide line. In deeper water, up to 10 m deep, plants form dense stands. It is a salt-requiring species, which only rarely is found in deltas and estuaries with diluted seawater (Den Hartog, 1970).

THALASSIA BANKS EX KOENIG

Thalassia is the most abundant genus of marine angiosperms in the tropical western Atlantic region. It occurs in a variety of loose substrates from mud to sand. While being relatively independent of water depth, plants require calm water for germination and growth. In quiet lagoons, a dense intertidal growth may occur, whereas in open water beds, these plants may be found only at depths down to 30 m. *Thalassia* plants tolerate salinity variations in the 1.0–4.8% range.

Thalassia plants develop erect, leafy shoots from a creeping rhizome buried at a depth of 5 to 10 cm in the substrate. New branches arise successively from near the apex of the rhizome. Erect shoots consist of a short stem bearing a small group of leaves with a sheathing base. Leaves contain a well-developed aerenchymatous tissue.

Plants are dioecious, and inflorescences are arranged within a spathe. Staminate flowers are long-pedicelled with a three petaloid perianth and six stamens. Pistillate flowers are nearly sessile, having a beaked six- to nine-celled ovary. The fruits are peculiarly stalked and oval in shape. They may float in the oceans for long distances and roll about in the surf before opening to discharge seeds. According to Phillips (1960), flowering occurs from May through July.

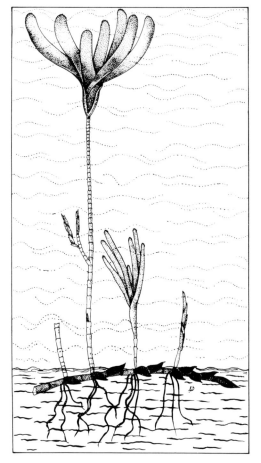

Fig. 14.3. *Thalassodendron ciliatum* (Forssk.) Den Hartog.

However, in view of a sparse production of flowers, vegetative reproduction is probably the main mode of reproduction of these plants.

HALOPHILA STIPULACEA (FORSSK.) ASCHERS.

This dioecious group of plants has thick horizontal rhizomes. Usually one root develops on each internode. Nodes are 1–4 cm long with leaves (1–1.5 cm wide and 5–8 cm long) arranged in pairs having distinct veins and serrulate margins.

Calcium carbonate precipitates are formed on leaf surfaces of most submerged plants which grow in alkaline water. Submerged halophytes are no exception and leaves of *Halophila, Zostera,* and *Cymodocea,* are encrusted with such precipitations.

Halophila occupies shallow water and firm sandy beaches in the Red Sea and in the western part of the Indian Ocean.

It is believed that *Halophila stipulacea* is one of the species which have invaded the Mediterranean via the Suez Canal. However, this has not been sufficiently investigated. *Halophila* was recorded in large masses in Rhodes in 1895 (Den Hartog, 1970). Thus, in the few years since the Canal was opened, this species could have reached Rhodes and spread widely in the Aegean Sea. On the other hand, such populations might have been part of an ancient East Mediterranean flora of *Halophila* that was established long ago.

ENHALUS L.C. RICH

This is a monotypic genus distributed in warm water of the tidal zone along the coasts around the Indian Ocean, Malaya, the Philippines, and various archipelagos of the Pacific (Den Hartog, 1970).

Plants have thick horizontal rhizomes covered with fiberlike remnants of leaf bases. Leaves are long (up to 150 cm) and wide (1–2 cm), and particularly distinct by septated air ducts. Flowering occurs during all the year, but mainly during the spring. Pistillate inflorescences are located on top of a spiraling stalk which contracts after pollination. Pollination occurs after detached staminate flowers collide with horizontally floating pistillate flowers during low tide. Dissemination occurs by fruits which may remain floating for long periods. Completing its development, the seed coat cracks and the embryo is released into the water. When the embryo germinates, it sinks to the sea bottom (Den Hartog, 1970). *Enhalus* has a submerged plumlike fruit with a salty taste which is eaten by the Australian aborigines.

Enhalus acoroides (= *E. Koenigii*) forms dense stands in warm water down to 4 m on sandy or muddy substrates. Distribution of this species is limited by water temperature during winter time. Plants frequently remain exposed on low tides.

ALTHENIA FILIFORMIS PETIT.

Even though it is not a marine plant, *Althenia filiformis* is another submerged halophyte. The plants are branched and bear alternating filiform leaves. Flowers are mono- or bisexual. The staminate flowers have a hyaline perianth and two to three stamens. Pistillate flowers bear a three-lobed perianth, and three distinct carpels. Distribution of *Althenia filiformis* is disjunctive appearing in saline ponds with stagnant water of the West Mediterranean and of South Africa. According to Willis (1966), it is one species which recently was subdivided into two subspecies.

PHYLLOSPADIX HOOK

These are leafy herbaceous plants with horizontal underground rhizomes and distinct internodes; one leaf and a few roots grow on each internode. Leaves are long, having a considerable leaf sheath. *Phyllospadix* plants are known to be adapted either to disturbed intertidal conditions or to deep water. Plants form large masses at or just below the low tide water mark, and are exposed generally to strong wave action. Plants are firmly fastened to rocky substrates by short, condensed rhizomes that often form a tough mat under a thin layer of shifting sand. Leaves are long, varying in width in the different species.

The plants are dioecious. Flowers are borne in two rows on the sides of a flattened spadix. Flowering stems vary in length. Inflorescences are enclosed by spathes. According to Dawson (1966), pistillate plants surpass the numbers of staminate ones by 12 to 1. The pollen grains move with the water to contact submerged stigmas. Pollination commonly occurs at the time of flood following a low tide. The pollen is extraordinary in being filamentous, about 5μ in diameter and 1000μ long. It is often released from the stamens only after the latter are exposed to the action of waves.

Young *Phyllospadix* plants gain a foothold in the surfy environment by an interesting mode of adaptation. Fruits have a pulpy exocarp, which is gradually worn away as the ripe, drupelike fruit is dashed out in the surf. This leaves a dark, horny mesocarp. The mesocarp has two projecting incurved armlike structures that bear inwardly pointing, stout-fringed bristles (Fig. 14.4). By means of these bristly arms, fruits are capable, upon occasion, of becoming fastened on branches of articulated corrallines growing on otherwise bare rocks. Fruits cling so tightly to these retuse bristles that seeds may germinate on the site, and send out a large number of stout roots from each segment of the developing rhizome. Root hairs are produced in abundance and attach themselves to whatever they touch.

ZOSTERA L.

Of the temperate region species of *Zostera*, only *Zostera marina* (eel grass) has a wide distribution area. Other species of *Zostera* are found locally along the coasts of Europe, eastern Asia, Australia, and South Africa. Upon germination, the young stem of *Zostera* elongates and pushes up the encompassed plumule; the cotyledonary sheath is ruptured and the leaves protrude. Consecutive growth involves development of a rhizome by elongation of internodes. At that stage, the first leaves are lost. Lateral buds and adventitious roots are then initiated at the nodes. Further growth, following a period of quiescence, involves extensive development of creeping rhizomes. Erect leafy branches develop, producing elongate fertile branches with twenty spadices or more, enclosed within spathes.

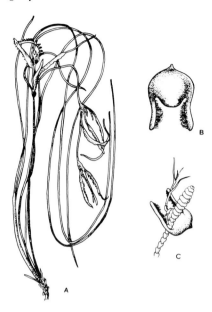

Fig. 14.4. *Phyllospadix torreyi.* (A) A mature plant with fruiting spadices; (B) a mature fruit with the exocarp worn away; (C) germination of a seed after the fruit was attached to a corralline. (After Dawson, 1966.)

Fertile shoots are ephemeral and perish at the end of the season, while vegetative prostrate shoots persist. After a few growth seasons, the rhizomes become fragmented.

Flowers are monosexual, consisting of either a pistil or a stamen. They are borne alternately in two rows in a series of about twelve on a spadix, and exhibit proterogyny, i.e., pistillate branches protruding first. Ripening of seeds is successive in the spadices with the lower ones ripening first. From 500 to 1000 seeds can be produced on a luxuriant plant in a season.

Dispersal of *Zostera* occurs by means of the fruit-bearing shoots. These become detached from the parent plants and are carried away by the sea currents and tides. The ripe fruits eventually drop off those branches and germinate (Fig. 14.5). Fruits of *Zostera* are susceptible to desiccation. Most seeds fall in the immediate vicinity of the mother plants, into the mud. Most of them are snatched and swallowed by fish and water fowl. Only a few become favorably buried in the substrate, and consequently germinate. *Zostera* beds are highly important for supporting fish populations because of their value as food as well as because of the epiphytic fauna and flora which they support (Arasaki, 1950; Den Hartog, 1970).

Zostera marina plants grow well on soft muck, as well as on hard sand or gravel (Tutin, 1942). Stands of *Zostera* greatly accelerate sedimentation of solids

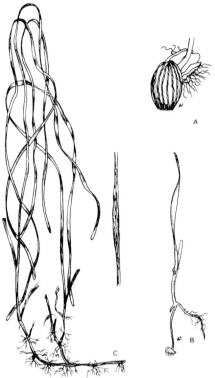

Fig. 14.5. *Zostera marina.* (A) Germinating seed; (B) a young seedling; (C) a fruiting spadix. (After Dawson, 1966.)

in tidal regions. However, when a stand is buried by sand the plants degenerate. Moreover, it takes a long time before stands which became covered by sand are recolonized (cf. Dexter, 1950).

Zostera marina tolerates salinity of marine water, as well as of brackish water. It grows in the Baltic Sea and in the estuaries of the Rhine and Meuse, as long as fluctuations in salinity are small (Den Hartog, 1970). The content of NaCl in *Zostera* leaves is about 6%, but over 80% of it can be leached out by consecutive rinsing with freshwater (Bacalbasa, 1951). Highest development of *Zostera* occurs in the Black Sea coasts in the high summer months, when plants develop the longest and most dense leaves (10–40 cm) (Bacalbasa, 1951). Growth of *Z. marina* plants in brackish water is slow and under such conditions plants of *Ruppia* sp. and *Potamogeton pectinatus* compete successfully with *Zostera marina.*

Plants stop growing when water temperatures fall below $10°$ C. Vegetative growth occurs in water with temperatures of $10°-15°$ C, while flowers are produced only if water temperature is above that range (Setchell, 1929).

Zostera plants generally occupy shallow sandy lagoons and bays, varying in depth. Usually depth is approximately 2–10 m, but sometimes *Zostera* plants may grow as deep as 50 m. In areas of their abundance, *Zostera* sp. and especially *Zostera marina* support a variety of marine animals comprising their winter food. Canada geese and black ducks depend upon *Zostera marina* to a great extent (Dawson, 1966) and a decline in populations of *Zostera* during various years (Tutin, 1938) resulted in reduction of numbers of wintering populations of the Atlantic brant.

Mangroves

Mangrove is a West Indian name given to a formation of trees and shrubs inhabiting the coasts of tropical or subtropical seas. This formation is comprised of a heterogenous group of plants which are mostly limited in their distribution to the tidal zone.

Mangroves do not represent a uniform taxonomic group, and include at least twelve genera in eight different families.

Avicenniaceae
  Avicennia sp.
Chenopodiaceae
  Suaeda monoica
Combretaceae
  Laguncularia sp.
  Lumnitzera sp.
Meliaceae
  Conocarpus sp.
  Xylocarpus sp.
Myrsinaceae
  Aegiceras sp.
Plumbaginaceae
  Aegialitis sp.
Rhizophoraceae
  Rhizophora sp.
  Bruguiera sp.
  Ceriops sp.
Sonneratiaceae
  Sonneratia sp.

Other species, e.g., *Scyphiphora hydrophyllacea, Scyphiphora malayana* of the Rubiaceae, *Acanthus ilicifolius* of the Acanthaceae, *Hibiscus tiliaceus* of the Malvaceae, and others, can also be considered as mangrove species. However, their distribution is very limited.

Mangroves are distributed along most of the warm water oceans, and are thus included in the following phytogeographical regions: the Caribbean region, the Brazilian region, the tropical rain forest region of western Africa, the coasts of the Savanna region of East Africa, and the Malaysian phytogeographical region.

Most of the highly developed mangrove forest is limited to the zone between the spring and the neap tides. These habitats usually comprise a narrow strip of shallow coastal habitats, deltas, estuaries, or lagoons, and they are highly affected by the climate, tides, substrate, and water salinity (Walter and Steiner, 1936; Chapman, 1960; Clarke and Hannon, 1967, 1969, Macnae, 1968).

Several factors characterize the mangroves and their habitats.

CLIMATE

Mangroves comprise a tropical phytogeographical element, and distribution of the various species is limited by temperature regimes. Regions in which the average temperature is below $19°C$, are beyond the ecological limits of mangrove plants. These plants also do not tolerate temperature fluctuations exceeding $10°C$, even in habitats with higher temperatures, and no mangroves develop satisfactorily in such regions (cf. Macnae, 1963). Although being tropical, certain species may still penetrate into bordering climatic regions. In places where mangroves are found beyond the tropical conditions, the water temperatures usually are constantly warm. Such habitats are found along the East African and the Red Sea coasts, where high radiation and shallow water create suitable temperatures. Still, populations of *Avicennia marina* var. *resinifera* can be found in New Zealand on sites which have occasional freezing temperatures (Chapman and Ronaldson, 1958). The northern margins of the mangrove distribution ranges are generally between the $29°$ and $30°N$ latitude. Again exceptions are found in places subjected to warmer local conditions. In North America, specimens of *Laguncularia racemosa* and *Rhizophora mangle* were recorded near the $30°N$ latitude (Uphof, 1937). On the coasts of the Red Sea, the limit of distribution of *Avicennia officinalis* (Fig. 14.6) is near $28°N$ latitude, probably due to incidental periods of low temperatures (cf. Ascherson, 1903). *Rhizophora* and *Ceriops* do not reach even these latitudes and specimens of these species are found in this region only south of the Tropic of Cancer. Mangroves are distributed down to $38°S$ in Australia and $44°S$ on Chatham Island near New Zealand.

TIDES

Mangroves are found in habitats which are periodically exposed to high variations in water levels. During the high spring tides, trees are mostly immersed in water with only the upper parts of their canopies protruding above sea level.

Fig. 14.6. General view of a stand of *Avicennia officinalis* in southern Sinai.

Since the substrate in many of the mangrove habitats is composed of heavy clays, it remains saturated even during the intertides (Walter and Steiner, 1936; Scholander *et al.*, 1955). Most of the root systems are thus continuously submerged. The periodic tidal levels (daily, monthly, and annual) affect the distribution of plants within the formation.

The inundated substrate is practically continuously anaerobic, thus exposing the plant roots to a reducing environment low in oxygen. Tides provide some specific ecological conditions by supplying the roots with oxygen containing fresh water. The tide also causes a periodic flush of the substrate with saline seawater. For most mangroves, these flushes are frequent. However, in regions affected by spring tides only, salts tend to accumulate during the intertides in the upper layers of the soil (Walter, 1936a,b).

SUBSTRATE

Most types of habitats occupied by mangroves have either sandy or heavy silt substrate. They contain large amounts of organic matter and in certain cases even some types of peat are formed under the prevailing anaerobic conditions. Mangrove peat beds, which were found in Florida beyond the tidal range (Penfound, 1952), were used as an evidence for changes in coast line level.

The pH of soil extracts taken under mangroves is a little lower (pH 7.6–8.2) than that of open seawater. This is caused by anaerobic decomposition of organic material, by the high content of iron sulfides, and by the lack of calcium carbonates. Thus, pH of the substrate fluctuates with the tides.

The high temperatures prevailing in these places, only accelerate the consumption by microorganisms of the small amount of oxygen that diffuses into the soil, making it practically oxygen-free. Unadapted plants would have suffocated rapidly under such anaerobic conditions. However, like other swamp plants (van Raalte, 1940), mangrove species also seem to possess adaptive mechanisms which enable them during each of the growth phases to withstand such conditions. Prevention of suffocation of seeds and young seedlings is achieved by the dissemination of plant propagules above the anaerobic mud. Such a viviparous system is most common in mangroves. Seedlings of *Rhizophora* float horizontally in seawater. Only after a month under suitable temperature conditions, will the root grow, and will eventually change the seedling's position into a vertical one. Such seedlings may remain floating in good conditions for over a year.

Avicennia does not exhibit a similar vivipary. However, seedlings falling into the mud are able to withstand long periods of inundation due to their anaerobic respiration (Walter, 1962). (See also Chapter 9.)

In established plants, the root system is aerated by means of the developing aerial roots, i.e., stilt roots and pneumatophores (see Fig. 13.6). Both types provide an efficient system for gas transport and exchange via the lenticels and the aerenchyma.

SALINITY

The seawater brought in by the tides contains on the average about 3.5% soluble salts and has an osmotic potential of about 24 bars. However, salinity remains more or less constant only in the limited lower range of the mangrove habitats. Even there salinity is constant only in a specific zone. As a great part of the mangroves are located on deltas or estuaries, the plants are periodically exposed either to sea- or to freshwater flushes. Plants growing under such conditions have to be adapted to the fluctuations in both osmotic and ionic environments.

Segregation between constantly and periodically saline habitats also results in plant segregation. Optimal growth of many mangrove species is in salty water (Stern and Voigt, 1959). Moreover, genera such as *Sonneratia* disappear in habitats which are frequently eluted with freshwater. However, *Rhizophora mangle* plants grown on 25% seawater produced more leaves and roots than plants grown in undiluted seawater (Egler, 1951; Pannier, 1959).

SUCCESSION OF MANGROVES

Succession of the various mangrove formations differs in various types of habitats. As a rule, plants of the genus *Avicennia* are capable of invading sandy habitats and form a pioneer community in the sere of sandy coasts (West, 1956; Macnae, 1963, 1968).

The buildup of a dense stand of *Avicennia* and the growth of a dense thicket of pneumatophores impede the tidal flow of water in such habitats, resulting in deposition of silt and fine clay minerals. A new shallow, muddy coastal habitat is thus formed. Consequently, additional species of plants are able to invade this newly built mangrove. *Avicennia* is also a light-requiring species which develops nicely on open sites. However, its best development is restricted on heavily populated and partly shaded habitats.

In certain sandy places, *Avicennia* forms pure stands on the seaward edge as well as on the landward edge. Between the two zones of *Avicennia,* a thicket of *Bruguiera* usually develops. Where the influence of freshwater is strong, *Hibiscus tiliaceus,* one of the so-called freshwater mangroves, is common.

On river banks and estuaries with a heavy clay substrate, *Rhizophora* instead of *Avicennia* dominates the pioneer communities. In habitats which are already shaded, a large number of *Bruguiera* plants appear. Light intensity inside the thickets of *Bruguiera* is too low for germination, establishment, and growth of *Avicennia* plants, but is optimal for propagation and growth of *Bruguiera* seedlings.

In northern Madagascar, *Lumnitzera racemosa* Willd. occupies the landward fringe of the mangrove formation. Roots of these plants are immersed in water, only during the highest spring tides (Macnae, 1963). *Lumnitzera* lacks pneumatophores and, thus, develops preferably either on sandy coasts not frequently inundated, or on elevated sites.

As mentioned previously, zonation of the various mangrove species depends on a few factors, i.e., on the tidal line, salinity levels, and on the substrate type. Plant distribution in an East African mangrove (Walter and Steiner, 1936) shows four distinct zones:

Zone a. This zone is inundated for the longest periods. The soil of this habitat is a heavy clay with a high organic content. Usually, this zone is inhabited by *Sonneratia,* which forms a monoculture.

Zone b. The substrate in this zone is similar to that of the first one mentioned, but immersion is of shorter duration. It is inhabited primarily by *Rhizophora* or *Bruguiera* trees. In habitats which are supplied by freshwater, *Rhizophora* rather than *Sonneratia* forms the seaward fringe.

Zone c. This zone is characterized by a well-drained substrate, but immersion is still of long duration. *Ceriops* is a typical plant of this zone.

Zone d. In this belt, the substrate is sandy and drained. The upper parts of this zone are inundated only by the biweekly tides, and are thus highly saline. The zone is characterized by *Avicennia* and *Lumnitzera.*

Succession in mangroves varies in different sites. *Rhizophora* forms the pioneer community in coastal swamps of North America. After the establishment of plants of this species on the mineral soil of the swamp and the consequent accumulation of organic matter, other species such as *Laguncularia racemosa* invade the deeper parts of the swamp, while species such as *Avicennia nitida* invade its shallower parts (Davis, 1940). *Rhizophora mangle* seedlings become established at, or below, mean low tide level, but not above it. It takes some 20–30 years before a thick stand of *Rhizophora* is developed (Davis, 1940).

Stands of *Avicennia* constitute a transition zone between swamp and marsh vegetation. In open areas, in between stands of *Avicennia,* other species such as *Batis maritima, Distichlis spicata,* and *Spartina alterniflora* become established. On drier sites of the landward edge of the *Avicennia* marshes in Florida, a belt of *Conocarpus erecta* is usually found.

Laguncularia racemosa is one of the mangrove species which has a wide ecological range, and may appear in *Rhizophora,* in *Avicennia,* or even in *Conocarpus* stands (Davis, 1940).

In Mexico, *Rhizophora mangle* plants characterize shore lines of lagoons. Succession in such stands results more from physical and chemical changes in habitat than from interactions among plants (Thom, 1967).

The biotic factor exerts an important influence on species distribution and plant succession in mangrove communities of the Sydney district (Clarke and Hannon, 1971). Besides being influenced by the regular conditions of the stand, successional changes in the mangrove belts in Australia are also affected by hurricanes. A mangrove belt may change after such storms into a tidal marsh or a *Casuarina* stand.

According to Kunkel (1966), succession in mangrove swamps of Liberia can develop along three lines: (1) On elevated and leached sites, *Rhizophora* is replaced by various Cyperaceae. Eventually, such sites will turn into stands of *Pandanus* or into *Elaeis guineensis* grass savanna. (2) On salty but dry sites *Rhizophora* is firstly replaced by *Avicennia* and only later gives way to grass formations and to savanna vegetation. (3) On saline and wet sites, *Rhizophora* is replaced by a salt marsh vegetation, changing into a freshwater marsh vegetation and eventually into a savanna.

The zones of *Bruguiera* or *Rhizophora* are frequently characterized by brackish water. In such habitats, in the vicinity of Lagos, Nigeria, an undergrowth of the fern *Acrostichum aureum* is commonly found.

A specific fauna, as well as a specific flora of algae are associated with mangrove trees. Their development is dependent, to a great extent, on the general climatic conditions of the habitat, as well as on the shade and humidity provided by the trees. The algal populations consist mostly of red algae (*Bostrychia, Caloglossa, Catenella,* and *Murrayella*). The fauna consists of the decapods (*Balanus* sp., *Sesarma* sp., and *Uca* sp.), amphipods (*Talorchestia* sp.), snails (*Littorina* sp., and *Cerithidea* sp.), fish (*Periophthalamus koelreuteri*) and others. The distribution of animals depends mostly on the presence of algae and trees (Macnae, 1968; Macnae and Kalk, 1962; Fishelson, 1971).

ADAPTATION TO AN ANAEROBIC SUBSTRATE

Various morphological and physiological means of adaptation are known in plants which grow in anaerobic habitats. Evidently, the capability of mangrove plants to survive in substrates with suboptimal aeration is related to their special system of aerial roots. The pneumatophores of *Avicennia* and *Sonneratia,* and stilt roots of *Rhizophora* and *Bruguiera* contain a highly developed aerenchyma, enabling rapid transport of gases.

As in all other aerenchymatous tissues, roots are rigid but elastic and have a low oxygen consumption (Williams and Barber, 1961). The pneumatic tissue has wide and continuous air ducts connecting the aerial roots both with the main root system and with the atmosphere. Access of the inner air spaces to the atmosphere is provided by a large number of lenticels. These lenticels are highly hydrophobic, a most important feature preventing water penetration into the air spaces during high tides. Penetration of air into the pneumatic tissues of the roots is certainly not a diffusion phenomenon only.

Westermaier (1900) attempted to explain the rapid flow of air into the roots by a "tidal pump." According to this hypothesis the hydrostatic pressure of water during high tide squeezes the soft and spongy roots. At low tide the roots which emerge out of the water expand to their normal size and suck in air. However, this hypothesis has never been supported experimentally. On the contrary, manometers attached by Scholander *et al.* (1955) to roots of mangrove plants showed that at high tide, suction, rather than pressure, develops inside the roots.

When roots are submerged during high tide, continuous consumption of oxygen results in a negative pressure in the roots (Scholander *et al.,* 1955). Such a negative pressure is not high enough to cause entry of water into the roots, but certainly causes suction of air during the intertides, as soon as the roots are exposed. Thus, periodic changes in tidal levels are followed by similar changes in air pressure and in the partial pressure of oxygen and of carbon dioxide inside the root aerenchyma.

ADAPTATION TO SALINITY

Similar to other halophytes, mangrove species also develop various mechanisms which enable them to survive in the highly saline seawater. Some of them secrete excess salts very efficiently through salt glands (see Chapter 8 on salt secretion). However, even salt-secreting plants do not absorb the external solution in its natural composition and concentration. Salt concentration in the sap of nonsecreting mangrove species (e.g., *Rhizophora* and *Sonneratia*) is about $\frac{1}{70}$ the salt concentration in seawater, i.e., almost ten times higher than the concentration of sap of normal plants (Scholander *et al.,* 1962). However, the concentration of sodium chloride in the bleeding sap of a secreting species, e.g., *Avicennia* roots, is only approximately $\frac{1}{7}$ the concentration of seawater. The salt concentration in the bleeding sap is usually constant and does not show diurnal fluctuations. Thus, most mangrove species are able to separate the entry of salt and water by a nonmetabolic ultrafiltration process (Scholander, 1968). Roots were found to be insensitive to chilling or to poisoning by DNP or by carbon monoxide. A similar filtering system was also found in the leaves, though permeability of leaf cells was disturbed by chloroform and ether. Such a response thus indicates that substances of lipoidal nature govern the process.

A negative hydrostatic pressure of at least 20 bars is thus needed to move water through the plant. This was, in fact, demonstrated by Scholander *et al.* (1962) for trunks of various mangrove species. All the mangrove species investigated had a negative xylem pressure of about 30–60 atm serving as the driving force for salt filtration (Scholander, 1967, 1968).

Lötschert and Liemann (1967) investigated the salt metabolism of *Rhizophora mangle* seedlings. It was suggested that a barrier to salt movement existed between the embryo and the mother plant. Chloride tended to increase with age in most tissues of the embryos, while the content of cations varied. Also an inverted osmotic gradient was found between the parent plant and hypocotyls of the viviparous seedlings (Walter and Steiner, 1936; Kipp-Goller, 1940). This means that the water and the nutrients needed for development of the seedlings would be supplied against normal osmotic gradients, due to participation of a tissue of a glandular type. In fact, such a tissue which may function in transport of nutrients was observed within viviparous mangrove seedlings (Haberlandt, 1895; Lötschert and Liemann, 1967). This was corroborated by Pannier (1962), who showed that such a specialized tissue showed high acid phosphatase activity, a characteristic typical of glandular tissues.

High Coast Halophytes

Effects of proximity of the sea on plants certainly do not terminate at the high tide level. Because of wave action and splash and spray of seawater, plants in such wind-swept habitats are subjected to salt injury. Salt-burnt plants can be detected as far as a few kilometers from the coast inland. Furthermore, wind-borne sand particles usually cause lesions to exposed plants. Thus, more rapid dehydration of such plants, as well as faster penetration of salt into their tissues, cause extension of the salt injuries. The further the plants are from the water line, and the higher they are on the coastal cliff, the less seawater spray will reach them. Such sites may thus be inhabited by salt-resistant as well as by salt-sensitive species. On sheltered and quiet coasts, such as the lochs of Scotland, fjords of Norway, or certain rocky bays of the Mediterranean, the hinterland terrestrial plants approach the high tide line (1–2 m) and in humid regions may almost reach the water line itself (cf. Gillner, 1952; Gimingham, 1964).

Zonation of plants in salt-beaten coastal habitats is very striking, and segregation of plant communities is abrupt. Whereas some plant communities may occupy a strip of only a few meters, the width of the distribution area of others can be measured in kilometers. Wide belts are usually found on shallow coasts or in deltas and river fans.

On the coast, various environmental parameters, of which salinity is only the most obvious one, change from the water line inward. One of the typical characteristics of many coasts is the constant effect of sea and land winds. Constancy of winds is known from many coasts around the world. In Israel, such winds alternate daily during the summer, showing minimal rates toward midnight and maximal rates in the early afternoon (cf. Eisikowitch, 1970). During the winter, winds are variable. Annual as well as daily fluctuations in temperature are also much more moderate on the coast than in inland habitats. A gradient in temperatures is, therefore, found from the beach inland. As a result, a gradual change in zonation of coastal plants regarding both species composition and plant density parallels the changes in the various ecological factors. Only plants with ability to withstand salinity, strong winds, root exposure, or cover by sand can inhabit the closest sites to the water line. Others occupy habitats which are at different distances from the high tide level.

ECOLOGICAL CONDITIONS AND PLANT DISTRIBUTION

Five major zones can usually be distinguished above the water line in coastal habitats. Each zone is characterized by specific ecological conditions and by typical plant associations (Fig. 14.7).

1. The beach
   a. Sea-washed boulders
   b. Sandy beach
   c. Shingle beach
2. Salt swamps and salt marshes
3. The cliff
4. The plateau on the cliff top
5. Sand dunes

The beach is defined as that strip of land which is stretched along the water line. It varies in width from a few meters in certain coasts to a few kilometers in others. Still, in other coasts, where a cliff drops directly into the sea, the beach may be absent altogether. The substrate on the beach is mostly sandy and is poor in nutrients. The texture of beach sand varies greatly and contains variable amounts of shell fragments.

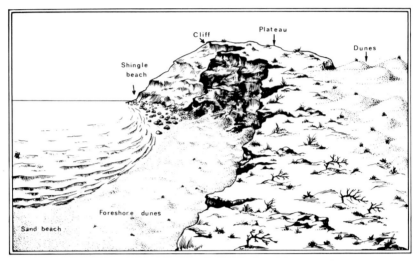

Fig. 14.7. A scheme of a high coast, with its various habitats.

The sea-washed boulders are usually covered by algal communities or by communities of submerged phanerogams. Only in the tropics may such habitats support swamp or terrestrial halophytes.

Although most coastal plants have a multiregional distribution pattern, the plant cover of beaches varies around the world from site to site. In most cases, it is not known whether the restriction of typical species, e.g., *Glaucium* sp., *Mesembryanthemum* sp., and *Limonium* sp., to coastal habitats occurs because of their special requirements, or if this is a result of hard competitive pressure in inland communities. Nevertheless, in each coast such plants certainly constitute a specific group. In the tropics and subtropical regions, beaches are inhabited by a number of species, such as, *Corchorus acutangulus, Spermacoce hispida, Launaea pinnatifida, Scaevola lobelia, Ipomoea pes-caprae, Neuracanthus sphaerostachys, Clerodendron inerme, Leucas aspera, Boerhavia diffusa,* and *Celosia argentea.* Most species are dwarf and prostrate. They have leathery leaves and are capable of producing adventitious roots. In India and West Africa, as well as in other places in the tropics, the low sandy coasts are occupied by *Ipomoea pes-caprae–Canavallia* associations. On stable sand coasts, or on those with heavier substrates, *Ipomoea* is replaced by species of *Sesuvium* and *Sporobolus* (Mullan, 1933; Walter, 1962).

In the Mediterranean, sandy beaches are characterized by hair-covered dwarf species of plants such as *Diotis maritima, Medicago marina, Lotus creticus, Convolvulus secundus, Launaea tenuifolia, Sporobolus arenarius, Cakile maritima, Pancratium maritimum,* and *Oenothera drummondii.* Only very few annuals are present in such formations.

On the sandy foreshores of northern Europe, the vegetation is dominated by plant associations of the Salsola-Minuartion peploides and of the Atriplicion litoralis alliances. Such habitats are characterized by species such as *Ammophila arenaria, Armeria maritima, Koeleria cristata, Silene maritima,* and *Tripleurospermum maritimum* (Gimingham, 1964). Such foreshores may be intermittently affected by tides. Seed germination and growth of plants tend to predominate late in the season.

Shingle beaches in northern Europe are mostly covered with lichens. Among the terrestrial higher plants *Atriplex glabriuscula, Silene maritima, Tripleurospermum maritimum,* and *Rumex crispus* are dominant (Braun-Blanquet and Tüxen, 1952; Gimingham, 1964).

Whenever sandy or shingle beaches are on top of a hardpan, salt marshes or salt bogs develop. Such marshes are common on the drier sites of the temperate regions and of subtropical coasts. They rarely appear in humid regions. Coastal swamps are occasionally submerged in seawater and are continuously wet. Plants characteristic of these habitats include: *Salicornia radicans, Glaux maritima, Suaeda maritima, Spartina alterniflora, Juncus gerardii,* and *Scirpus maritimus.* Occasionally, also *Plantago maritima, Armeria maritima,* and *Silene maritima* can be found.

Sand particles blown from the beach tend to accumulate in small heaps at the base of the coastal cliff. Such sand consists mostly of coarse grained particles which are too large to be blown over the cliff. In such habitats above the beach or coastal marsh levels, different types of halophytes can be found. These consist mostly of plants which can tolerate splashes of seawater on their shoots, but cannot withstand high salt concentrations in their root zone. A combination of a few species of plants, such as *Ammophila arenaria, Carex arenaria, Plantago maritima,* and *Festuca rubra* is typical in such sites.

Usually, high rocky coasts are exposed periodically or continuously to the action of a rugged sea. These coasts are characterized by an ascending cliff covered by a prostrate wind-swept vegetation. Since plants are anchored on rock debris or weathered material and depend upon them for their growth requirements, coastal cliff vegetation varies with exposure of geological rock formations. Thus, different plants inhabit cliffs made of siliceous rocks, granites, calcareous rocks, dolomites, sandstones, etc. In eroded niches or rock pockets, stability and depth of the soil or degree of rock hardness determine the type of plants present. Among the most common species in such habitats in Europe are *Plantago maritima, Festuca rubra, Scilla verna, Sedum roseum,* and *Primula vulgaris.* In the Mediterranean, species such as *Agropyron junceum* or *Ononis stenophylla* dominate.

Coastal habitats are more saline under arid climates than under humid ones. Thus, plants on the high coasts of the temperate as well as tropical regions are less halophytic than those of arid regions.

As the dominating factors in coastal habitats are wind and salinity, plant cover changes accordingly. A typical example of such changes on a coastal cliff in Barbados is that reported by Randall (1970b) (Fig. 14.8).

Evidently, only the salt-tolerant species, *Sesuvium,* was able to survive on the foreshores or on the lower parts of the cliff. Even on higher elevations on the cliff, where a few additional species appeared, *Sesuvium* remained the major component. Only in sheltered niches, or on the plateau on top of the cliff, additional inland species were also able to grow.

On hard limestones of the Mediterranean coasts, salt-tolerant ecotypes of the Batha and Garigue species dominate (*Poterium spinosum, Teucrium polium,*

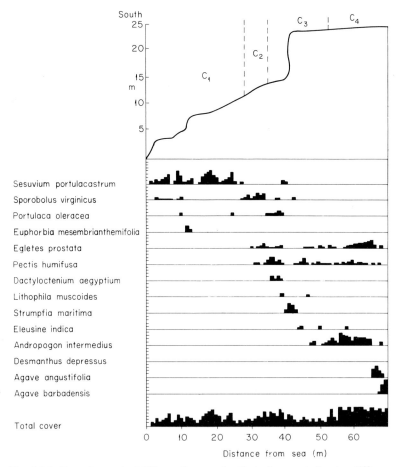

Fig. 14.8. Type transect at Olivers Cave on the Barbados coast showing differences in cover and composition of species. Four zones (C_1–C_4) are distinguished. Each division = 5% cover. (After Randall, 1970b.)

Thymus capitatus, etc.). On soft psammoliths of the same region, plants such as *Agropyron junceum, Crithmum maritimum, Ononis stenophylla,* or *Thymus capitatus* comprise the main plant cover.

On top of the cliff, but still under the heavy impact of sea winds, other species and other types of vegetation units can be found. Although the habit of such plants remains compact and dwarf, the number of species increases considerably, and includes an appreciable number of annuals. Usually on a stable substrate and on rich and deep soils, vegetation is climatogenic. On psammoliths or other dry and poor sites, plants with lower ecological requirements such as *Limonium* sp., *Retama roetam, Atractylis flava,* or *Echium angustifolium* appear in great numbers. Desert plants frequently invade such sites on the southern and eastern coasts of the Mediterranean.

Vegetation on the lee side of the high coast is usually much richer than on the sea side. On hard rock or in habitats with deep alluvial soil, vegetation is climatogenic. However, on sandy sites this part of the coast may be characterized by drifting sand dunes. Consequently, plant cover is typically psammophilous and consists of relatively few species only. Hinterland sand dunes in Europe are characterized by *Ammophila arenaria, Elymus arenarius, Corynephorus canescens, Carex arenaria,* and *Agropyron junceum.* In the East Mediterranean, such dunes are inhabited by *Ammophila arenaria,* but also by *Diotis maritima, Lotus creticus, Salsola kali,* etc. In the tropics, plants of the coastal dunes are characterized by *Ipomoea pes-caprae, Boerhavia diffusa,* etc.

Most plants of such habitats form vigorous stands on sand dunes for a short period only. When the dunes are drifting and new sands accumulate on the top of, and in front of the present stand, plants are able to produce new systems of adventitious roots and to continue to show vigorous growth. However, when dunes become stabilized, the old stands of plants degenerate (Marshall, 1965).

Adaptation of Epilitoral Halophytes to Their Environment

Inhibition of plant growth on exposed coasts is a well known phenomenon and was reported in the late 19th century (Focke, 1871; Borgreve, 1890; Gerhardt, 1900). However, only Wells and Shunk (1938) and Oosting (1945) have shown that such suppression of growth was caused at least partly by wind-borne salt spray. Correlations between spray intensity, location of plants near the coast, and their adaptability to salt spray were reported for many regions (Oosting and Billings, 1942; Oosting, 1945; Zohary and Fahn, 1952; Boyce, 1954). Similar correlations were observed when vicarious pairs of ecotypes of coastal and inland habitats were compared.

Coastal plants usually can avoid salt damage by two different mechanisms: (a) morphological mechanisms: salt droplets are prevented from reaching live tissues because of plant form (prostrate) or structure (efficient cuticle, thick hair cover, or a sclerenchymatous sheath); (b) physiological mechanisms: salt spray reaches the live tissues and salts penetrate the cells. However, as the cells are highly salt tolerant, the cytoplasm remains undamaged. Usually, plants exhibit both morphological and physiological modes of adaptation.

The selective advantage of the prostrate forms in coastal habitats is self-evident. Not only are tall vegetative parts of the plants salt burnt, but their chances of being pollinated and producing seeds are reduced to nil. Prostrate forms and high protoplasmic salt tolerance are encountered in coastal ecotypes irrespective of the species to which they belong.

The capacity of trees to survive in coastal habitats is of special interest and has many practical aspects. Only very few species of trees can survive seacoast conditions, and even those are primarily located on protected sites, or under conditions where regular leaching occurs. Most trees found near the coasts are salt burnt. Some are completely damaged and die when reaching a certain size, whereas others become salt damaged on their wind side (Fig. 14.9). Growth is thus unidirectional and trees obtain the so-called flag shape. Even tree species such as *Tamarix*, which can resist high salt concentrations in the soil and which secrete salts on their leaf surface, are badly damaged by exposure to salt spray.

The behavior of *T. aphylla* in coastal habitats is of special interest. This species is known to be a fast growing one in the inland habitats of arid regions (Zohary and Waisel, 1956; Friedman and Waisel, 1964). However, on the

Fig. 14.9. Flag-shaped tree on an exposed coast of the Dead Sea, Israel. (Photo by Z. Kuller.)

exposed coasts, growth and survival of this species were poor. The same is also true for *Tamarix meyeri,* which grows on highly saline soils, but cannot survive salt spray on its top.

As in other halophytes, also coastal plants have a typical structure which is believed to be of adaptive value. This was discussed in detail in Chapter 13. Such characteristics (succulence, water tissues, hair cover, blue-green color, etc.) observed in coastal plants are largely genetically fixed. Leaves of coastal ecotypes of *Atractylis flava, Pallenis spinosa, Artemisia monosperma, Lotus villosus,* etc., are thicker, have a larger surface area, fewer stomata, and fewer veins per unit area than their respective inland ecotypes (Waisel, 1960d).

Sonchus oleraceus, Euphorbia peplis, or *Eryngium maritimum* are among the few plant species which repel the salt droplets by their highly hydrophobic cuticles (Fig. 14.10). Spray droplets thus slide from leaf surfaces without penetrating the tissues or causing them any damage.

Other species of coastal plants can be included among the salt-enduring halophytes. Höfler and Weixl-Hofmann (1939) and Repp (1939) showed that coastal plants absorb high quantities of salts, and have higher osmotic values than comparable noncoastal glycophytes. Biebl (1953) who investigated a few plant species of the English coast showed that tissues of coastal plants had higher osmotic potentials than tissues of glycophytes, and that tissues of coastal plants survived longer immersion periods in concentrated salt solutions.

Fig. 14.10. Water droplets on the hydrophobic cuticle of *Sonchus oleraceus* (1) and *Euphorbia peplis* (3) and of the less hydrophobic cuticle of *Capsicum annuum* (2).

   Not much is known of the specific toxicities caused by the spray. In certain species, toxicity is induced by chloride and in others by sodium or magnesium. In still other species, sodium is the active ingredient and toxicity probably involves the metabolism of potassium. In such plants, recovery from the effects of spray is observed under heavy fertilization with potash.

   Ecological conditions on exposed seacoasts delay several aspects of plant development, such as flower production, pollination, and seed production. Most coastal plants flower during the summer when salt spray intensity is lowest and the activity of pollinators is greatest. Late flowering probably is caused by the relatively slow rise in temperature of coastal habitats in the spring and by slow growth of plants. No evidence was obtained that late flowering was generally due to different photoperiodic requirements of coastal plants. The same physiological behavior pertains to coastal ecotypes. The behavior of coastal ecotypes is discussed further in Chapter 16.

   On the basis of flowering characteristics, coastal plants may generally be divided into two groups (Table 14.1): (a) night flowering; (b) day flowering. Flowers of the first group, e.g., *Pancratium maritimum* and *Oenothera drummondii,* open at dusk when wind velocity and salt spray are at their minimum. Flowers of such plants remain open throughout the night. *Pancratium maritimum* flowers at night, also because it is totally dependent on hawk moths, such as, *Chaerocampa, Macroglossum,* and *Daphnis,* for pollination and seed production. Plants of this species, which flower beyond the migration season of the hawk moths produce no seeds. Flowers of *Pancratium,* which are pollinated

Table 14.1
FLOWERING (TIME AND SEASON) OF VARIOUS COASTAL HALOPHYTES IN ISRAEL[a]

Species	Flowering season	Hours of flowering
Pancratium maritimum	July 15–Oct. 25	17:30–Noon
Oenothera drummondii	April 1–Dec. 15	17:00–Noon
Silene succulenta	Mar. 10–Sept. 10	16:00–10:00
Glaucium flavum	April 15–Oct. 1	4:00–14:00
Ipomoea stolonifera	April 1–Nov. 1	6:00–14:00
Convolvulus secundus	April 15–Nov. 1	6:00–14:00
Statice sinuata	Mar. 15–Aug. 15	6:00–16:00
Mesembryanthemum crystallinum	May 1–Nov. 1	9:00–15:00
Crithmum maritimum	Aug. 1–Nov. 1	–
Lotus creticus	May 1–Sept. 1	–
Urginea maritima	Aug. 1–Nov. 1	4:00–18:00
Statice oleifolia[b]	June 1–Nov. 1	6:00–18:00

[a] After Eisikowitch, 1970.
[b] Apomictic.

during the night, wilt in the morning. Unpollinated flowers remain open a few hours longer and wilt during the following day (Eisikowitch, 1970).

Flowers of plants of the second group, e.g., *Ipomoea stolonifera, Convolvulus secundus,* and *Glaucium flavum,* open in the early morning hours and are pollinated by day insects before the wind increases in velocity.

Flowers of most species of coastal plants in Israel are short-lived, i.e., they are open for short periods only (Eisikowitch, 1970). Most of the flowers remain open for less than 24 hours and thus are not injured by long exposure to salt spray. Plants in which flowers remain open for longer periods have other means of salt avoidance. Some of such species, e.g., *Statice oleifolia* are apomictic. Others, such as *Statice thouinii*, have the salt-sensitive parts of the flowers, i.e., the stigmas and stamens inserted deep inside the floral tube, thus being protected from sea spray.

Bees are among the best pollinators under coastal conditions, due to their ability to remain active even under windy conditions. Yet even in such habitats their pollinating activity is restricted primarily to flowers near the ground.

Morning flowering plants are exposed to spray for relatively long periods. Flowering branches of such plants are prostrate. As wind velocity near the flowers is low, the activity of pollinators is not disturbed and flowers are pollinated.

Flowers of most day-flowering halophytic species show high salt tolerance in comparison to glycophytic species and the stigmata and pollen of the former are not affected even when exposed to relatively high concentrations of NaCl. *Mesembryanthemum crystallinum* is an exception in this respect, as it lacks salt tolerance. This species flowers during the day (approximately 9 A.M.– 3 P.M. .) but since it is bud pollinated, the inhibitory effects of external environmental factors on pollination and fertilization are minimal. Bud pollination, i.e., pollination within the closed bud before the flower opens is also common in *Glaucium flavum*. However, under suitable conditions, this species can still be cross-pollinated (Eisikowitch, 1970).

―――――――――――――15

Ecological Notes on Some Terrestrial Halophytes

Introduction

Ecological conditions which prevail in terrestrial salines are complex. Thus, it often seems best to define salines by the nature of plants inhabiting them rather than by their physical or chemical conditions. Certain species of halophytes require a combination of saline and waterlogged soils, and disappear when the water level in their root zone drops 30 cm below the soil surface. Yet other species thrive only under such conditions. Whereas certain species require constant environmental conditions, others need seasonal alternation. Thus, appearance of a given species can provide information about growth conditions of a site even without touching the soil. The autecology of terrestrial halophytes has been inadequately studied and more research is needed on the biology of individual species. Nevertheless, in this chapter an attempt was made to summarize the available information on common terrestrial halophytes. More attention was paid to some species, mostly to those which have been investigated lately. Other species were treated only briefly because the data on their ecology has been discussed earlier by other investigators (cf., Chapman, 1960, etc.).

Aeluropus litoralis (Willd.) Parl.

This species is a perennial procumbent grass (Gramineae) of wet saline habitats (Fig. 15.1). Its distribution area covers the Mediterranean, Irano-Turanian, Saharo-Arabian, and Sudanian phytogeographical regions. *Aeluropus* is a thermophilic species which grows and flowers during the summer. The spikelets are nearly sessile, crowded into an ovate or oblong one-sided spike and the leaves are robust, 3–5 cm long, and covered with salt crystals. The plant has an extensive system of rhizomes and spreads vegetatively by long, above-ground runners. Short-noded branches develop during the summer from underground buds.

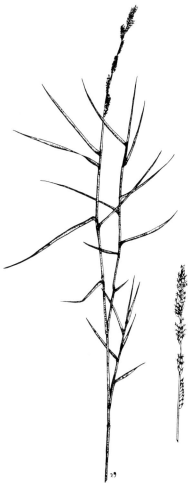

Fig. 15.1. *Aeluropus litoralis.* A flowering branch.

Germination of *Aeluropus* seems to be best under salt-free conditions. An increase in salt concentration of the medium causes a decrease in germination. The highest salt concentration in which germination is still possible is 0.4 *M* NaCl. Caryopses of *Aeluropus* germinate only in light; the optimal temperature for germination being 25°–30°C.

Cuttings root in a wide range of saline conditions, and growth of developing roots is rapid. As practically no seedlings of *Aeluropus* were found in many developed stands, vegetative reproduction under field conditions seems to have an advantage over germination, and secures the establishment of that species in saline habitats.

Plants of *Aeluropus litoralis* are capable of adding new growth in a wide range of saline media, up to 0.4 *M* NaCl. Growth of seedlings seems to be optimal at a salt concentration of approximately 0.1 *M*. Accelerated growth of the seedlings under saline conditions results from combined effects of sodium and chloride ions.

Plants of *Aeluropus litoralis* survived even after transfer into NaCl solutions as concentrated as 0.6 *M*. Salt resistance of osmotically adapted plants was even higher, and plants survived exposure to 1.1 *M* NaCl solutions.

Aeluropus litoralis is a salt-secreting species. Salty brine droplets are secreted from glands located on both sides of the longitudinal furrows between the leaf bundles. Quality and quantity of various ions contained in the growth medium affect the composition and concentration of the secreted fluid. The total amount of secreted sodium, as well as its concentration in the secreted solution, increase proportionally with sodium concentration in the growth medium. Sodium and chloride are secreted by an active and highly selective process, with sodium secreted in preference to potassium or calcium. However, retention of potassium and calcium by leaf tissues of *A. litoralis* is much higher than that of sodium. In addition to mineral salts, the secreted solution contains various organic compounds. Salt secretion maintains a low salt content in *Aeluropus* leaves, prevents accumulation to toxic levels, and contributes to adaptation of the plants to saline habitats.

Aeluropus repens (Desf.) Parl. is another species of this genus which inhabits the east Mediterranean under similar although somewhat more saline conditions. Recently these two species were considered as being one.

For further details the reader is referred to Chapman (1960), Zohary (1962), Gruenberg-Fertig (1966), Pollak (1967), Pollak and Waisel (1968, 1970, 1972).

Armeria maritima (Mill.) Willd.

These are narrow-leaved shrubs of the Plumbaginaceae. The genus includes some forty species over the northern hemisphere, but only few are true halophytes. Although distribution of *Armeria* in the northern hemisphere is circumglobal, this species is missing from the southern hemisphere and from warm climates. *Armeria maritima* is one of the temperate region species which occupy salines as well as nonsaline habitats. It is morphologically variable, and at least four varieties were defined in Europe (var. *maritima*, var. *elongata*, var. *intermedia*, and var. *salina*) and another four in North America (var. *labradorica*, var. *sibirica*, var. *purpurea*, and var. *californica*). There are also distinct groups of plants with respect to their pollination biology.

Flowers are arranged in dense heads and are accompanied by a few whorls of membranaceous bracts.

Plants have a relatively low salt resistance, and growth is negatively correlated with increases in salinity. Their tolerance to inundation is low, and *Armeria* usually appears on well-drained soils.

For further details the reader is referred to Iversen (1936), Lawrence (1947), Baker (1948, 1949), and Chapman (1960).

Arthrocnemum Moq.

Arthrocnemum plants are branched chamaephytes of the Chenopodiaceae with succulent leaves and stems (Fig. 15.2). The genus consists of some twelve species of wet and saline habitats in the Mediterranean and Irano-Turanian phytogeographical regions, as well as in certain coasts of western Europe, North

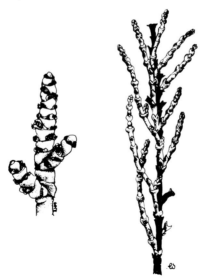

Fig. 15.2. *Arthrocnemum glaucum.* Flowering and vegetative branches.

America, and Australia. *Arthrocnemum* flowers during the late autumn. Flowers are bisexual and arranged in groups of three. They comprise a fleshy perianth with few lobes, two stamens, and two stigmas. Seeds are small (<1 mm) with a black hard seed coat. *Arthrocnemum* plants usually have a bluish-green appearance and attain sizes of up to 1 m in height. Rooting of lateral branches is very common.

Arthrocnemum glaucum (Del.) Ung. germinates in salt solutions of up to 0.5 *M.* No great differences were found between germination in isoosmotic concentrations of NaCl or of $CaCl_2$, and it seems that germination is controlled by osmotic potential of the medium. Plants of this species show only small seasonal variations in growth. A major growth flush occurs in the summer, but plants which are exposed to continuously inundated habitats show practically uninterrupted growth. The plants are highly salt resistant as long as the medium is wet and growth is best in the presence of NaCl.

Arthrocnemum plants have high osmotic potentials and typically are restricted to slightly inundated soils or flooded depressions.

For further details the reader is referred to Wood (1937), Simmoneau (1952), Waisel (1958), Chapman (1960), and Zohary (1962).

Aster tripolium L.

This is a species of semihemicryptophytic perennial herbs of the Compositae. The plants are leafy, glabrous, and erect, with stems 30–60 cm long. The leaves are alternate and succulent, with three distinct veins. Each plant has numerous inflorescences, 1–1.5 cm in diameter. The number of flowers produced is related to plant vigor. Disc florets are yellow, whereas ray florets have a bright blue, light pink, or white color. In some varieties, the ray florets are missing. The plant is pollinated by insects of various groups. Fruits are hairy and equipped with a grayish-white pappus.

Plants have short, but thick rhizomes with well-developed aerenchyma. The aerenchyma is also developed in roots and more so in those plants growing on inundated sites, than in those of well-aerated sites.

Aster tripolium is a common species in European coastal salines, but is also found in North Africa, and in Asia around the Black Sea, the Caspian Sea, and Lake Baikal. *Aster tripolium* appears in Europe and central Asia also in inland salines. In central Asia, it reaches an altitude of 400 m. It is highly frost-resistant, and plants survive even in Lapland. This species probably consists of a number of ecological forms and at least one distinct variety (var. *discoideus*) was described.

Aster tripolium does not seem to be restricted to any particular type of soil, whether saline, sodic, or organic. It frequently appears on sandy soils, but also on rocks and cliffs.

Seed production is abundant, and sometimes 450–1600 seeds produced per plant were reported. Germination of achenes of *A. tripolium* is lowered by NaCl concentrations above 1% and is stopped by a 2% NaCl solution. Root growth of seedlings is slow and, in the earliest stages of growth, the seedlings are liable to be washed away.

The plants are susceptible to grazing and become dwarfed by heavy pasturing. Where plants are overshaded, they become etiolated and usually do not flower. *Aster tripolium* flowers in the second season only. At that time, branches in the axils of the old leaves end their period of dormancy and elongate. Adventitious roots appear on these shoots, and eventually, when contact with the parent plant is broken by decay of the connecting tissues, the young plants grow independently. Finally a whole cluster of young shoots arises, so that it is impossible to ascertain which was the original plant. It is assumed that the life-span of an individual plant is about 4 to 5 years.

For further details, the reader is referred to Iversen (1936), Milton (1939), Van Eijk (1938), Clapham *et al.* (1942), Feekes (1943), and Chapman (1960).

Atriplex L.

Annual, as well as perennial herbs and shrubs of the Chenopodiaceae comprise the genus *Atriplex* which contains some 120 species. Most of them are halophytes, although some are xerophytes or ruderals. Range of distribution of the genus *Atriplex* covers the temperate and warm regions of America, Europe, Asia, Africa, and Australia, with the latter being one of the centers of its speciation.

Atriplex halimus L. is a large-sized shrub which may attain a height of 3 m (Fig. 15.3). Branches have a white-yellow periderm, and leaves are rhomboid, or triangular, grayish-green. They are covered with vesiculated hairs, containing high concentrations of NaCl and oxalate. Flowers are arranged in panicles which have very small leaves or lack leaves altogether. The staminate flowers are located mainly at the upper part of the panicle and the pistillate ones on the lower parts. Pistillate flowers are enclosed between two bracts which are approximately 5 mm wide and long. Seeds are dark brown 1–2 mm long. Flowering occurs in the East Mediterranean during early summer.

Atriplex halimus is a semihalophyte rather than a halophyte. Nevertheless, under normal conditions, it exhibits a considerable positive response to addition of NaCl. Moreover, even increased salinity, up to 100 m*M* did not exert an inhibitory effect on growth. Similar behavior is also known for *A. nummularia, A. polycarpa,* and *A. inflata.*

When young, *Atriplex* plants make good fodder. The old leaves contain too much sodium, chloride, and oxalate, to enable continuous feeding.

For further details the reader is referred to Wood (1925), Black (1954),

Fig. 15.3. *Atriplex halimus.* Flowering and fruit-bearing branches; surface-view of a fruit-bract.

Ashby and Beadle (1957), Beadle *et al.* (1957), Koller (1957), Chapman (1960), Zohary (1962), Gruenberg-Fertig (1966), Osmond (1967), Blumenthal-Goldschmidt and Poljakoff-Mayber (1968), Rosenblum and Waisel (1969), Chatterton and McKell (1969), Springfield (1970), Gale *et al.* (1970), Mozafar and Goodin (1970), Jones (1970), Jones and Hodgkinson (1970), Jones *et al.* (1970), and Anderson (1970).

Halimione portulacoides (L.) Aellen.

Halimione plants are low perennial shrubs of the Chenopodiaceae. Branches are procumbent, angular, and grayish; leaves are opposite, fleshy, elliptical or lanceolate, and grayish with a shiny appearance. Flowers are loosely located on spikelike branches. Bracteoles attain a length of 3–4 mm enveloping the seeds. Flowers in the East Mediterranean appear during June and July.

The genus *Halimione* consists of three species of which *H. portulacoides* is the most important. It is generally found in saline and wet habitats throughout the Mediterranean, Irano-Turanian and West Euro-Siberian regions and was introduced into North America and South Africa, where it is now widespread.

Halimione portulacoides consists of at least two morphologically distinct ecotypes: var. *parvifolia* and var. *latifolia.* They differ in size and ecological requirements; var. *parvifolia* being characteristic of sandy soils. Other varieties are also known.

The species occurs over a wide range of soil types on inundated sites as well as those with good drainage. However, long exposure to waterlogged conditions inhibits plant growth. *Halimione* plants suffer also under coastal conditions from a too frequent tidal submersion. Growing plants are temperature-sensitive. Nevertheless, *H. portulacoides* is frost-resistant and during winter tolerates prolonged freezing without injury. Flowering occurs during late summer and autumn, depending on temperature. This species is anemophilic. Seed germination is best in fresh water, but growth is stimulated by low NaCl concentrations.

Halimione plants make good grazing material, even though the plants are damaged by heavy grazing.

For further details the reader is referred to Chapman (1937, 1950, 1960), Orshan and Zohary (1955), Zohary (1962), Gruenberg-Fertig (1966), and Kappen (1969).

Halogeton C. A. Mey.

These plants are annuals or small chamaephytes of the Chenopodiaceae. The genus includes a few species in the Irano-Turanian (southwest Siberia and northwestern China) and Saharo-Arabian regions.

Leaves are succulent, semicylindrical in shape, and terminate with a stiff bristle. Inflorescences are covered with cotton-woollike hairs, in which flowers are embedded.

Plants have a high oxalate content (5–25% of their dry weight) and thus are poisonous to sheep and cattle.

Halogeton glomeratus (M.B.) Mey is a salt-requiring halophyte and exhibits better growth under slightly saline conditions than under salt-free ones. It was introduced into the United States around 1930, and since then has spread throughout the semidesert and saline lands of the mid and northwest. It is found on saline–alkaline soils at altitudes of up to 2000 m.

For further details the reader is referred to Burge and O'Harra (1952) and Williams (1960).

Juncus L.

Juncus plants are tall perennials forming thick tussocks (Fig. 15.4). However, while tussocks of some species, e.g., *J. acutus* L. or *J. arabicus* (Aschers. et Buch.) Adams have a circular growth pattern, *J. maritimus* Lam. forms a loose tussock

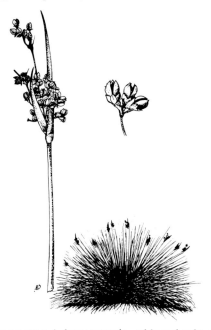

Fig. 15.4. *Juncus acutus.* A dense tussock, and branches bearing fruits.

with radiating rhizomes. Leaves are long and pointed at their end. Inflorescences are compact. The perianth consists of six leaves. The flower has six stamens and a twisted yellowish or reddish stigma. The plant is anemophilic. The capsule is yellow turning into dark brown on ripening.

First flowering takes place in most species of *Juncus* only 2 or 3 years after seed germination. The minute seeds produced per plant can amount to millions. Germination, under suitable conditions, is fast and is completed in less than 12 hours. Seeds require light for satisfactory germination. Fresh seeds are highly salt tolerant, but tolerance decreases with age.

The genus consists of several species of circumglobal distribution in subtropical and temperate regions. Most species are inhabitants of brackish water, but tend to avoid permanently inundated habitats with stagnant water, and concentrate in fringes of salines. Species with wide distribution areas in the arid regions, e.g., *Juncus arbicus* or *Juncus acutus,* are able to penetrate also into highly saline sites.

Most species are polymorphic, and because of their wide distribution have developed a number of ecotypes. Some of those ecotypes have now been classified as distinct taxonomic entities.

The various species of *Juncus* take an important part in the succession of saline habitats, and under certain conditions form a long prevailing community (sere climax). *Juncus* plants are used for fodder, but grazing quality varies among species. While *Juncus gerardii* is grazed to a certain degree, species such as *Juncus acutus* or *Juncus arabicus* are too tough and prickly for most domestic grazing animals, and are eaten only by camels.

For further details the reader is referred to Jones and Richards (1954), Waisel (1958), Tadmor *et al.* (1958), Chapman (1960), and Teal and Teal (1969).

Limonium Mill.

These are perennial herbs or nano-chamaephytes of the Plumbaginaceae. The genus includes some 180 species, and has a wide distribution. Leaves are mostly confined to a rosette. Flowers are arranged in panicles or spikes. The sepals are mostly membranaceous (Fig. 15.5). They are species of high saline inundated sites and in the East Mediterranean are accompanied by *Salicornia fruticosa* and *Halimione portulacoides. Limonium mexicanum* Blake of the western hemisphere can tolerate occasional submersion in seawater, probably because of the presence of a well-developed aerenchyma in petioles and roots. In contrast, *L. carolinianum* (Walt.) Britton cannot survive submersion.

Root aerenchyma in this genus seems to be of importance in protecting plants against waterlogging. The development of successive rosettes of leaves elevates plants above the general level of marsh soil and enables them to avoid inundation.

Though propagules of *L. vulgare* Mill and *L. humile* Mill are abundantly dispersed by the sea, germination is better in brackish or fresh water. Such conditions are also needed for seedling establishment and plant growth. Both species have a high potential of vegetative reproduction.

Growth is generally inhibited by salinity, although various species of *Limonium* have been defined as obligatory halophytes. In *L. carolinianum,* the leaf size depends on salt concentration, and is smaller with increasing salinity. All species of *Limonium* are adversely affected by mowing or grazing.

All species of the genus possess typical salt glands on their leaves and stems and are salt secretors. Salt secretion enables such plants to limit salt concentration within their leaves, at least to a certain level.

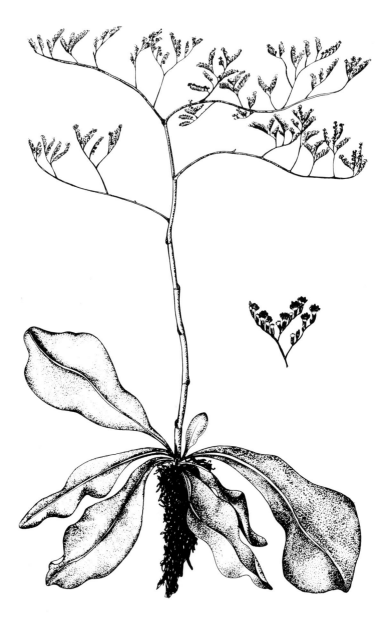

Fig. 15.5. *Limonium latifolium.* A flowering plant.

For further details the reader is referred to De Fraine (1916), Blake (1916), Tsopa (1939), Arisz *et al.* (1955), Chapman (1960), and Boorman (1967, 1968).

Nitraria retusa (Forssk.) Aschers.

This is an evergreen spiny shrub of the Zygophyllaceae. Plants of *Nitraria* lack a single major stem. The plants have two types of branches; brachyblasts and dolichoblasts. The dolichoblasts end in a spine, while brachyblasts produce the flowering branches. Leaves are succulent.

Nitraria retusa is distributed over the Irano-Turanian, Saharo-Arabian and Sudanian phytogeographical regions. The species probably originated in central Asia. A vicarious species (*Nitraria schoberi*) is found in similar habitats in eastern Asia.

Nitraria retusa is a desert plant which seldom appears in exposed coastal salines. It either forms a pure stand or grows together with *Tamarix tetragyna* and *Zygophyllum album. Nitraria* plants have a deep and branched root system.

Nitraria retusa in Israel shows four distinct periods in its annual growth cycle.

(a) The period of minimal activity (September–December). Abscission of leaves occurs with no apparent new growth. Plants are mostly covered with thickened leaves, the result of the swelling of mesophyll mucus cells.

(b) The period of renewed activity (December–February). Plants appear spiny and bear small fleshy leaves arranged in groups of two to three. Leaves are shed readily. The short, lateral, mostly naked branches end in a spiny point. Slow sprouting begins after the first rains.

(c) The period of maximal activity (February–June). During this period, thin leaves appear on young, as well as on older branches. Brachyblasts bearing two to seven leaves grow in the axils of those leaves. New leaves frequently appear in May; the older leaves then increase in size, but their thickness changes only slightly. At the end of this period, the plants are covered with dense foliage, and have an extensive transpiring surface.

(d) The period of diminishing activity (June–September). During this period, leaves reach their maximal size, with the outer ones attaining the length of 2–2.5 cm and the central ones 0.5–1.0 cm. Shedding of outer leaves and thickening of the inner ones begins.

Changes in leaf surface and leaf moisture content were found in *Nitraria* during the year. During periods of activity, plants are densely covered with leaves and transpire a substantial amount of water. Also during periods of minimal activity, plants show high transpiration per unit of surface. However, since they bear only few leaves during this period, transpiration of the whole plant is greatly reduced. Water balance is thus regulated by reduction of transpiring surface.

Osmotic potentials of *Nitraria retusa* plants are high in comparison with other xeromorphs. Annual oscillations in the osmotic value are also great.

Under field conditions, the temperature of *Nitraria retusa* leaves during the day exceeds that of the air by $3°-6°C$ and *Nitraria* plants are thus regarded as belonging to the "over-temperature" type.

The endocarp of *Nitraria* fruits contains germination inhibitors. Such substances cannot be leached out with water and are temperature stable. The activity of the inhibitors can be overcome by subjecting the fruits of *Nitraria* to cycles of moistening and drying.

Nitraria seeds remain viable at least for 2 years. However, rates of seed germination decrease with aging. Seeds are insensitive to light, and germination proceeds similarly in the light and dark. Germination occurs over a wide range of temperature conditions, i.e., between $10°-35°C$. Seeds are highly salt tolerant and in spite of a decline in germination rates, the upper limit of salt concentration which still permits germination is $0.65 M$ NaCl. The response of germinating seeds to salinity is conditioned by temperature, being highest at optimal temperatures. Seeds of *Nitraria* may remain immersed for long periods (over 2 months) in solutions of high salt concentrations without losing viability. Although germination is prevented under such conditions, seeds will germinate when moved to less saline conditions.

Seedlings of *Nitraria retusa* were capable of growing in a wide range of saline conditions; up to $0.6 M$ NaCl or more. Growth of seedlings was optimal in $0.1-0.2 M$ range of NaCl solutions, being higher as seedlings aged.

The content of sodium in stems and roots increased and potassium content decreased with increasing NaCl concentration of the medium. Sodium content of leaves was higher than that of roots and stems. It remained more or less constant over a wide range of NaCl solutions.

Osmotically adapted plants of *Nitraria retusa* tolerated a high concentration of sodium chloride in their growth medium. Such plants survived in a NaCl solution of $1.8 M$. Leaves of *Nitraria retusa* plants growing in nonsaline media were easily infected and damaged.

Nitraria retusa plants are also sensitive to anaerobic conditions, and were severely injured when their roots were inundated for long periods. Under such conditions, shedding of leaves occurred and was greater in young, vigorously growing plants than in old ones.

Plants of *Nitraria retusa* are capable of dark CO_2 fixation.

For further details, the reader is referred to Seybold (1930), Zohary (1944), Shmueli (1948), Reese (1958), Lange (1959), Kassas and Zahran (1962, 1967), and Keren (1970).

Phragmites communis Trin.

Phragmites is a perennial grass of the Gramineae which grows mostly in fresh but also in brackish and saline water. Plants have long and branched canes and long and narrow leaves. Flowering occurs in the autumn. Caryopses are light and hairy, and are disseminated by wind to great distances. Vegetative reproduction by runners is rapid and its extensive system of rhizomes may extend for distances of 50 m or more. Growth occurs mostly during the summer. *Phragmites* has circumglobal distribution and can be found in swamps, salines, and sand-covered wet soils. Variations in fertility exist in different populations. However, this characteristic does not seem to be connected with habitat conditions.

No morphological differences, and very few differential physiological characteristics, were encountered between *Phragmites communis* plants taken from glycophytic and halophytic populations in Israel. The two populations examined seemed to be at very early stages of ecotypic differentiation.

Caryopses of *Phragmites* germinate in a wide range of saline media (0–0.5 *M* NaCl). High germination percentages (>90%) were obtained for caryopses of the saline ecotype in treatments containing up to 0.4 *M* NaCl and only slightly lower (70%) in the 0.5 *M* NaCl treatment. Germination percentages obtained for the glycophytic ecotype were similar in treatments below 0.3 *M* NaCl. However, only 20% of the caryopses germinated in the 0.4 *M* NaCl medium and none in the 0.5 *M* treatment.

Seedlings of both populations grew much better in liquid media than in soil or sand culture. Plants of the halophytic and glycophytic populations exhibited similar rates of growth, relative growth, sodium uptake, and ion distribution in various organs. Root growth of seedlings of the saline ecotype was optimal in treatments containing low concentrations of NaCl. Root growth of both populations was equally low in salinity treatment in the 0.1–0.5 *M* range. Fertility and responses to inundation were also similar in halophytic and glycophytic populations.

For further details, the reader is referred to Luther (1950), Bittmann (1953), Bakker (1957a,b), Chapman (1960), van den Tooren (1964, 1966), Rechav (1967), and Waisel and Rechav (1972).

Prosopis farcta (Banks et Sol.) Eig.

This is a spiny hemicryptophyte of the Mimosaceae, belonging to the Irano-Turanian chorotype. However, plants penetrate in cultivated fields into the Mediterranean and in saline spots, also into the Saharo-Arabian regions.

Prosopis farcta plants have an extensive system of roots which grow as deep as 15 m. The deep root system enables the plants to obtain appreciable amounts of water throughout the year. Leaves are bipinnate, having a low osmotic potential. Flowers are yellow, arranged in cylindrical spikes. Fruits are hard legumes each containing a few seeds. Seeds have a hard seed coat and will germinate as soon as the seed coat is cracked.

Prosopis farcta is a thermophilic species, and in most places grows and flowers during the summer months only. During winter, as temperatures drop, the plants shed their leaves, and branches die back down to the soil surface. In a few habitats, where winter temperatures are not limiting, e.g., near the Dead Sea, Israel, plants remain active throughout the year and attain the form of a shrub.

Ecotypic differentiation in *Prosopis farcta* is well known, and was investigated with particular attention given to saline ecotypes.

Salt concentration in the germination medium has been negatively correlated with seed germination. A medium containing 0.4–0.5 M NaCl stopped germination altogether. There was but a little difference between halophytic and glycophytic populations, although germination of seeds from the halophytic population tended to be influenced less by the presence of NaCl.

Irrigation with NaCl solutions hindered growth of plants of both populations. However, at low NaCl concentrations, growth of the glycophytic population was inhibited much more.

No differences in sodium absorption mechanisms were found between the two populations of *P. farcta*. The absorption mechanisms seem to be similar to those known for crop plants. Rates of absorption from a 10 mM NaCl solution were about 8 μmoles/gm fresh weight/hour. An absorption curve of a saturation type was obtained when absorption was plotted against concentration. The temperature ratio ($30°/20°$ C) of sodium uptake was in the metabolic range. At a concentration of 0.5 mM, the ratio was 1.6, whereas at 5 mM it was 2.2. At low external concentrations (0.5 mM), sodium accumulated only in roots and hypocotyls and was excluded from upper parts of the shoots. At high sodium concentrations in the medium (10 mM NaCl), the barrier restricting upward movement of sodium was overcome and accumulation of sodium occurred in the stem and petioles. At low concentrations, sodium reached the leaflets of plants of the glycophytic population, and in lesser amounts those of the halophytic plants.

Retranslocation of sodium from the stem to the cotyledons occurred before the cotyledons abscissed. Such an "evading mechanism" which excludes sodium from sensitive leaf tissues seems to contribute to survival of *Prosopis* in saline habitats.

Salt resistance of most organs was similar in plants of both halophytic and glycophytic ecotypes. Only viability of cotyledons differed for the two populations; with those of halophytic plants showing higher tolerance.

As only minute differences were found between different populations of *P. farcta* it was assumed that differentiation into distinct ecotypes in the species is only at its early stages.

For further details, the reader is referred to Eig (1946), Shmueli (1948), Zohary and Orshansky (1949), Waisel (1960d), Zohary (1962), and Eshel and Waisel (1965).

Salicornia L.

Salicornia plants are succulent low herbs. Leaves are opposite; reduced to a sheath around the stem. Flowers are embedded in cavities along upper parts of the branches. The flower consists of a fleshy perianth, two stamens, and two stigmas. There are some fifty species of *Salicornia* in saline but moist habitats of temperate and subtropical regions.

Salicornia is one of the best known salt-requiring halophytic genera which show optimal growth in saline media. Distribution of *Salicornia* plants is restricted to wet and saline habitats, but plants do not tolerate prolonged inundation. This is especially critical during the period of reproduction. Seedlings of *Salicornia perennis* Mill. and *Salicornia herbacea* L. require a few days free of tidal submersion for anchorage of germinating seedlings and their establishment. Propagation of *Salicornia* plants in coastal marshes of the temperate regions is limited to neap tide periods.

Salicornia plants tolerate high salt concentrations with OP above 50 atm, provided that soil moisture content is high as well. *Salicornia herbacea* is a pioneer species on newly emerged salt flats, and in the Mediterranean is accompanied by *Salicornia fruticosa.* There is some confusion in the taxonomy of *Salicornia herbacea* L. and names such as *S. europaea* L. and *S. stricta* Dum. seem to be synonymous.

For further details, the reader is referred to Johnson and York (1915), Montfort (1927), Wiehe (1935), Iversen (1936), Van Eijk (1939), Repp (1939), Purer (1942), Shmueli (1948), Orshan and Zohary (1955), Ball and Tutin (1959), Chapman (1960), and Zohary (1962).

Spartina Schreb.

These are large-sized or small grasses of the Gramineae, typical of waterlogged and anaerobic soils of coastal salt marshes. The genus consists of a group of species (*S. alterniflora* Loisel, *S. anglica* C. E. Hubbard, *S. gracilis* Trin. *S. patens* (Ait.) Muhl., and *S. stricta* (Ait.) Roth), each being dominant in certain parts of the world. In regions bordering distribution areas of two or more species, hybrids can be formed (e.g., *S. townsendii* H. et J. Groves). Some species of *Spartina* are tall and have long and robust leaves (*S. alterniflora*) whereas others (*S. patens*) are short and fine.

Propagation of *Spartina* plants occurs vegetatively from underground rhizomes in established stands, as well as from seeds in new sites. Germination of at least some species, e.g., *S. anglica,* is inhibited by light. Long submersion under muddy seawater may thus be of stimulatory value for germination of

Spartina seeds by reducing the inhibitory light intensity. Seeds are also temperature-sensitive and germinate poorly below 4°C. Seed production of *S. townsendii* is not ample, with about 100 seeds per cane. Seed production apparently varies with locality. As many as 100 tillers per square meter were found in Germany in a *Spartina* stand. Vitality of the seeds is retained for up to 14 weeks. Germination percentages vary greatly, i.e., between 1 and 90%, when seeds are fresh, but rates decrease rapidly on drying of seeds. Germination of *S. patens* caryopses is best in alternating temperature 18°- 35°C.

Seeds of *Spartina alterniflora* could not withstand desiccation and at moderate temperatures (25°C) lost viability within 40 days. Viability was retained for up to 8 months, when seeds were stored in seawater at 6°C. Germination was inhibited by salinity, probably because of its osmotic effect. Maximal limit of salt tolerance for germination was between 6–8% NaCl. Germination of seeds taken from short, medium, and tall clones responded similarly to storage and temperature. Growth of seedlings taken from the same sources was also similar, exhibiting higher rates in 0.5–1% NaCl solutions.

Spartina seedlings have a basal rosette of leaves, each with a bud in its axil. Buds are borne on nodes separated by short internodes. They may develop into tillers or rhizomes. Seedlings may reach 20 cm after one growth season. The lowest buds on the parent stock give rise to rhizomes which develop into tillers or new short rhizomes. Phyllotaxy in *Spartina* is one-half, and buds are opposite. Under favorable conditions, 20 years are sufficient to change a few scattered clumps of *Spartina* to a continuous meadow. Poole Harbour, England, became colonized within 18 years, and similar results have been reported for the Schleswig-Holstein coast in Germany.

Spartina plants are well adapted to tidal regions by having an extensive, deep, and well-aerated anchoring root system. Plants also have a superficial layer of fine absorbing roots. *Spartina* demonstrates an exceptionally high capacity to withstand immersion. Hubbard (1969), reported that *Spartina* plants survived submersion in clear seawater for over 4½ months. *Spartina anglica* plants may survive periods of immersion up to 23½ hours in neap tide cycles.

Spartina anglica is photoperiodically responsive. It flowered under long days (16 hours), but remained vegetative under short photoperiods (10 hours). The time of flowering in the field was preconditioned during the preceding autumn.

Spartina plants grow best on muddy soils and have only low competitive ability with other species on sandy substrates. *Juncus gerardii, Distichlis spicata,* and *Phragmites* seem capable of competing with *S. townsendii,* but not *Scirpus maritimus* and *Puccinellia* spp.

Like most salt marsh species, *Spartina* plants exhibit reduced growth with increasing salinity. *Spartina alterniflora* is restricted to low marsh conditions, because of its moderate salinity and high iron requirements. If grown either in media with a low iron content, or in freshwater, the plants become chlorotic.

Neither *Distichlis spicata,* nor *S. patens,* species with which *S. alterniflora* competes along its upper ecotone, show any ill effects when grown under conditions poor in iron.

Spartina alterniflora seems to be salt-obligatory, while *S. patens* is a salt-facultative halophyte. Best growth of *Spartina townsendii* seedlings was at 0.5–1.0% NaCl concentration when grown in soil, but in salt-free media when grown in water cultures.

Spartina plants are among the cold-requiring species and growth of the various species is successful only in regions with cold winter temperatures.

Repeated mowing stimulated growth of *S. alterniflora.* It caused uniform growth, higher plant density, and early flowering. Adverse effects of mowing were found for *S. patens.*

For further details see Sutherland and Eastwood (1916), Oliver (1925), Taylor (1939), Purer (1942), König (1948), Chater and Jones (1957), Caldwell (1957), Chapman (1960), Ranwell (1961, 1964), Bird and Ranwell (1964), Walter (1968), Goodman (1969), Teal and Teal (1969), Hubbard (1969, 1970), Marchant and Goodman (1969), and Mooring *et al.* (1971).

Suaeda Forssk. ex Scop.

This is a group of halophytic species of the Chenopodiaceae which occupy the saltiest niches in wet salines. The origin of some species is boreal (*S. maritima*), but of others (e.g., *Suaeda monoica*) is Sudanian–tropical.

Suaeda maritima (L) Dum. plants are decumbent glabrous annuals, which greatly resemble *S. fruticosa.* Plants are 10–30 cm tall. They are widely distributed in the northern hemisphere, both in coastal and inland habitats. The species consists of a few ecotypes and varieties. It is restricted to wet habitats, and has a wide range of salt tolerance. The American populations of *S. maritima* have relatively low salt tolerance and do not survive in media containing more than 1% NaCl. Salt content of *S. maritima* tissues is high and leaves may contain NaCl up to 14.5% of their dry weight.

The related species, *S. australis* and *S. novaezelandiae* are found in Australia and New Zealand.

Suaeda monoica Forssk. ex J. F. Gmel. plants are large shrubs or small trees that may sometimes attain the height of 4 m and have one trunk. Leaves are oblong, somewhat flat on their upper and lower sides and succulent (Fig. 15.6). Plants flower during the mid summer months, and seeds ripen in October–November. Flowers are yellow-green and seeds (~1 mm) have a hard black seed coat.

Suaeda monoica is a Saharo-Arabian Sudanian species, occupying salines in warm regions. In certain tropical sea coasts, it forms one of the mangrove inner

Fig. 15.6. *Suaeda monoica.* A vegetative branch (left) and flower clusters (right).

belts. It is a thermophilic species that cannot withstand even the mild winter temperatures of the East Mediterranean.

Seeds of *Suaeda monoica* germinate readily, the optimal temperature for germination being about 25°C. Sodium chloride solutions up to 0.1 M did not inhibit or stimulate seed germination. Salt solutions above this concentration reduced both percentage as well as rates of germination. The upper limit of salinity which permitted germination was 0.6 M. The limiting factor for germination under high salt concentrations seemed to be mainly osmotic.

Suaeda monoica is a salt-requiring halophyte and its growth is negatively affected by a salt-free medium. The optimal concentration of NaCl for its growth varies between 0.05 to 0.1 M NaCl depending on the plant age. Yellowing, curling, and wilting of leaves were noticed when plants were grown on a NaCl-free Hoagland's solution. Growth was maintained for long periods in NaCl concentrations up to 0.4 M. When NaCl concentration was increased to 0.6 M, growth continued for a short period only and then stopped. Signs of wilting appeared when higher concentrations of NaCl were given to the plants.

The mineral content of the various tissues of *Suaeda monoica* is extremely high. Ash content as high as 43% of dry leaves was recorded. Uptake of sodium into various organs depended on external NaCl concentration and on duration of exposure. The sodium content of roots was relatively low. Most of the sodium absorbed by roots was translocated into the stem and concentrated in the cotyledons and young leaves.

Potassium concentration in various organs of *S. monoica* depends on its external concentration, as well as on the sodium concentration. Under sodium deficiency, potassium substitutes for sodium in the tissues. Antagonism between sodium and calcium was also shown. Low calcium content in the leaves and a high calcium content in the roots and stems were associated with high sodium content in leaves and low sodium content in other organs.

In Israel, *Suaeda* species display only small seasonal changes. During the early summer, *Suaeda monoica* as well as *S. palaestina* Eig et Zoh. and *S. fruticosa* Forssk. ex J. F. Gmel. develop large succulent leaves. Later in the summer, small and thin leaves develop. Various anatomical differences exist between the two types of leaves.

Suaeda monoica is a deep-rooted species. Despite its ability to reduce transpiration and despite the large water reserves contained in its leaves, the plant must receive a constant supply of water for maintaining satisfactory growth. The root systems of *S. palaestina* and *S. fruticosa* are much shallower than those of *S. monoica* and are affected more by fluctuations in soil water content and salinity.

For further details, the reader is referred to Schratz (1934), Repp (1939), Chapman (1947, 1960), Shmueli (1948), Mackay and Chapman (1954), Turner (1956), Binet (1962, 1963), Binet and Bouchard (1964), Ungar and Capilupo (1968), Ovadia (1969), and Waisel and Ovadia (1972).

Tamarix L.

Tamarix is a deciduous or evergreen tree or shrub. Leaves are small, engulfing the stem. Flowers are bisexual with four to five petals and sepals, four to ten stamens, and three to four stigmas.

The genus *Tamarix* includes some seventy-five species in a few complexes. The taxonomy of these species is extremely confused. Some of them are highly halophytic, while others grow mainly in freshwater. All of them have salt glands. Some species are spring flowering (Vernales) whereas others flower only in autumn (Aestivales). Some species of *Tamarix*, e.g., *T. jordanis* Boiss. and *T. nilotica* (Ehreb.) Bge., are biseasonal (Vernales-Aestivales). The flowering of others is restricted to a short and specific season (*T. aphylla*).

Some species of *Tamarix* have a wide distribution area, while others are endemic to small localities. The Middle East comprises one of the centers of speciation and distribution of the genus *Tamarix*. However, distribution of the genus is practically circumglobal in semiarid and temperate regions.

Most species of *Tamarix* have a hydrohalophytic character. Some of them are psammophytes, whereas the remainder are phreatophytes.

The highly halophytic species of *Tamarix* in Israel, *T. tetragyna* Ehrenb. and *T. meyeri* Boiss. (four stamen tamarisks) have long spikes of rose flowers and can be used as ornamentals. Plants of those species can tolerate even 10% sodium chloride in the soil solution (Zohary, 1956). Also *T. passerinoides* Del ex. Desv. and *T. amplexicaulis* Ehrenb. (ten stamen tamarisks) are highly salt resistant, but because of their cold sensitivity are restricted to warm climates, like the Dead Sea region.

All species of *Tamarix* transpire profusely, and are among the halophytic species with highest rates of transpiration.

Tamarix aphylla (L.) Karst. is a tree of moderate size. Crown habit is spherical, greatly varying in size; old trees reach a height of 8–11 m. Twigs are mostly greenish-gray and leafless. Stem diameter at breast height may reach 40–60 cm. The bark is grayish-brown.

Tamarix aphylla has a very wide range of distribution in North Africa and in western Asia. It is also found in South Africa together with *T. austro-africana* Schinz and other *Tamarix* species which are closely related taxonomically.

Tamarix aphylla was considered a Saharo-Sindian species penetrating into the Sudano-Deccanian region (Eig, 1931–1932). However, its occurrence together with many Sudanian species, its high temperature requirements, its phenological behavior and rhythm of cambial activity, and its capacity to reproduce from seeds only under certain conditions of summer rains, suggest that it is tropical—more exactly of Sudanian origin.

These trees produce large numbers of minute seeds annually which are equipped with a hairy pappus and are disseminated over long distances. Still, because of rapid loss of seed viability and slow growth of seedlings, reproduction by seeds is rare. Vegetative propagation of *T. aphylla* is rapid. Salinity negatively affects its reproduction.

Owing to its easy vegetative propagation, the tree has been widely distributed since ancient times, and in recent years has been extensively used for afforestation of arid semisaline habitats all over the world. Nevertheless, it has a narrow amplitude of temperature tolerance, and cannot withstand severe frosts.

Tamarix aphylla grows well in sandy desert regions with an annual precipitation as low as 100 mm. Its presence depends on edaphic and on topographical conditions. The trees occur over a wide range of different soils including saline ones.

The root system of *T. aphylla* is very extensive. Shallow horizontal roots up to 34 m long, were observed on sand dunes; the vertical root system may reach a depth of over 10 m.

In comparison with other species of *Tamarix,* transpiration rates of *T. aphylla* are rather low, especially if one considers its rate of growth. It is very sensitive to dehydration.

Growth of *T. aphylla* trees is inhibited by salinity at concentrations as low as $0.1M$ NaCl. Growth is stopped altogether in a medium containing approximately $0.5M$ NaCl.

Like all the Tamaricaceae, *T. aphylla* is also a salt secretor. Salts are secreted even when trees are grown on nonsaline soils, but the salt composition differs. Sodium and chloride are secreted more readily than other ions.

Although being a salt-secreting plant, *T. aphylla* cannot tolerate heavy salt spray on its twigs. Other species of the genus seem to thrive better in coastal habitats.

Usually *T. aphylla* is not grazed. However, when forage is in short supply, it is grazed by camels. Grazing or pruning results in abundant growth of new branches. A similar effect is seen in trees which were strongly injured by locusts.

For further details, the reader is referred to Shmueli (1948), Zohary (1956), Merkel and Hopkins (1957), Fahn (1958), Waisel (1959a, 1960a, b, c, 1961), Decker *et al.* (1962), Wilgus and Hamilton (1962), Horton (1964), Gary and Horton (1965), Waisel and Friedman (1965), and Howard and Horton (1965).

———————————————16

Ecotypic Differentiation

General Remarks

In each environment, pressure is exerted for selection of some of the characteristics of the inhabiting plants. Such selective pressure is more intensive as the general conditions for plant growth become limiting. Thus, under conditions of extreme cold, drought, or salinity, only the most adapted plants can survive.

Adaptation of plants to any environment induces morphological, anatomical, phenological, and physiological changes. In some cases, adaptation may be temporary and phenotypical, and frequently expressed in some parts of the plant only. In other cases, the selected characteristics are genetically fixed, thus forming ecotypes.

Ecotypes are closely related taxonomic entities of the same species whose genotypes differ only in the frequency of a few alleles. Their distribution overlaps the occurrence of one, or of very few environmental factors. Vicarious ecotypes form the first stage in differentiation of higher taxonomic units. Investigation of many vicarious pairs of ecotypes may thus give valuable information on trends in adaptability of plants and responses to natural selection.

Theoretically, phenotypical adaptation is sufficient for enabling plants to evade and survive the pressure of natural selection without genotypical changes (Clements *et al.,* 1950). Nevertheless, numerous investigations into the constitution of various species of plants of different ecological groups showed that only few phenotypical adaptations—ecads—exist. Most of the populations investigated went through the action of natural selection to form genetically fixed ecotypes (Turesson, 1922a, b, 1931, 1936; Clausen *et al.,* 1940, 1945, 1948; Stebbins, 1950; Clausen, 1951; Waisel, 1959a, 1960d; McNaughton, 1966). With regard to halophytes, it should be noted that the first modern definition of ecotypes was based on certain populations of coastal halophytes (Turesson, 1922a).

Morphological Adaptation

As the proximity to the sea is the dominant ecological factor under coastal conditions, formation of similar ecotypes along coasts is frequently encountered in similar although isolated habitats. This was found in various plants, disregarding their taxonomical composition or geographical location. For example, Clausen *et al.* (1948) described an independent differentiation of similar coastal ecotypes of *Achillea millefolium* along the western coast of North America. Such ecotypes, formed from California to Alaska, were all prostrate, had a compact growth form, and wide and succulent leaves.

Among ecotype pairs of seven different coastal species investigated in Israel, only in one species (*Sonchus oleraceus* L.) were there no physiological differences between coastal and inland ecotypes. All other species (*Echium angustifolium* Mill., *Atractylis flava* Desf., *Nigella arvensis* L. *Pallenis spinosa* (L) Cass., *Poterium spinosum* L., and *Verbascum sinuatum* L.) exhibited ecotypic differentiation. Plants of the coastal ecotypes were morphologically distinct (Fig. 16.1) and grew normally on exposed coasts, while plants of the inland ecotypes were salt-burned under such conditions. The inland ecotype of *Echium angustifolium* escaped damage on the coast by forming a prostrate ecad.

Using *Nigella arvensis* as an example, the positive selective value of prostrate forms in coastal habitats was experimentally shown by Waisel (1959b). It was demonstrated that the erect specimens of unadapted inland populations were not only salt-burned, but were unable to flower, produce seeds, and consequently to secure their perpetuation on the coast. Under the same conditions of an exposed coastal cliff, the adapted prostrate native populations were undamaged by prevailing strong winds and by intensive salt spray; they flowered and produced seeds normally. Artificial hybrids between the two

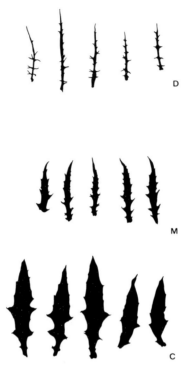

Fig. 16.1. Leaves of *Atractylis flava*. D, Desert ecotype; M, inland ecotype; C, coastal ecotype. (After Waisel, 1960d.)

ecotypes showed intermediate characteristics (Fig. 16.2). It is interesting to note that usually the sensitivity of flowers to salt spray is far higher than that of the vegetative organs. Perennial plants may thus survive under such conditions even though their reproduction is prevented. A prostrate form can thus explain the success of many coastal plants to obtain a permanent hold in such habitats. Prostrateness can be substituted in some cases by dwarfness. For example, the inflorescence of the coastal *Echinops viscosus* ssp. *litoralis* remains near the ground level and is protected from intensive salt spray, while that of the inland population is over 1 m tall. Natural selection discards tall plants, but is not selective against the various means which enable plants to remain close to the ground. A few clones, which differ in height, were also distinguished in *Spartina alterniflora*. However, while Stalter and Batson (1969) described such differences as being inherent, Mooring *et al.* (1971) concluded that those forms were ecophenes (ecads) and not true ecotypes.

Various species of coastal plants were shown to be genetically diverse not only on a physioecological level but also on a morphological basis. *Matricaria*

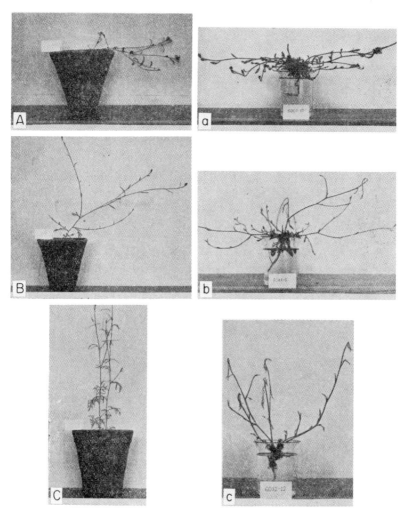

Fig. 16.2. Effects of exposure to coastal conditions on the behavior of different ecotypes of *Nigella arvensis*. (A) and (a), coastal ecotype; (B) and (b), hybrids; (C) and (c), inland ecotype. Capital letters: grown in the inland experimental plot in Jerusalem. Small letters: grown in the coastal experimental plot in Mikhmoret. (After Waisel, 1959b.)

inodora var. *maritima* and *Armeria vulgaris* var. *maritima* were defined as saline ecotaxonomic units within those species (Turesson, 1922a, b). Both ecotypes exhibited a prostrate compact form and a tendency for succulence. The appearance of the saline ecotype of *Armeria* is related to saline conditions, being absent on the nonsaline east coast of Sweden. On the southern coast, both the saline and nonsaline ecotypes intermingle. Also other succulent varieties develop from nonsucculent species on salty sites, e.g., *Tetragonolobus siliquosus* var. *maritimus, Plantago major* var. *carnosa,* and *Plantago coronopus* var. *maritima.*

Ecotypic differentiation in coastal habitats is obvious and coastal halophytic ecotypes show divergent modes of adaptation. Mostly, plants are salt avoiders because of their dense and low form. Some of these ecotypes are coated with a highly water-repellent cuticle. Others are protected against salt spray by different means. Nevertheless, in spite of such morphological measures of defense, even leaves of various coastal ecotypes artificially sprayed with seawater showed some damage. However, as a rule, such plants were more resistant to salt spray than their vicarious glycophytic ecotypes (Boyce, 1951; Waisel, 1960d).

Ecotypic differentiation occurs not only in coastal halophytes, but also in plants from other saline habitats. Langlois (1961) reported ecotypical differentiation in *Salicornia herbacea. Salicornia* plants, grown from seeds which were collected on different sites, exhibited distinct morphological and physiological characteristics. On a morphological basis, those plants had previously been classified into three varieties: (a) *Salicornia herbacea* L. var. *stricta* Dumort.; (b) *S. herbacea* L. var. *patula* Moss; (c) *S. herbacea* L. var. *appressa* Dumort.

Halimione protulacoides also consists of at least a few established varieties (var. *latifolia;* var. *parvifolia,* and var. *angustifolia*) segregated into distinct habitats. The morphological difference among such varieties were genetically fixed (Chapman, 1937).

Phenological Adaptations

Halophytes tend to be somewhat late flowering. In part, this is the result of exposure to either coastal or inundated habitats, where temperatures rise slowly during the spring. In part, time of flowering is genetically controlled.

Under inland garden conditions, plants of coastal ecotypes always flowered later than their respective inland ecotypes (Waisel, 1960d). For example, plants of the coastal ecotype of *Atractylis flava* flowered 20 days later than plants of the Mediterranean ecotype and 1 month later than plants of the desert ecotype (Fig. 16.3). Flowering periods of plants from desert and coastal ecotypes overlapped only during very few days. Thus, time of flowering constitutes a barrier for gene exchange between the two ecotypes and comprises one of the first stages in speciation.

Anatomical Adaptations

Differences in structure were also described for various ecotypes of a few species of plants in Israel (Fig. 16.4). Plants of the coastal ecotypes usually had

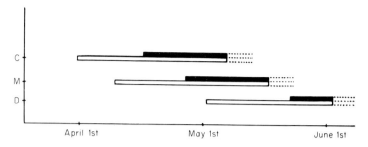

Fig. 16.3. The phenology of three ecotypes of *Atractylis flava*. C, Coastal ecotype; M, inland Mediterranean ecotype; D, desert ecotype. Open bars, flowering; full bars, fruiting; dots, scattered flowering and fruiting. (After Waisel, 1960d.)

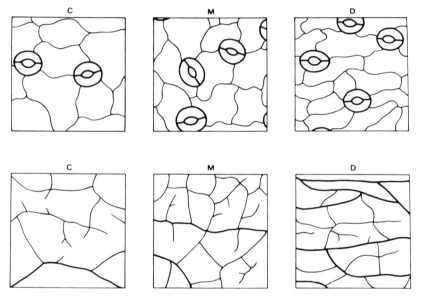

Fig. 16.4. Leaf anatomy of the coastal (C), inland (M), and desert (D) ecotypes of *Atractylis flava*. Upper group, surface view of the lower epidermis; lower group, venation of mature leaves. (After Waisel, 1960d.)

fewer veins and fewer stomata per unit leaf area, even when plants were grown side by side in the same habitats. These are genetically fixed characteristics and such populations thus can be defined as ecotypes.

Physiological Adaptations

Resistance to salt spray of coastal ecotypes of *Artemisia monosperma,* *Poterium spinosum,* and *Atractylis flava* was higher than that of inland ecotypes of the same species. The latter showed greater leaf burning when sprayed with seawater, and died earlier than plants of coastal ecotypes (Waisel, 1960d). As in other cases of extreme conditions, also in this case, the younger leaves were more tolerant than the older ones.

Differences in salt resistance were found between various populations of *Atriplex polycarpa*. Growth of certain ecotypes was even accelerated when plants were given saline treatments (Chatterton and McKell, 1969). Similarly, relatively higher growth rates were observed in the halophytic ecotype of *Prosopis farcta* than in desert and Mediterranean ecotypes, when grown under saline conditions. The higher rates of growth were attributed primarily to fast initial development of seedlings.

Because of their exposure to intensive salt spray, tall plants growing in coastal habitats must be adapted on the physiological level. In this respect, the behavior of trees in coastal habitats is of special interest and has important economic implications. The ability to grow on exposed coasts, to withstand wind, salt spray, and sand blow, and serve as an efficient windbreak was shown to differ among various species and ecotypes of *Tamarix* sp.

Growth rates of seven clones of *Tamarix* were investigated by Waisel and Friedman (1965). Three species were tested: (1) *T. gallica* ssp. *nilotica*—a species which grows naturally in various habitats along the coastal plain in Israel; (2) *T. aphylla*—a fast-growing species in sandy habitats, and (3) *T. meyeri*—native halophytic species of coastal salines. One population from each of the two species, *T. aphylla* and *T. meyeri* and plants originating from five successful specimens of *T. gallica* were compared.

High intra- and interspecific variations were observed. Growth rates and survival ability of one clone of *Tamarix gallica* were higher than in all other clones tested (Fig. 16.5). The average height of plants of the latter clone was about 60% greater than that of other clones. *Tamarix gallica* also consisted of varieties with a higher survival capability. Three years after planting, 73% of the plants of one of its clones survived, whereas survival of all other clones ranged between 23.1 and 46.2%.

Ecotypic differentiation was also found in *Zostera marina, Ruppia rostellata*, and *Aster tripolium*. Plants grown from salt-rich seeds, grew better in saline than in nonsaline media and tolerated high concentrations of salts.

Salt-resistant forms have evolved in *Typha angustifolia* L. and *Typha latifolia* L. on salt-affected soils (McMillan, 1959; McNaughton, 1966). Similar results were obtained for *Phragmites communis* (Waisel and Rechav, 1972). Seeds taken from a halophytic ecotype of reed exhibited high salt tolerance and were able to germinate in concentrations of NaCl which inhibited germination of caryopses of a glycophytic ecotype (Fig. 16.6). Also seedlings' growth of such ecotypes was differently affected by salinity. While growth of seedlings of the glycophytic populations was inhibited by increasing salinity, growth of halophytic plants yielded an optimum type of curve. (Fig. 16.7).

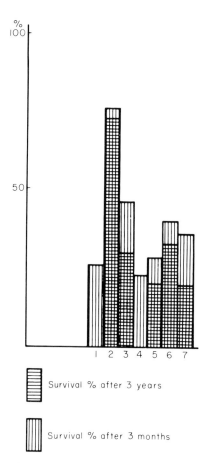

Fig. 16.5. Survival of *Tamarix* trees grown on a coastal cliff in Mikhmoret, Israel. Columns left to right: 1–5, different ecotypes of *Tamarix gallica*; 6, *Tamarix meyeri*; 7, *Tamarix aphylla*. (After Waisel and Friedman, 1965.)

An investigation into the nutritional behavior of parallel ecotypes for comparison of their response to sodium also seemed to be of great interest. However, one of the difficulties in such investigations is the scarcity of suitable ecotypic pairs. For instance, only three species (*Prosopis farcta, Alhagi maurorum,* and *Phragmites communis*) which include both halophytic and glycophytic ecotypes, were found in the flora of Israel (Waisel, 1960d). No significant differences were observed between the two populations of *Prosopis farcta* (Eshel and Waisel, 1965), and only minor differences were found in *Phragmites* (Waisel and Rechav, 1972) in their sodium, potassium, or calcium metabolism. Other modes of adaptation to saline conditions must be sought for these plants.

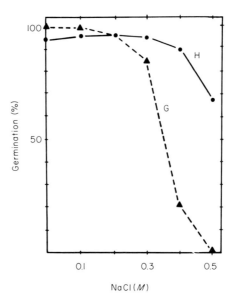

Fig. 16.6. Effects of salinity on germination of a halophytic ecotype (H) and a glycophytic one (G) of *Phragmites communis.* (After Waisel and Rechav, 1972.)

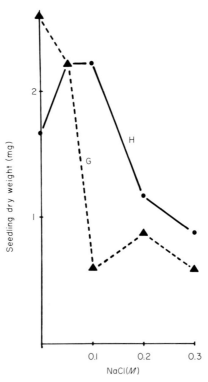

Fig. 16.7. Effects of salinity on growth of seedlings of a halophytic ecotype (H) and a glycophytic ecotype (G) of *Phragmites communis.* (After Waisel and Rechav, 1972.)

Differences in responses to salts of two closely related species—*Agropyron elongatum,* a salt-tolerant species, and *A. intermedium,* a salt-sensitive species (Dewey, 1960)—were claimed by Elzam and Epstein (1969a, b). Uptake of sodium and chloride was at least three times higher in the salt-tolerant *A. elongatum* than in the salt-sensitive *A. intermedium.* Potassium uptake was highly reduced in *A. elongatum* in the presence of sodium but in the salt-sensitive species, *A. intermedium,* it was unaffected. The two species differed also in response to divalent ions. In the salt-tolerant species, *A. elongatum,* magnesium and calcium had similar effects on the sodium uptake mechanism. However, effects of magnesium on *A. intermedium* were less than those of calcium. In both species higher environmental salinity caused an increase in sodium and chloride contents but did not affect the potassium content. Calcium content decreased under similar conditions.

As such trends in sodium and chloride metabolism are known to occur in other halophytes as well, those differences seem to be of adaptive value and selected for at very early stages of species development.

References

Abd el Malek, Y., and Rizk. S. G. (1963). Bacterial sulphate reduction and the development of alkalinity. *J. Appl. Bacteriol.* **26**, 7–26.

Adams, D. A. (1963). Factors influencing vascular plant zonation in North Carolina salt marshes. *Ecology* **44**, 445–456.

Adriani, M. J. (1945). Sur la phytosociologie, la synecologie et le bilan d'eau de halophytes. Comm. SIGMA 88a, Groningen: Wolters **5**, 1–217.

Adriani, M. J. (1956). Der Wasserhaushalt der Halophyten. *Encycl. Plant Physiol.* **3**, 902–914.

Adriani, M. J. (1958). Halophyten. *Encycl. Plant Physiol.* **4**, 709–736.

Ahi, S. M., and Powers, W. L. (1938). Salt tolerance of plants at various temperatures. *Plant Physiol.* **13**, 767–789.

Ahmad, R. (1968). The mechanism of salt tolerance in *Suaeda fruticosa* and *Haloxylon recurvum. Plant Soil* **28**, 357–362.

Ahmad, R., and Hewitt, E. J. (1971). Studies on the growth and phosphatase activities in *Suaeda monoica. Plant Soil* **34**, 691–696.

Allen, M. B., and Arnon, D. I. (1955). Studies on nitrogen fixing blue-green algae. II. The sodium requirement of *Anabaena cylindrica. Physiol. Plant.* **8**, 653–659.

Altman, A. (1969). Chloride uptake by the roots of citrus root stocks. Ph.D. Thesis, Hebrew Univ. Jerusalem.

Alvim, P. de T. (1960). Moisture stress as a requirement for flowering of coffee. *Science* **132**, 354.

Amen, R. D., Carter, G. E., and Kelly, R. J. (1970). The nature of seed dormancy and germination in the salt marsh grass *Distichlis spicata. New Phytol.* **69**, 1005–1013.

Anderson, D. J. (1967). Studies on structure in plant communities. V. Pattern in *Atriplex vesicaria* communities in South Eastern Australia. *Aust. J. Bot.* **15**, 451–458.

Anderson, D. J. (1970). Analysis of pattern in *Atriplex vesicaria* communities from the Riverine plain of New South Wales *In* "The Biology of *Atriplex*" (R. Jones, ed.), pp. 63–68. Div. Plant Industry, C.S.I.R.O., Canberra, Australia.

Anderson, V. G. (1945). Some effects of atmospheric evaporation and transpiration on the composition of natural waters in Australia. *J. Aust. Chem. Inst. Proc.* **12**, 41–68, 83–98.

Anderson, W. P., and Allen, E. (1970). A comparison of potassium translocation in excised roots and intact maize seedlings. *Planta* **93**, 227–232.

Anderson, W. P., and House, C. R. (1967). A correlation between structure and function in the root of *Zea mays. J. Exp. Bot.* **18**, 544–555.

Anderson, W. P., and Reilly, E. J. (1968). The effects of temperature on the exudation process of excised primary roots of *Zea mays. J. Exp. Bot.* **19**, 648–657.

Anderson, W. P., Aikman, D. P., and Meiri, A. (1970). Excised root exudation, a standing gradient osmotic flow. *Proc. Roy. Soc.* **B174**, 445–458.

Arasaki, S. (1950). Studies on the ecology of *Zostera marina* and *Zostera nana. Bull. Jap. Soc. Sci. Fish.* **16**, 70.

Archibald, E. E. A. (1948). Plant populations. 1. A new application of Neyman's contagious distribution. *Ann. Bot. (London)* **12**, [N.S.] 221–235.

Arisz, W. H. (1956). Significance of the symplast theory for transport in the root. *Protoplasma* **46**, 5–62.

Arisz, W. H. (1963). Influx and efflux by leaves of *Vallisneria spiralis.* I. Active uptake and permeability. *Protoplasma* **57**, 5–26.

Arisz, W. H. (1964). Influx and efflux of electrolytes. *Acta Bot. Neerl.* **13**, 1–58.

Arisz, W. H., and Sol, H. H. (1956). Influence of light and sucrose on the uptake and transport of chloride in *Vallisneria* leaves. *Acta Bot. Neerl.* **5**, 218–246.

Arisz, W. H., Camphuis, I. J., Heikens, H., and van Tooren, A. J. (1955). The secretion of the salt glands of *Limonium latifolium* Ktze. *Acta Bot. Neerl.* **4**, 322–338.

Arnold, A. (1955). Die Bedeutung der Chlorionen für die Pflanze, insbesondere physiologische Wirksamkeit. "Botanische Studien," Pt. 2. Fischer, Jena.

Ascherson, P. (1903). Der nördlichste Fundort der Mangrove in Aegypten. *Bot. Z.* **61**, 235–238.

Ashby, W. C., and Beadle, N. C. W. (1957). Studies in halophytes. Salinity factors in the growth of Australian salt bushes. *Ecology* **38**, 344–352.

Atkinson, M. R., and Polya, G. M. (1967). Salt-stimulated adenosine triphosphate from carrot, beet and *Chara australis. Aust. J. Biol. Sci.* **20**, 1069–1086.

Atkinson, M. R., Findlay, G. P., Hope, A. B., Pitman, M. G., Saddler, H. D. W., and West, K. R. (1967). Salt regulation in the mangroves *Rhizophora mucronata* Lam. and *Aegialitis annulata* R.Br. *Aust. J. Biol. Sci.* **20**, 589–599.

Augenfeld, J. M. (1969). The role of Na$^+$, K$^+$ activated, ouabain-sensitive ATPase in the response of *Artemia salina*, L. to salinity changes. *Life Sci.* **8**, 973–978.

Ayers, A. D. (1952). Seed germination as affected by soil moisture and salinity. *Agron. J.* **44**, 82–84.

Ayers, A. D., and Hayward, H. E. (1948). A method for measuring the effects of soil salinity on seed germination with observations on several crop plants. *Soil Sci. Soc. Amer. Proc.* **13**, 224–226.

Baalman, R. J. (1965). Vegetation of the salt plains wildlife refuge, Jet. Oklahoma. Ph.D. Thesis, Univ. of Oklahoma, Norman, Oklahoma, 129pp.

Bacalbasa, N. (1951). Posibilitatea valorificarrii ierbii de mare *(Zostera)* in R.P. Romana. *Bull. Inst. de Cer. Pisc.* **1**, 25–33.

Bagley, J. M. (1967). The salinity problem in reuse of water. *In* "Water Supplies for Arid Regions" (L. Gardner and L. E. Meyers, eds.). Contrib. No. 10 of the Committee on Desert and Arid Zones Research AAAS and the University of Arizona, Arizona.

Baker, H. G. (1948). Dimorphism and monomorphism in *Plumbaginaceae*. I. A survey of the family. *Ann. Bot. (London)* **12**, 207–219.

Baker, H. G. (1949). *Armeria* studies. *Carnegie Inst. Wash. Yearb.* **48**, 103–105.

Bakker, D. (1957a). Voorlopige resultaten van proven, in 1956 op "Het Spijk" genomen, met riet, rietgras en wiezen. *Flavo-Berichten B. No.* **3**, 1–6.

Bakker, D. (1957b). Het Riet in de Noordoostpolder. *Van Zee tot Land No.* **21**, 5–25.

Ball, P. W., and Tutin, T. G. (1959). Notes on annual species of *Salicornia* in Britain. *Watsonia* **4**, 193–205.

Barbier, G., and Chabannes, J. (1951). Accumulation du sodium dans les racines des plants. *C.R. Acad. Sci. (Paris)* **232**, 1372–1374.

Barbour, M. G. (1969). Age and space distribution of the desert shrub *Larrea divaricata*. *Ecology* **50**, 679–685.

Baumeister, W., and Schmidt, L. (1962). Über die Rolle des Natriums im pflanzlichen Stoffwechsel. *Flora (Jena)* **152**, 24–56.

Beadle, N. C. W. (1952). Studies in halophytes. I. Germination of the seeds and establishment of the seedlings of five species of *Atriplex* in Australia. *Ecology,* **33**, 49–62.

Beadle, N. C. W., Whalley, R. D. B., and Gibson, J. B. (1957). Studies in halophytes. II. Analytic data on the mineral constituents of three species of *Atriplex* and their accompanying soils in Australia. *Ecology* **38**, 340–344.

Beauchamp, R. S. A. (1953). Sulphates in African inland waters. *Nature (London)* **171**, 769–771.

Béchet, M. A., and Binet, P. (1964). Croissance de *Triglochin maritimum* L. après refrigeration des semences imbibées. *Physiol. Veg.* **2**, 25–34.

Becker, H. (1913). Über die Keimung verschiedenartiger Früchte und Samen bei derselben Spezies. *Beih. Bot. Centralbl.* **29**, 21–143.

Beiler, A. (1939). Untersuchungen über die Kohlensäureassimilation der Strand und Dünen pflanzen. *Jahr. Wiss. Bot.* **87**, 356–407.

Beljakoff, E. (1930). Über den Einfluss der Temperatur auf die Kohlensäureassimilation bei zwei Klimatischen Pflanzenrassen. *Planta* **11**, 727–764.

Benecke, W. (1930a). Zur Biologie der Strand und Dünen flora. I. Vergleichende Versuche über die Salztoleranz von *Ammophila arenaria* Link., *Elymus arenarius* L. und *Agropyrum junceum* L. *Ber. Deut. Bot. Ges.* **48**, 127–139.

Benecke, W. (1930b). Kulturversuche mit *Aster tripolium*. *Z. Bot.* **23**, 745–766.

Benzioni, A., Itai, Ch., and Vaadia, Y. (1967). Water and salt stress, kinetin and protein synthesis in tobacco leaves. *Plant Physiol.* **42**, 361–365.

Berger-Landefeldt, U. (1933). Die Hydratur einiger Halophyten in Ihrer abhangigkeit von der Substrat-Konzentration. *Beih. Bot. Centralbl.* **51**, 697–710.

Bernstein, L. (1961). Osmotic adjustment of plants to saline media. I. Steady state. *Amer. J. Bot.* **48**, 909–918.

Bernstein, L. (1963). Osmotic adjustment of plants to saline media. II. Dynamic phase. *Amer. J. Bot.* **50**, 360–370.

Bernstein, L., and Ayers, A. D. (1953). Salt tolerance of five varieties of carrot. *Proc. Amer. Soc. Hort. Sci.* **61**, 360–366.

Bernstein, L., and Hayward, H. E. (1958). Physiology of salt tolerance. *Ann. Rev. Plant Physiol.* **9**, 25–46.

Bernstein, L., Brown, J. W., and Hayward, H. E. (1956). The influence of rootstock on growth and salt accumulation in stone fruit trees and almonds. *Proc. Amer. Soc. Hort. Sci.* **68,** 86–95.

Berry, W. L. (1970). Characteristics of salts secreted by *Tamarix aphylla. Amer. J. Bot.* **57,** 1226–1230.

Berry, W. L., and Thomson, W. W. (1967). Composition of salt secreted by salt glands of *Tamarix aphylla. Can. J. Bot.* **45,** 1774–1775.

Bharucha, F. R., and Shirke, V. S. (1947). A study in the important catabolic changes in the seedlings of *Avicennia officinalis* L. *J. Univ. Bombay* **15,** 1–14.

Bickenbach, K. (1932). Zur Anatomie und Physiologie einiger Strand und Dünenpflanzen. Beitrage zum Halophytenproblem. *Beitr. Biol. Pflanz.* **19,** 334–370.

Biebl, R. (1953). Seewasserresistenz von Meerstrandpflanzen. *Photogr. Forsch.* **5,** 174–192.

Biebl, R., and Kinzel, H. (1965). Blattbau und Salzhaushalt von *Laguncularia racemosa* (L.) Gaertn. und andere Mangrovebäume auf Puerto Rico. *Oesterr. Bot. Z.* **112,** 56–93.

Biebl, R., and Weissenbock, G. (1968). Vergleichende Untersuchungen der Salz-Trocken und Strahlensukkulenz bei Wild und Kulturpflanzen. *Oesterr. Bot. Z.* **115,** 229–254.

Binet, P. (1960). Cultures sans sol de plantes halophiles; Les enseignements de quelques essais preliminaires. *Bull. Soc. Linn. Normandie* **1,** 28–39.

Binet, P. (1962). La repartition du sodium et du potassium chez *Suaeda macrocarpa* Moq. *Physiol. Plant.* **15,** 428–436.

Binet, P. (1963). Le sodium et le potassium chez *Suaeda vulgaris. Physiol. Plant.* **16,** 615–622.

Binet, P. (1964). Action de la temperature et de la salinité sur la germination des graines de *Plantago maritimum* L. *Bull. Soc. Bot. Fr.* **111,** 407–411.

Binet, P., and Bouchard, J. (1964). Vue d'ensemble sur la croissance et la nutrition minerale de *Suaeda maritima* Dum. *Bull. Soc. Fr. Physiol. Veg.* **10,** 227–241.

Bird, E. C. F., and Ranwell, D. S. (1964). *Spartina* salt marshes in southern England. *J. Ecol.* **52,** 355–366.

Bittmann, E. (1953). Das Schilf (*Phragmites communis* Trin.) und seine Verwendung in Wasserbau. *Angew. Pflanz. Soz.* **7,** 1–47.

Björkman, O., Nobs, M. A., and Hiesey, W. M. (1969a). Growth photosynthetic and biochemical responses of contrasting *Mimulus* clones to light intensity and temperature. *Carnegie Inst. Stanford Calif. Ann. Rept. 1968–1969,* pp. 614–620.

Björkman, O., Gauhl, E., and Nobs, M. A. (1969b). Comparative studies of *Atriplex* species with and without β-carboxylation photosynthesis and their first generation hybrid. *Carnegie Inst. Stanford Calif. Ann. Rept. 1968–1969,* pp. 620–633.

Black, R. F. (1954). The leaf anatomy of Australian members of the genus *Atriplex.* I. *Atriplex vesicaria* Heward and *A. nummularia* Lindli. *Aust. J. Bot.* **2,** 269–286.

Black, R. F. (1956). Effects of NaCl in water culture on the ion uptake and growth of *Atriplex hastata* L. *Aust. J. Biol. Sci.* **9,** 67–80.

Black, R. F. (1958). Effects of sodium chloride on leaf succulence and area of *Atriplex hastata* L. *Aust. J. Bot.* **6,** 306–321.

Black, R. F. (1960). Effects of NaCl on the ion uptake and growth of *Atriplex vesicaria* Heward. *Aust. J. Biol. Sci.* **13,** 249–266.

Blake, S. F. (1916). *Statice* in North America and Mexico. *Rhodora* **18,** 53–66.

Bleasdale, A. (1957). Afforestation of catchment areas. The physicist's approach to problems of water loss from vegetation. *Empire Forestry Rev.* **36,** 59–68.

Blumenthal-Goldschmidt, S., and Poljakoff-Mayber, A. (1968). Effects of substrate salinity on growth and submicroscopic structure of leaf cell of *Atriplex halimus* L. *Aust. J. Bot.* **16,** 469–478.

Bocquier, G. (1964). The presence of tropical solothized solonetz features in the Chad basin. *8th Intern. Congr. Soil Sci. Abst. Papers* **5**, 103–106.

Bodrogközy, Gy. (1958). Synökologische Auswertung des Einflusses verschiedener Behandlungen auf der Lepidio-Puccinelietum Limosae Kalk und Sodahalitger Boden. *Acta Agron. (Budapest)* **8**, 343–376.

Bodrogközy, Gy. (1962). Die Standortökologischen Verhältnisse der halophilen Pflanzengesellschaften des Pannonicum. I. Untersuchungen an den Solontschak-Szikböden der Südlichen Kiskunsag. *Acta Bot.*, **8**, 1–37.

Bonting, S. L., and Caravaggio, L. L. (1966). Studies on $Na^+ - K^+$ activated adenosinetriphosphatase. XVI. Its absence from the cation transport system of *Ulva lactuca. Biochim. Biophys. Acta* **112**, 519–523.

Bonython, C. W. (1958). The influence of salinity upon the rate of natural evaporation. *Proc. Canberra Symp.* UNESCO Arid Zone Res. *Climatol. Microclimatol.* **11**, 65–71.

Boorman, L. A. (1967). Biological flora of the British Isles: *Limonium vulgare* Mill. and *L. humile* Mill. *J. Ecol.* **55**, 221–232.

Boorman, L. A. (1968). Some aspects of the reproductive biology of *Limonium vulgare* Mill. and *Limonium humile* Mill. *Ann. Bot. (London)* **32**, 803–824.

Borgreve, A. (1890). Zur wirkung des Seewindes auf den Waldwuchs. *Forstl.* **6**, 28–32.

Bove, J. M., Bove, C., Whatley, F. R., and Arnon, D. I. (1963). Chloride requirements for oxygen evolution in photosynthesis. *Z. Naturforsch.* **18**, 683–688.

Bowen, G. D. (1968). Chloride efflux along *Pinus radiata* roots. *Nature (London)* **218**, 686–687.

Bowen, G. D., and Rovira, A. D. (1966). Microbial factor in short-term phosphate uptake studies with plant roots. *Nature (London)* **211**, 665–666.

Bowen, G. D., and Rovira, A. D. (1967). Phosphate uptake along attached and excised wheat roots measured by an automatic scanning method. *Aust. J. Biol. Sci.* **20**, 369–378.

Bowen, G. D., and Rovira, A. D. (1968). The influence of micro-organisms on growth and metabolism of plant roots. *In* "Root Growth" (W. J. Whittington, ed.), Proc. 5th Easter School in Agric. Sci. Univ. Nottingham 1968, pp. 170–199. Butterworth, London.

Bowen, H. J. M. (1966). "Trace Elements in Biochemistry." Academic Press, New York.

Bower, C. A., and Wadleigh, C. H. (1948). Growth and cationic accumulation by four species of plants as influenced by various levels of exchangeable sodium. *Soil Sci. Soc. Amer. Proc.* **13**, 213–223.

Bowling, D. J. F. (1966). Active transport of ions across sunflower roots. *Planta* **69**, 377–383.

Bowling, D. J. F., and Spanswick, R. M. (1964). Active transport of ions across the root of *Ricinus communis J. Exp. Bot.* **15**, 422–427.

Bowling, D. J. F., Macklon, A. E. S., and Spanswick, R. M. (1966). Active and passive transport of the major nutrient ions across the root of *Ricinus communis. J. Exp. Bot.* **17**, 410–416.

Boyce, S. G. (1951). Salt hypertrophy in succulent dune plants. *Science* **114**, 544–545.

Boyce, S. G. (1954). The salt-spray community. *Ecol. Mongr.* **24**, 26–67.

Boyer, J. S. (1965). Effects of osmotic water stress on metabolic rates of *cotton* plants with open stomata. *Plant Physiol.* **40**, 229–234.

Braun-Blanquet, J. (1931). Zur Frage der "Physiologischen Trockenheit" der Salzböden. *Ber. Schweiz. Bot. Ges.* **40**, 31–39.

Braun-Blanquet, J., and Tüxen, R. (1952). Irische Pflanzengesellschaften. *In* "Die Pflanzenwelt Irlands" (W. Ludi, ed.), Vol. 25, pp. 224–420. Veröff. Geobot. Inst. Rübel.

Breazeale, J. F. (1930). Maintenance of moisture equilibrium and nutrition of plants at and below the wilting percentage. *Bull. Ariz. Agr. Exp. Sta.* **29**, 137–177.

Breazeale, J. F., and Crider, F. J. (1934). Plant association and survival and build-up of moisture in semi-arid soils. *Bull. Ariz. Agr. Exp. Sta.* **53**, 95–123.

Breazeale, E. L., and McGeorge, W. T. (1949). A new technique for determining the wilting percentage of soil. *Soil Sci.* **68**, 371–374.

Briggs, G. E., and Robertson, R. N. (1957). Apparent free space. *Ann. Rev. Plant Physiol.* **8**, 11–30.

Briggs, G. E., Hope, A. B., and Robertson, R. N. (1961). "Electrolytes and Plant Cells," 217 pp. Blackwell, Oxford.

Briggs, L. J., and Shantz, H. L. (1912) The wilting coefficient for different plants and its indirect determination. *Bull. Bur. Plant Ind. U.S. Dept. Agr.* p. 230.

Brix, H. (1962). The effect of water stress on rate of photosynthesis and respiration in tomato plants and loblolly pine seedlings. *Physiol. Plant.* **15**, 10–20.

Bromefield, S. M. (1954). The reduction of iron oxide by bacteria. *J. Soil Sci.* **5**, 129–139.

Brown, J. C., and Ambler, J. E. (1969). Characterization of boron deficiency in soybeans. *Physiol. Plant.* **22**, 177–185.

Brown, R. (1963). Cellular differentiation in the root. *Symp. Soc. Exp. Biol.* **17**, 1–17.

Brown, R., and Broadbent, D. (1950). The development of cells in the growing zones of the root. *J. Exp. Bot.* **1**, 249–263.

Brownell, P. F. (1965). Sodium as an essential micronutrient element for a higher plant (*Atriplex vesicaria*). *Plant Physiol.* **40**, 460–468.

Brownell, P. F. (1968). Sodium as an essential micronutrient for some higher plants. *Plant Soil* **28**, 161–164.

Brownell, P. F., and Wood, J. G. (1957). Sodium as an essential element for *Atriplex vesicaria* Heward. *Nature (London)* **179**, 635–636.

Broyer, T. C., Carlton, A. B., Johnson, C. M., and Stout, P. R. (1954). Chlorine as a micro-nutrient. *Plant Physiol.* **29**, 526–532.

Brunner, C. (1909). Beiträge zur vergleichenden Anatomie der Tamaricaceae. *Jahrb. Wiss. Anst. Hamburg,* **27**, 89–162.

Bunch, R. L., and Ettinger, M. B. (1964). Water quality depreciation by municipal use. *J. Water Pollut. Control Fed.,* **36**, 1411–1414.

Bureau, P., and Roederer, R. (1961). Sols gypseux au sud de Tunisie. Association Française pour l'Etude du Sol. *Bull. Notice Spéc.* pp. 150–176.

Burge, L. M., and O'Harra, J. L. (1952). *Halogeton* in Nevada—Animal poisoning and control studies. *Nev. State Dept. Agr. Bull. No. 4.*

Burkser, V. V. (1951). Aerogeochemical investigations in the steppe and forest steppe region. *Ukr. Khim. Zh.* **17**, 472–476.

Burvin, A. (1963). The ecology of a few plant communities in the saline of Sedom. M.Sc. Thesis, The Hebrew University, Jerusalem.

Caldwell, P. A. (1957). The spatial development of *Spartina* colonies growing without competition. *Ann. Bot. (London)* **21**, [N.S.] 203–214.

Caldwell, J. (1961). Further evidence of polar movement of nutrients in plants. *Nature (London)* **190**, 1028–1029.

Campbell, C. J., and Strong, J. E. (1964). Salt gland anatomy in *Tamarix pentandra.* *Southwest Nat.* **9**, 232–238.

Campbell, G. S., Zollinger, W. D., and Taylor, S. A. (1966). Sample changes for thermocouple psychrometers; construction and some applications. *Agron. J.* **58**, 315–318.

Canning, R. E., and Kramer, P. J. (1958). Salt absorption and accumulation in various regions of roots. *Amer. J. Bot.* **45**, 378–382.

Cassidy, N. G. (1971). Cyclic salt and plant health. *Endeavour* **30**, 82–86.

Catsky, J. (1962). Water saturation deficit in the wilting plant. The preference of young leaves and the translocation of water from old into young leaves. *Biol. Plant.* **4**, 306–314.

Chang, C. W., and Dregne, H. E. (1955). Effect of exchangeable sodium on soil properties and on growth and cation content of alfalfa and cotton. *Soil Sci. Soc. Amer. Proc.* **19**, 29–35.

Chapman, V. J. (1937). A note upon *Obione portulacoides* (L.) Gaert. *Ann. Bot. (London)* **1**, [N.S.] 305–310.

Chapman, V. J. (1938). Studies in salt marsh ecology. Sections I–III. *J. Ecol.* **26**, 144–179.

Chapman, V. J. (1940a). Studies in salt marsh ecology. Sections VI–VII. Comparison with marshes on the east coast of North America. *J. Ecol.* **28**, 118–152.

Chapman, V. J. (1940b). Succession on the New England salt marshes. *Ecology* **21**, 279–282.

Chapman, V. J. (1942). The new perspective in the halophytes. *Quart. Rev. Biol.* **17**, 291–311.

Chapman, V. J. (1947). *Suaeda maritima* (L.) Dum. in Biological Flora of the British Isles. *J. Ecol.* **35**, 293–302.

Chapman, V. J. (1950). *Halimione portulacoides* (L.) Aell. in Biological Flora of the British Isles. *J. Ecol.* **38**, 214–222.

Chapman, V. J. (1954). The influence of salts upon the terrestrial halophytes. *Proc. 8th Intern. Bot. Congr. Paris*, pp. 194–200.

Chapman, V. J. (1958). Salt marshes and ecological terminology. *Vegetatio* **8**, 215–234.

Chapman, V. J. (1960). "Salt Marshes and Salt Deserts of the World," 392 pp. Leonard Hill Books Ltd., London.

Chapman, V. J. (1962). Respiration studies of mangrove seedlings. II. *Bull. Mar. Sci. Gulf Carib.* **12**, 245–263.

Chapman, V. J. (1968). Vegetation under saline conditions. *In* "Saline Irrigation for Agriculture and Forestry" (H. Boyko, ed.), pp. 201–216. Dr. W. Junk Publ., The Hague.

Chapman, V. J., and Ronaldson, J. W. (1958). The mangrove and salt marsh flats of the Auckland isthmus, *N.Z. Dep. Sci. Ind. Bull.* **125**.

Chater, E. H., and Jones, H. (1957). Some observations on *Spartina townsendii*. H. and J. Groves in the Dovey Estuary. *J. Ecol.* **45**, 157–167.

Chatterton, N. J., and McKell, C. M. (1969). *Atriplex polycarpa*. I. Germination and growth as affected by sodium chloride in water cultures. *Agron. J.* **61**, 448–450.

Chatterton, N. J., Goodin, J. R., and Duncan, C. (1971). Nitrogen metabolism in *Atriplex polycarpa* as affected by substrate nitrogen and NaCl salinity. *Agron. J.* **63**, 271–274.

Chaudari, G. V. (1968). Effects of soil salinity on germination and survival of some steppe plants in Washington. *Ecology* **49**, 465–471.

Chaudhri, I. I., Shah, B. H., Naqvi, N., and Mallik, I. A. (1964). Investigations on the role of *Suaeda fruticosa* in the reclamation of saline and alkaline soils in west Pakistan. *Plant Soil* **21**, 1–7.

Clapham, A. R. (1936). Over-dispersion in grassland communities and the use of statistical methods in plant ecology. *J. Ecol.* **24**, 232–251.

Clapham, A. R., Pearsall, W. H., and Richards, P. W. (1942). *Aster tripolium* L. in biological flora of the British Isles. *J. Ecol.* **30**, 385–395.

Clarke, L., and Hannon, N. J. (1967). The mangrove swamp and salt marsh communities of the Sydney district. I. Vegetation, soils and climate. *J. Ecol.* **55**, 753–771.

Clarke, L., and Hannon, N. J. (1969). The mangrove swamp and salt marsh communities of the Sydney district. II. The holocoenotic complex with particular reference to physiography. *J. Ecol.* **57**, 231–234.

Clarke, L., and Hannon, N. J. (1971). The mangrove swamp and salt marsh communities of the Sydney district. IV. The significance of species interaction. *J. Ecol.* **59**, 535–553.

Clausen, J. (1951). "Stages in Evolution of Plant Species." Cornell Univ. Press, New York.

Clausen, J., Keck, D. D., and Hiesey, W. M. (1940). Experimental studies on the nature of species. I. The effect of varied environment on western North American plants. *Carnegie Inst. Wash. Publ. No. 520.*

Clausen, J., Keck, D. D., and Hiesey, W. M. (1945). Experimental studies on the nature of species. II. Plant evolution through amphiploidy, with examples from the Madiinae. *Carnegie Inst. Wash. Publ. No. 564.*

Clausen, J., Keck, D. D., and Hiesey, W. M. (1948). Experimental studies on the nature of species. III. Environmental responses of climatic races of *Achillea. Carnegie Inst. Wash. Publ. No. 581.*

Clements, F. E., Martin, E. V., and Long, F. L. (1950). "Adaptation and Origin in the Plant World." Chronica Botanica, Waltham, Massachusetts.

Collander, R. (1930). Permeabilitätsstudien an *Chara ceratophylla. Acta Bot. Fenn.* **6**, 3–20.

Collander, R. (1941). Selective absorption of cations by plants. *Plant Physiol.* **16**, 691–720.

Conard, H. S., and Galligar, G. C. (1929). Third survey of a Long Island salt marsh. *Ecology* **10**, 326–336.

Conway, E. J. (1942). Mean geochemical data in relation to oceanic evolution. *Proc. Roy. Irish Acad.* **48**, 119–159.

Cooil, B. J., de la Fuente, R. K., and de la Pena, R. S. (1965). Absorption and transport of sodium and potassium in squash. *Plant Physiol.* **40**, 625–633.

Correns, C. W. (1937). Die Sedimente des äquatorialen atlantischen Ozeans. *Meteor* **3**, 274–277.

Cory, S., and Finch, L. R. (1967). Further studies on the incorporation of phosphate into nucleic acid of normal and boron deficient tissue. *Phytochemistry* **6**, 211–215.

Coupland, R. T. (1950). Ecology of the mixed prairie in Canada. *Ecol. Monogr.* **20**, 271–315.

Crafts, A. S., and Broyer, T. C. (1938). Migration of salts and water into xylem of the roots of higher plants. *Amer. J. Bot.* **25**, 529–535.

Craigie, J. S. (1963). Dark fixation of C^{14} bicarbonate by marine algae. *Can. J. Bot.* **41**, 317–325.

Crocker, W., and Barton, L. V. (1953). "Physiology of Seeds." Chronica Botanica, Waltham, Massachusetts.

Crozier, W. D., and Seely, B. K. (1952). A second chloride cloud in California. *Bull. Amer. Meteorol. Soc.* **33**, 117–118.

Dainty, J. (1963). The polar permeability of plant cell membranes to water. *Protoplasma* **57**, 220–228.

Dalby, D. H. (1963). Seed dispersal in *Salicornia pusilla. Nature (London)* **199**, 197–198.

Damadian, R. (1969). Ion exchange in *Escherichia coli.* Potassium binding proteins. *Science* **165**, 79–81.

Dan, J., Koyumdjisky, H., and Yaalon, D. H. (1962). Principles of proposed classification for the soils of Israel. *Trans. Intern. Soil Conf. Sect. IV and V*, pp. 1–13.

Danielli, J. F. (1943). Theories of cell permeability. *In* "Permeability of Natural Membranes" (H. Dawson and J. F. Danielli, eds.), pp. 310–340. Cambridge Univ. Press, Cambridge.

Dansereau, P. (1947). Zonation et succession sur la Restinga de Rio de Janeiro. I. Halosere. *Contrib. Inst. Biol. Gen. Zool. Univ. Montreal* **20**, 448–477.

Davis, J. H., Jr. (1940). The ecology and geologic role of mangroves in Florida. Papers from Tortugas Laboratory. *Carnegie Inst. Wash. Publ. No. 517*, pp. 305–412.

De Vlaming, V., and Proctor, V. W. (1968). Dispersal of aquatic organisms: Viability of seeds recovered from the droppings of captive kildeer and mallard ducks. *Amer. J.*

Bot. **55**, 20–26.

Dawson, E. Y. (1966). "Marine Botany." Holt, Rinehart, and Winston, New York.

De Bary, A. (1877). "Vergleichende Anatomie der Vegetationsorgane." W. Engelmann, Leipzig.

de Broyn, J. A. (1966). The *in vitro* germination of pollen of *Setaria sphacelata.* II. Relationships between boron and certain cations. *Physiol. Plant.* **19**, 322–327.

Decker, J. P. (1961). Salt secretion by *Tamarix pentandra* Pall. *Forest Sci.* **7**, 214–217.

Decker, J. P., Gaylor, G. W., and Cole, F. D. (1962). Measuring transpiration of undisturbed tamarisk shrubs. *Plant Physiol.* **37**, 393–397.

de Fraine, E. (1912). The anatomy of the genus *Salicornia. Linn. J. Bot.* **41**, 317–346.

de Fraine, E. (1916). The morphology and anatomy of the genus *Statice* as represented at Blakeney Point. I. *Statice binervosa,* G. E. Smith and *Statice bellidifolia* D.C. (= *S. reticulata*). *Ann. Bot. (London)* **30**, 239–282.

Delf, E. M. (1911). Transpiration and behaviour of stomata in halophytes. *Ann. Bot.* **25**, 485–505.

de Lobelius, M. (1570). Stirpium Adversaria Nova.

den Hartog, C. (1964). The taxonomy of the sea-grass genus *Halodule* Endl. (Potamogetonaceae). *Blumea* **12**, 289–312.

den Hartog, C. (1970). "The Sea Grasses of the World." North Holland, Amsterdam.

de Saussure, N. T. (1804). "Recherches chemique sur la Vegetation." Paris.

de Sigmond, A. A. J. (1927). The classification of alkali and salty soils. *Proc. 1st Intern. Congr. Soil Sci.* **1**, 330–334.

de Sigmond, A. A. J. (1938). "The Principles of Soil Science," 362 pp. T. Murby, London.

Dewey, D. R. (1960). Salt tolerance of twenty-five strains of *Agropyron. Agron. J.* **52**, 631–635.

De Wit, C. T. (1960). On competition. *Versl. Landbouwk. Onderzoek.* **66**, 1–81.

De Wit, C. T., Tow, P. G., and Ennik, G. C. (1966). Competition between legumes and grasses. *Versl. Landbouwk. Onderzoek.* **687**, 1–30.

Dexter, R. W. (1950). Restoration of the *Zostera* faciation at Cape Ann, Massachusetts. *Ecology* **31**, 286–288.

Dixon, M., and Webb, E. C. (1957). "Enzymes." Longmans, Green, London.

Dodds, J. J. A., and Ellis, R. J. (1966). Cation stimulated adenosine triphosphatase activity in plant cell walls. *Biochem. J.* **101**, 3.

Dodoens, R. (1563). Cruijde Boeck Vander Loe Drukkerij Mehelen.

Donald, C. M. (1963). Competition among crop and pasture. *Advan. Agron.* **15**, 1–118.

Dorph-Petersen, K., and Steenbjerg, F. (1950). Investigations of the effect of fertilizers containing sodium. *Plant Soil* **2**, 283–300.

Downton, W. J. S., and Tregunna, E. B. (1968). Carbon dioxide compensation—its relation to photosynthetic carboxylation reactions, systematics of the Graminae and leaf anatomy. *Can. J. Bot.* **46**, 207–215.

Dregne, H. E. (1968). Surface materials of desert environments. *In* "Deserts of the World." (W. G. McGinnies, B. J. Goldman, and P. Paylore, eds.), pp. 287–377. Univ. Arizona Press, Tucson, Arizona.

Eardley, A. J. (1962). Gypsum dunes and evaporite history of the Great Salt Lake Desert. *Utah Geol. Mineral Surv. Spec. Stud. 2.*

Eaton, F. M. (1927). The water requirement and cell sap concentration of Australian saltbush and wheat as related to the salinity of the soil. *Amer. J. Bot.* **14**, 212–226.

Eaton, F. M. (1942). Toxicity and accumulation of chloride and sulfate salts in plants. *J. Agr. Res.* **64**, 357–399.

Eaton, F. M., and Horton, C. R. (1940). Effect of exchange sodium on the moisture equivalent and wilting coefficient of soils. *J. Agr. Res.* **61**, 401–425.

Egler, F. E. (1951). Southeast saline Everglades vegetation, Florida and its management. *Vegetatio* **3**, 213–265.

Ehlig, C. F. (1960). Effect of salinity on four varieties of table grapes grown in sand culture. *Proc. Amer. Soc. Hort. Sci.* **76**, 323–335.

Eig, A. (1931–1932). Les éléments et les groupes phytogeographiques auxiliaires dans la flora Palestinienne. *Repertorium Specierum Novarum Regni Vegetabilis, Beih.* **63**, 1–200; (2) 1–120.

Eig, A. (1946). Synopsis of the phytosociological units of Palestine. *Palestine J. Bot. Jerusalem Ser* **3**, 183–246.

Eisikowitch, D. (1970). The ecology of pollination of the seashore plants in Israel. Ph.D. Thesis, Tel-Aviv Univ. Israel.

Elazari-Volcani, B. (1940). Studies on the Microflora of the Dead Sea. Ph.D. Thesis, Hebrew University of Jerusalem, Israel.

El Sheikh, A. M., Ulrich, A., and Broyer, T. C. (1967). Sodium and rubidium as possible nutrients for sugar beet plants. *Plant Physiol.* **42**, 1202–1208.

Elzam, O. E., and Epstein, E. (1969a). Salt relations of two grass species differing in salt tolerance. I. Growth and salt content at different salt concentrations. *Agrochimica* **13**, 187–195.

Elzam, O. E., and Epstein, E. (1969b). Salt relations of two grass species differing in salt tolerance. II. Kinetics of the absorption of K, Na and Cl by their excised roots. *Agrochimica* **13**, 196–206.

Epstein, E. (1961). The essential role of calcium in selective cation transport by plant cells. *Plant Physiol.* **36**, 437–444.

Epstein, E. (1966). Dual pattern of ion absorption by plant cells and by plants. *Nature (London)* **212**, 1324–1327.

Epstein, E. (1969). Mineral metabolism in halophytes. *In* "Ecological Aspects of the Mineral Nutrition of Plants" (I. H. Rorison, ed.), pp. 345–355. Blackwell, Oxford.

Epstein, E., and Hagen, C. E. (1952). A kinetic study of the absorption of alkali cations by barley roots. *Plant Physiol.* **27**, 457–474.

Epstein, E., and Leggett, J. E. (1954). The absorption of alkaline earth cations by barley roots: Kinetics and mechanism. *Amer. J. Bot.* **41**, 785–791.

Epstein, E., and Stout, P. R. (1952). The micronutrient cations iron, manganese, zinc and copper: their uptake by plants from the adsorbed state. *Soil Sci.* **72**, 47–65.

Epstein, E., Rains, D. W., and Elzam, O. E. (1963). Resolution of dual mechanisms of potassium absorption by barley roots. *Proc. Natn. Acad. Sci. U.S.* **49**, 684–692.

Eriksson, E. (1958). The chemical climate and saline soils in the arid zone. *UNESCO, Arid Zone Res. Climatol. Rev. Res.* **10**, 147–180.

Eshel, A. (1971). Patterns of sodium uptake along corn roots. M.Sc. Thesis, Tel-Aviv Univ., Tel-Aviv.

Eshel, A., and Waisel, Y. (1972). Variations in sodium uptake along primary roots of corn seedlings. *Plant Physiol.* **49**, 585–589.

Eshel, Y. (1966). Ecotypic differentiation in *Prosopis farcta* (Banks et Sol.) Eig. M.Sc. Thesis, Tel-Aviv Univ., Tel-Aviv.

Eshel, Y., and Waisel, Y. (1965). The salt relations of *Prosopis farcta* (Banks et Sol.) Eig. *Israel J. Bot.* **14**, 50.

Etherton, B. (1963). Relationship of cell transmembrane electropotential to potassium and sodium accumulation ratios in oat and pea seedlings. *Plant Physiol.* **38**, 581–585.

Etherton, B. (1967). Steady state sodium and rubidium effluxes in *Pisum sativum* roots. *Plant Physiol.* **42**, 685–690.

Etherton, B., and Higinbotham, N. (1960). Transmembrane potential measurements of cells of higher plants as related to salt uptake. *Science* **131**, 409–410.

Evenari, M. (1962). Plant physiology and arid zone research. *UNESCO, Arid Zone Res. The Problems of the Arid Zone* **18**, 175–195.

Evenari, M., and Richter, R. (1937). Physiological–ecological investigations in the Wilderness of Judaea. *J. Linn. Soc. (London)* **51**, 333–381.

Fahn, A. (1958). Xylem structure and annual rhythm of development in trees and shrubs of the desert. I. *Tamarix aphylla, T. jordanis* var. *negevensis, T. gallica,* var. *maris-mortui. Trop. Woods* **109**, 81–94.

Fahn, A. (1963). The fleshy cortex of articulated Chenopodiaceae. *J. Indian Bot. Soc.* **42A**, 39–45.

Fahn, A. (1964). Some anatomical adaptations of desert plants. *Phytomorphology* **14**, 93–102.

Fahn, A. (1967). "Plant Anatomy." Pergamon Press, London.

Fahn, A., and Arzee, T. (1959). Vascularization of articulated Chenopodiaceae and the nature of their fleshy cortex. *Amer. J. Bot.* **46**, 330–338.

Feekes, W. (1936). De ontwikkeling van de natuurlijke vegetatie in de Wieringermeer Polder–de eerste groote droogmakerij van de Zuiderzee. *Ned. Kruidk. Arch.* **46**, 1–295.

Feekes, W. (1943). De Piamer Kooiwaard en Makkumerwaard. *Ned. Kruidk. Arch.* **53**, 288–330.

Ferchau, H. A. (1959). Ecotypic variability of *Tsuga canadensis* photosynthesis and related processes. *Diss. Abstr.* **19**, 2440.

Ferguson, J. (1952). The rate of natural evaporation from shallow ponds. *Aust. J. Sci. Res.* **A5**, 315–330.

Fireman, M., and Hayward, H. E. (1952). Indicator significance of some shrubs in the Escalente Desert, Utah. *Bot. Gaz. (Chicago)* **114**, 143–155.

Fischer, J., and Hodges, T. K. (1969). Monovalent ion stimulated adenosine triphosphatase from oat roots. *Plant Physiol.* **44**, 385–395.

Fishelson, L. (1971). Ecology and distribution of the benthic fauna in the shallow waters of the Red Sea. *Marine Biol.* **10**, 113–133.

Flowers, S. (1934). The vegetation of the Great Salt Lake region. *Bot. Gaz. (Chicago)* **95**, 353–418.

Focke, W. O. (1871). Untersuchungen über die Vegetation des nordwest-deutschen Tieflands. *Naturwiss. Ver. Bremen* **2**, 405–411.

Frey-Wyssling, A. (1935). "Die Stoffausscheidung der Höheren Pflanzen." Verlag von Julius Springer, Berlin.

Fried, M., and Shapiro, R. E. (1961). Soil-plant relationships in ion uptake. *Ann. Rev. Plant. Physiol.* **12**, 91–112.

Friedman, J., and Waisel, Y. (1964). Contribution to the arboreal flora of Israel: *Tamarix aphylla* (L) Karst. *La-Yaaran* **13**, 156–161.

Fukuda, Y. (1937). Die Anpassungsfahigkeit der Pflanzen bezuglich des osmotischen Druckes. I. Vorliebe der Halophyten fur NaCl. *Bot. Mag. (Tokyo)* **51**, 445–456.

Gaastra, P. (1959). Photosynthesis of crop plants as influenced by light, carbon dioxide, temperature and stomatal diffusion resistance. *Mededel. Landbouwhogesch. Wageningen* **59**, 1–68.

Gabrielsen, E. K., and Larsen, P. (1935). Über den Kohlenstoffhaushalt der terrestrischen Halophyten. *Medd. Kgl. Danske Videnskabs Selskab Biol.* **11**, 1–48.

Gale, J., and Poljakoff-Mayber, A. (1970). Interrelations between growth and photosynthesis of salt bush (*Atriplex halimus* L.) grown in saline media. *Aust. J. Biol. Sci.* **23**, 937–945.

Gale, J., Naaman, R., and Poljakoff-Mayber, A. (1970). Growth of *Atriplex halimus* L. in sodium chloride salinated culture solutions as affected by the relative humidity of the air. *Aust. J. Biol. Sci.* **23**, 947–952.

Ganguly, S. N., Sanyal, T., Sircar, P. K., and Sircar, S. M. (1970). A new gibberellin in the leaves of *Sonneratia apetala* Ham. *Chem. Ind.* **25**, 832–833.

Ganmore-Neumann, R. (1970). The mechanism of sodium uptake by excised barley roots. Ph.D. Thesis, Hebrew Univ., Jerusalem.

Ganmore-Neumann, R., and Waisel, Y. (1972). The mechanisms of sodium uptake by excised barley roots. In press.

Gardner, R. (1945). Some soil properties related to the sodium salt problem in irrigated soils. *U.S. Dept. Agr. Tech. Bull.* **902**, 28 pp.

Garrahan, P. J., and Glynn, I. M. (1967). The incorporation of inorganic phosphate into adenosine triphosphate by reversal of the sodium pump. *J. Physiol. (London)* **192**, 237–256.

Gary, H. L., and Horton, J. S. (1965). Some sprouting characteristics of five stamen tamarisk. *U.S. Forest Service Res. Note* **39**, 1–7.

Gates, C. T., and Bonner, J. (1959). The response of young tomato plant to a brief period of water shortage. IV. Effects of water stress on the ribonucleic acid metabolism of tomato leaves. *Plant Physiol.* **34**, 49–55.

Gauch, H. G., and Dugger, W. M. (1954). The physiological action of boron in higher plants. *Univ. Md. Exp. Sta. Bull. A,* **80**, 1–43.

Gauch, H. G., and Wadleigh, C. H. (1951). The salt tolerance and chemical composition of Rhodes and Dallis grasses grown in sand culture. *Bot. Gaz. (Chicago)* **112**, 259–271.

Gavalas, N. A., and Demetriades, S. D. (1964). Calcium distribution in relation to the chlorosis in boron deficient olive leaves. *Ann. Inst. Phytopatol. Benaki,* **7**, 37–46.

Gerhardt, J. (1900). "Handbuch der deutschen Dünenbaues." 438 pp. P. Parey, Berlin.

Gibb, D. C. (1957). The free-living forms of *Ascophyllum nodosum* (L.) Le Jol. *J. Ecol.* **45**, 49–84.

Giese, A. C. (1957). "Cell Physiology." Saunders, Philadelphia, Pennsylvania.

Gillham, M. E. (1957). Coastal vegetation of Mull and Iona in relation to salinity and soil reaction. *J. Ecol.* **45**, 757–778.

Gilliland, H. B. (1952). The vegetation of eastern British Somaliland. *J. Ecol.* **40**, 91–124.

Gillner, V. (1952). Die Gurtelung der Strandwiesen und der Wasserstandswechsel an der Westkuste Schwedens. *Sv. Bot. Tidksr.* **46**, 393–428.

Gimingham, C. H. (1964). Maritime and sub-maritime communities. *In* "The Vegetation of Scotland" (J. H. Burnett, ed.), pp. 67–142. Oliver & Boyd, Edinburgh and London.

Gingrich, J. R., and Russell, M. B. (1957). A comparison of effects of soil moisture tension and osmotic stress on root growth. *Soil Sci.* **84**, 185–194.

Ginzburg, C. (1964). Ecological anatomy of roots. Ph.D. Thesis, Hebrew University, Jerusalem.

Glawion, H. (1939). Staub und Staubfälle in Arosa. *Beitr. Phys. Freien Atmosphere* **25**, 1–43.

Goldberg, A. A. (1958). Contributions to the study of Nari in Israel and especially in the Eastern Esdraelon Valley. Ph.D. Thesis, The Hebrew University, Jerusalem.

Golley, F., Odum, H. T., and Wilson, R. F. (1962). The structure and metabolism of a Puerto-Rican red mangrove forest in May, 1962. *Ecology* **43**, 9–19.

Goodall, D. W. (1960). Quantitative effects of intraspecific competition: An experiment with mangolds. *Bull. Res. Counc. Israel* **8D**, 181–194.

Goodin, J. R., and Mozafar, A. (1970). Quantitative histochemistry of oxalate in vesiculated hairs of *Atriplex halimus* L. *Histochemie* **21**, 366–368.

Goodman, P. J. (1969). *Spartina* Schreb. in Biological Flora of the British Isles. *J. Ecol.* **57**, 285–287.

Goodman, P. J., Braybrooks, E. M., and Lembert, J. M. (1959). Investigations into "die-back" in *Spartina townsendii* agg. I. The present status of *Spartina townsendii* in Britain. *J. Ecol.* **47**, 651–677.

Gorham, A. V., and Gorham, E. (1955). Iron, manganese, ash and nitrogen in some plants from salt marsh and shingle habitats. *Ann. Bot. (London)* **19**, [N.S.] 571–577.

Green, J. (1959). Geochemical table of the elements for 1959. *Bull. Geol. Soc. Amer.* **70**, 1127–1184.

Greenway, H. (1962a). Plant response to saline substrates. I. Growth and ion uptake of several varieties of *Hordeum* during and after sodium chloride treatment. *Aust. J. Biol. Sci.* **15**, 16–38.

Greenway, H. (1962b). Plant response to saline substrates. II. Chloride, sodium and potassium uptake and translocation in young plants of *Hordeum vulgare* during and after a short sodium chloride treatment. *Aust. J. Biol. Sci.* **15**, 39–57.

Greenway, H. (1968). Growth stimulation by high chloride concentration in halophytes. *Israel J. Bot.* **17**, 169–177.

Greenway, H., and Klepper, B. (1969). Relation between anion transport and water flow in tomato plants. *Physiol. Plant.* **22**, 208–219.

Greenway, H., and Osmond, C. B. (1970). Ion relations, growth and metabolism of *Atriplex* at high external electrolyte concentrations. *In* "The Biology of *Atriplex*" (R. Jones, ed.), pp. 49–56. Div. Plant Industry C.S.I.R.O., Canberra, Australia.

Greenway, H., and Pitman, M. G. (1965). Potassium retranslocation in seedlings of *Hordeum vulgare. Aust. J. Biol. Sci.* **18**, 235–247.

Greenway, H., and Thomas, D. A. (1965). Plant response to saline substrates. V. Chloride regulation in the individual organs of *Hordeum vulgare* during treatment with sodium chloride. *Aust. J. Biol. Sci.* **18**, 505–524.

Greenway, H., Gunn, H., and Thomas, D. A. (1966). Plant response to saline substrates. VIII. Regulation of ion concentrations in salt-sensitive and halophytic species. *Aust. J. Biol. Sci.* **19**, 741–756.

Greenway, H., Hughes, P. G., and Klepper, B. (1969). Effect of water deficit on phosphorus nutrition of tomato plants. *Physiol. Plant.* **22**, 199–207.

Greig-Smith, P. (1952). The use of random and contiguous quadrats in the study of the structure of plant communities. *Ann. Bot. (London)* **16**, [N.S.] 293–316.

Gruenberg-Fertig, I. (1966). List of Palestine plants with data on their geographical distribution. Ph.D. Thesis, The Hebrew University, Jerusalem.

Gruener, N., and Neumann, J. (1966). An ion-stimulated adenosine triphosphatase from bean roots. *Physiol. Plant.* **19**, 678–682.

Grunwald, C. (1966). Calcium uptake by potato tuber mitochondria. I. Isolation and identification of ^{45}Ca complexes. *Physiol. Plant.* **19**, 335–347.

Haberlandt, G. (1895). Über die Ernährung der Keimlinge und die Bedeutung des Endosperms bei viviparen Mangrovepflanzen. *Ann. Jardin Bot. Buitenzorg* **12**, 91–114.

Hackett, D. P. (1961). Effects of salts on DPNH oxidase activity and structure of sweet potato mitochondria. *Plant Physiol.* **36**, 445–452.

Halket, A. C. (1911). Some experiments on absorption by aerial parts of certain salt-marsh plants. *New Phytol.* **10**, 121–139.

Halket, A. C. (1915). The effect of salt on the growth of *Salicornia. Ann. Bot. (London)* **29**, 143–154.

Halket, A. C. (1928). The morphology of *Salicornia*–an abnormal plant. *Ann. Bot. (London)* **42**, 523–530.

Handley, R., Ramon Dios Vidal, and Overstreet, R. (1960). Metabolic and non-metabolic uptake of sodium in roots of *Zea mays. Plant Physiol.* **35**, 907–912.

Handley, R., Metwally, A., and Overstreet, R. (1963). Effects of Ca upon metabolic and nonmetabolic uptake of Na and Rb root segments of *Zea mays. Plant Physiol.* **40**, 513–520.

Handley, R., Metwally, A., and Overstreet, R. (1965). Divalent cations and the permeability to Na of the root meristem of *Zea mays. Plant Soil* **22**, 200–206.

Hannon, N., and Bradshaw, A. D. (1968). Evolution of salt tolerance in two co-existing species of grass. *Nature (London)* **220**, 1342–1343.

Hansson, G., and Kylin, A. (1969). ATPase activities in homogenates from sugarbeet roots, relation to Mg^{++} and $(Na^+ + K^+)$ stimulation. *Z. Pflanzenphysiol.* **60**, 270–275.

Harmer, P. M., and Benne, E. J. (1945). Sodium as a crop nutrient. *Soil Sci.* **60**, 137–148.

Harmer, P. M., Benne, E. J., Laughlin, W. M., and Key, C. (1953). Factors affecting crop response to sodium applied as common salt on Michigan muck soil. *Soil Sci.* **76**, 1–17.

Harper, J. L., and Litav, M. (1967). A method for studying spatial relationships between the root systems of two neighbouring plants. *Plant Soil* **26**, 389–392.

Harris, F. S. (1915). Effect of alkali salts in soils on the germination and growth of crops. *J. Agr. Res.* **5**, 1–53.

Harris, F. S., and Pittman, D. W. (1918). Soil factors affecting the toxicity of alkali. *J. Agr. Res.* **15**, 287–319.

Harris, F. S., and Pittman, D. W. (1919). Relative resistance of various crops to alkali. *U.S. Utah Agr. Exp. Sta. Bull. No. 168.*

Harris, J. A., and Lawrence, J. V. (1917). The osmotic concentration of the sap of the leaves of mangrove trees. *Biol. Bull.* **32**, 202–211.

Harris, J. A., Gortner, R. A., Lawrence, W. F., and Valentine, A. T. (1924). The osmotic concentration, specific electrical conductivity and chloride content of tissue fluids of the indicator plants of Tooele Valley. *Utah J. Agr. Res.* **27**, 893–924.

Hasson-Porath, E., and Poljakoff-Mayber, A. (1971). Content of adenosine phosphate compounds in pea roots grown in saline media. *Plant Physiol.* **47**, 109–113.

Hassouna, M. G., and Wareing, P. F. (1964). Possible role of rhizosphere bacteria in the nitrogen nutrition of *Ammophila arenaria. Nature (London)* **202**, 467–469.

Hatch, M. D., and Slack, C. R. (1966). Photosynthesis by sugar cane leaves. A new carboxylation reaction and pathway of sugar formation. *Biochem. J.* **101**, 103–111.

Hatch, M. D., and Slack, C. R. (1970). Photosynthetic CO_2 fixation pathways. *Ann. Rev. Plant Physiol.* **21**, 141–162.

Hayward, H. E. (1956). Plant growth under saline conditions. *UNESCO, Arid Zone Res. Utilization of Saline Water* **4**, 37–71.

Hayward, H. E., and Bernstein, L. (1958). Plant growth relationship on salt affected soils. *Bot. Rev.* **24**, 584–635.

Hayward, H. E., and Long, E. M. (1941). Anatomical and physiological responses of the tomato to varying concentrations of sodium chloride, sodium phosphate and nutrient solutions. *Bot. Gaz. (Chicago)* **102**, 437–462.

Hayward, H. E., and Wadleigh, C. H. (1949). Plant growth on saline and alkali soils. *Advan. Agron.* **1**, 1–158.

Heiman, H., and Ratner, R. (1961). The influence of K on the uptake of Na by plants under saline conditions. *Bull. Res. Counc. Israel* **10A**, 55–62.

Helder, R. J. (1956). The loss of substances by cells and tissues (salt glands). *Encycl. Plant Physiol.* **2**, 468–486.

Helder, R. J., and Boerma, J. (1969). An electron microscopical study of the plasmodesmata in the roots of young barley seedlings. *Acta Bot. Neerl.* **18**, 99–107.

Hendrickson, W. H., and Vité, J. P. (1960). The pattern of water conduction and tracheidal alignment in Douglas Fir. *Contrib. Boyce Thompson Inst.* **20**, 353–362.

Henis, Y., and Eren, J. (1963). Preliminary studies on the microflora of a highly saline soil. *Can. J. Microbiol.* **9**, 902–904.

Henkel, P. A. (1954). "Salt Resistance of Plants and the Ways of Directing its Rise." USSR Academy of Science, Moscow.

Henkel, P. A., and Shakhov, A. A. (1945). The ecological significance of the water regime of certain halophytes. *J. Bot. USSR* **30**, 154–166.

Henkel, P. A., and Strogonov, B. P. (1961). Physiologie of plants consuming saline water. *UNESCO, Arid Zone Res. Salinity Problems in the Arid Zones* **14**, 145–151.

Hewitt, E. J. (1963). The essential nutrient elements: Requirements and interactions in plants. *In* "Plant Physiology" (F. C. Steward, ed.), Vol. III, pp. 137–360, Academic Press, New York.

Hiatt, A. J. (1970). An anomaly in potassium accumulation by Barley roots. III. Effect of calcium concentration and ^{86}Rb labeling. *Plant Physiol.* **45**, 411–414.

Hiatt, A. J., and Lowe, R. H. (1967). Loss of organic acids, amino acids, K and Cl from barley roots treated with metabolic inhibitors. *Plant Physiol.* **42**, 1731–1736.

Higinbotham, N. (1968). Cell electropotentials and ion transport in higher plants. *Proc. Intern. Symp. Transport Distribution Matter Cells of Higher Plants Schloss Reinhardsbrunn.* Oct. 1968. pp. 14–19.

Higinbotham, N. (1970). Movement of ions and electrogenesis in higher plant cells. *Amer. Zoologist* **10**, 393–403.

Higinbotham, N., Etherton, B., and Foster, R. J. (1964). Effect of external K, NH_4, Na, Ca, Mg, and H ions on the cell membrane potential of *Avena coleoptile*. *Plant Physiol.* **39**, 196–203.

Higinbotham, N., Etherton, B., and Foster, R. J. (1967). Mineral ion contents and cell transmembrane electropotential of pea and oat seedling tissue. *Plant Physiol.* **42**, 37–46.

Higinbotham, N., Pratt, M. J., and Foster, R. J. (1962). Effects of calcium, indolacetic acid, and distance from stem apex on potassium and rubidium absorption by excised segments of etiolated pea epicotyl. *Plant Physiol.* **37**, 203–214.

Hilgard, E. W. (1906). "Soils, Their Formation, Properties, Composition and Relations to Climate and Plant Growth," 593 pp. McGraw-Hill, New York.

Hill, A. E. (1967). Ion and Water transport of *Limonium*. II. Shoot circuit analysis. *Biochim. Biophys. Acta* **135**, 461–465.

Hill, A. E. (1970a). Ion and water transport in *Limonium* III. Time constants of the transport system. *Biochim. Biophys. Acta* **196**, 66–72.

Hill, A. E. (1970b). Ion and water transport in *Limonium* IV. Delay effects in the transport process. *Biochim. Biophys. Acta* **196**, 73–79.

Hill, T. G. (1908). Observations on the osmotic properties of the root hairs of certain salt marsh plants. *New Phytol.* **7**, 133–142.

Hillel, D. (1965). "Water Movement in Soils and Their uptake by Plant Roots." National Univ. Inst. of Agr. Rehovoth, Israel. (In Hebrew.)

Hillel, D. (1971). "Soil and Water: Physical Principles and Processes." Academic Press, New York.

Hoagland, D. R. (1948). "Lectures on the Inorganic Nutrition of Plants." Chronica Botanica, Waltham, Massachusetts.

Hodges, T. K., and Vaadia, Y. (1964). The kinetics of chloride accumulation and transport in exudating roots. *Plant Physiol.* **39**, 490–494.

Höfler, K., and Weixl-Hofmann, H. (1939). Salzpermeabilität und Salzresistenz der Zellen von *Suaeda maritima*. *Protoplasma* **32**, 416–422.

Hoffman, G. J., and Splinter, W. E. (1968). Water potential measurements of an intact system. *Agron. J.* **60**, 408–413.

Hoffmann, P., and Sachert, H. (1967). Der Einfluss von Harnstoff auf die Entwicklung von *Salicornia brachystachya* G. F. W. Meyer—ein Beitrag zum Halophytenproblem. *Ber. Deut. Bot. Ges.* **80**, 437–446.

Hofstra, G., and Hesketh, J. D. (1969). Effects of temperature on the gas exchange of leaves in the light and dark. *Planta* **85**, 228–237.

Honda, S. I., and Robertson, R. N. (1956). Studies in the metabolism of plant cells. XI. The Donnan equilibration and the ionic relations of plant mitochondria. *Aust. J. Biol. Sci.* **9**, 305–320.

Horovitz, C. T., and Waisel, Y. (1970a). Cation-activated ATPase and salt uptake by plants with different salt resistance. *Proc. 18th Intern. Hort. Congr.* p. 89.

Horovitz, C. T., and Waisel, Y. (1970b). Different ATPase systems in glycophytic and halophytic plant species. *Experientia* **26**, 941–942.

Horton, J. S. (1964). Notes on the introduction of deciduous tamarisk. *U.S. Forest. Serv. Res. Note No. 16.*

Horton, J. S., Mounts, F. C., and Craft, J. M. (1960). Seed germination and seedling establishment of phreatophyte species. *Rocky Mt. Forest Range Exp. Sta. Paper No.* **48**, 1–26.

House, C. R., and Findlay, N. (1966). Water transport in isolated maize roots. *J. Exp. Bot.* **17**, 344–354.

Howard, L. G., and Horton, J. S. (1965). Some sprouting characteristics of five stamen tamarisk. *Rocky Mt. Forest Range Exp. Stat. U.S. Forest Serv. Res. Note RM 39.*

Hubbard, J. C. E. (1969). Light in relation to tidal immersion and the growth of *Spartina townsendii* (s.l.). *J. Ecol.* **57**, 795–804.

Hubbard, J. C. E. (1970). Effects of cutting and seed production in *Spartina anglica*. *J. Ecol.* **58**, 329–334.

Hulett, G. K., and Tmanek, G. W. (1961). Effects of some environmental factors on germination of salt Cedar (*Tamarix pentandra*) Pall. *Trans. Kansas Acad. Sci.* **64**, 96–101.

Hunt, C. B. (1960). The Death Valley salt pan; Study of evaporites. *U.S. Geol. Surv. Prof. Paper* **400-B**, 456–458.

Hunter, A. S., and Jungen, J. A. (1952). The response of sugar beets to fertilizers, spacing, and irrigation on eastern Oregon soils. *J. Amer. Soc. Sugar Beet Technol.* **7**, 180–188.

Hutton, J. T. (1958). The chemistry of rain water with particular reference to conditions in southeastern Australia. *UNESCO Arid Zone Res. Climatol. Microclimatol. Proc. Canberra Symp.* **11**, 285–294.

Hyder, S. Z., and Greenway, H. (1965). Effects of Ca^{++} on plant sensitivity to high NaCl concentrations. *Plant Soil* **23**, 258–260.

Hylmö, B. (1953). Transpiration and ion absorption. *Physiol. Plant.* **6**, 333–405.

Iljin, M. M. (1946). Some results gained by studying the formation of flora in the deserts of Central Asia. "Materials on the History of the Flora and Vegetation of the USSR," Vol. 2, pp. 196–256. (In Russian with English Summary.)

Iljin, W. S. (1932a). Anpassung der Halophyten an Konzentrierte Salzlösungen. *Planta* **16**, 352–366.

Iljin, W. S. (1932b). Zusammensetzung der Salze in der Pflanze auf verschiedenen Standorten. Kalkpflanzen. *Beih. Bot. Centralbl.* **50**, 95–137.

Iljin, W. S. (1953). Causes of death of plants as a consequence of loss of water: Conservation of life in desiccated tissues. *Bull. Torrey Bot. Club* **80**, 166–177.

Ingelsten, B. (1966). Absorption and transport of sulphate by wheat at varying mannitol concentrations in the medium. *Physiol. Plant.* **19**, 563–579.

Ingham, G. (1950). Effect of materials absorbed from the atmosphere in maintaining soil fertility. *Soil Sci.* **70**, 105–212.

Itai, C. (1967). Shoot and root interaction under different water regimes. Ph.D. Thesis, Hebrew Univ., Jerusalem.

Itai, C., Richmond, A., and Vaadia, Y. (1968). The role of cytokinins during water and salinity stress. *Israel J. Bot.* **17**, 187–195.

Iversen, J. (1936). "Biologische Pflanzentypen als Hilfsmittel in der Vegetationsforschung." Dissertation. Medd. fra Skalling laboratoriet, Copenhagen.

Jackson, E. A. (1957). Soil features in arid regions with particular reference to Australia. *J. Aust. Inst. Agr. Sci.* **25**, 196–208.

Jackson, P. C., and Taylor, J. M. (1970). Effects of organic acids on ion uptake and retention in barley roots. *Plant Physiol.* **46**, 538–542.

Jackson, P. C., Taylor, J. M., and Hendricks, S. B. (1970). Entry of Organic Acid Anions into roots. *Proc. Natl. Acad. Sci. (U.S.)* **65**, 176–183.

Jacobson, L., Overstreet, R., Carlson, R. M., and Chastain, J. A. (1957). The effect of pH and temperature on the absorption of potassium and bromide by barley roots. *Plant Physiol.* **32**, 658–662.

Jacoby, B. (1964). Function of bean roots and stems in sodium retention. *Plant Physiol.* **39**, 445–449.

Jacoby, B. (1965). Sodium retention in excised bean stems. *Physiol. Plant.* **18**, 730–739.

James, L. E., and Kyhos, D. W. (1961). The nature of the fleshy shoot of *Allenrolfea* and allied genera. *Amer. J. Bot.* **48**, 101–108.

Janitzky, P., and Whittig, L. D. (1964). Mechanisms of formation of Na_2CO_3 in soils. II. Laboratory study of biogenesis. *J. Soil Sci.* **15**, 145–157.

Jarvis, P. G., and Jarvis, M. S. (1963). Effects of several osmotic substrates on the growth of *Lupinus albus* seedlings. *Physiol. Plant.* **16**, 485–500.

Jennings, D. H. (1967). Electrical potential measurements, ion pumps and root exudation—a comment and a model explaining cation selectivity by the root. *New Phytol.* **66**, 357–369.

Jennings, D. H. (1968a). Microelectrode experiments with potato cells: A re-interpretation of the experimental findings. *J. Exp. Bot.* **19**, 13–18.

Jennings, D. H. (1968b). Halophytes, succulence and sodium in plants—a unified theory. *New Phytol.* **67**, 899–911.

Jenny, H., and Ayers, A. D. (1939). Influence of the degree of saturation of soil colloids on the nutrient intake by roots. *Soil Sci.* **48**, 443–459.

Joffe, J. S. (1949). "The ABC of Soils." Pedology Publ. New Brunswick, New Jersey.

Johnson, D. S., and York, H. H. (1915). The relation of plants to tide levels. *Carnegie Inst. Wash. Publ. No.* 206, 161 pp.

Johnson, H. S., and Hatch, M. D. (1969). The C_4-dicarboxylic acid pathway of photosynthesis. Identification of intermediates end products and quantitative evidence for the route of carbon flow. *Biochem. J.* **114**, 127–134.

Jones, E. W. (1955). Ecological studies on the rain forest of southeastern Nigeria. IV. The plateau forest of Okomu Forest Reserve *J. Ecol.* **43**, 564–594.

Jones, R. (ed.) (1970). "The Biology of *Atriplex*." Div. Plant Industry, C.S.I.R.O., Canberra, Australia.

Jones, R., and Hodgkinson, K. C. (1970). Root growth of rangeland chenopods: morphology and production of *Atriplex nummularia* and *Atriplex vesicaria*. *In* "The Biology of *Atriplex*." (R. Jones, ed.), pp. 77–85. Div. of Plant Industry, C.S.I.R.O., Canberra, Australia.

Jones, R., Hodgkinson, K. C., and Rixon, A. J. (1970). Growth and productivity in rangeland species of *Atriplex*. *In* "The Biology of *Atriplex*." (R. Jones, ed.). pp. 31–42. Div. Plant Industry, C.S.I.R.O., Canberra, Australia.

Jones, V., and Richards, P. W. (1954). *Juncus acutus* L. in biological flora of the British Isles. *J. Ecol.* **42**, 639–650.

Joshi, G., Dolan, T., Gee, R., and Saltman, P. (1962). Sodium chloride effect on dark fixation of CO_2 by marine and terrestrial plants. *Plant Physiol.* **37**, 446–449.

Junge, C. (1956). Recent investigations in air chemistry. (Cited by Eriksson, 1958).

Junge, C. E. (1958). Atmospheric chemistry. *Advan. Geophys.* **4**, 1–108.

Kadman-Zahavi, A. (1955). Notes on the germination of *Atriplex rosea* L. *Palestine J. Bot. Jerusalem Ser.* **6**, 375–378.

Kahane, J., and Poljakoff-Mayber, A. (1968). Effects of substrate salinity on the ability for protein synthesis in pea roots. *Plant Physiol.* **43**, 1115–1119.

Kappen, L. (1969). Frostresistenz einheimischer Halophyten in Beziehung zu ihrem Salz-Zucker und Wassergehalt im Sommer und Winter. *Flora (Jena)* **158**, 232–260.

Karmarkar, S. M., and Joshi, G. V. (1969). Effect of sand culture and sodium chloride on growth, physical structure and organic acid metabolism in *Bryophyllum pinnatum*. *Plant Soil* **30**, 41–48.

Karschon, R. (1958). Leaf absorption of wind-borne salt and leaf scorch in *Eucalyptus camaldulensis* Dehn. *Ilanoth* **4**, 5–25.

Kassas, M., and Zahran, M. A. (1962). Studies on the ecology of the Red Sea coastal land. *Bull. Soc. Geog. d'Egypte* **35**, 123–175.

Kassas, M., and Zahran, M. A. (1967). On the ecology of the Red Sea littoral salt marsh, Egypt. *Ecol. Monogr.* **37**, 297–316.

Kavanau, J. L. (1966). Membrane structure and function. *Fed. Proc.* **25**, 1096–1107.

Katz, J., and Wood, H. G. (1963). The use of $^{14}CO_2$ yields from glucose-1- and -6-^{14}C for the evaluation of the pathways of glucose metabolism. *J. Biol. Chem.* **283**, 517–523.

Kearney, T. H., and Schofield, C. S. (1936). The choice of crops for saline land. *U.S. Dept. Agr. Cir.* **404**, 24 pp.

Kearney, T. H., Briggs, L. J., Shantz, H. L., McLane, J. W., and Piemeisel, R. L. (1914). Indicator significance of vegetation in Tooele Valley, Utah. *J. Agr. Res.* **1**, 365–417.

Keith, L. B. (1958). Some effects of increasing soil salinity on plant communities. *Can. J. Bot.* **36**, 79–90.

Keller, B. (1925). Halophyten und Xerophytenstudien. *J. Ecol.* **13**, 224–261.

Keller, B. (1926). Die Vegetation auf den Salzböden der russichen Halbwüsten und Wüsten. *Z. Bot.* **18**, 113–137.

Keller, B. (1929). Die Einsalzung die Transpiration und Salzanhäufung bei *Atriplex verruciferum* T. *Bot. Optyn. Stant. Voronezh.* **1**, 75–78.

Keller, P., and Henis, Y. (1970). The effect of sodium chloride on some physiological groups of micro-organisms inhabiting a highly saline soil. *Israel J. Agr. Res.* **20**, 71–75.

Kelley, W. P. (1951). "Alkali Soils, Their Formation, Properties, and Reclamation." Reinhold, New York.

Kende, H. (1964). Preservation of chlorophyll in leaf sections by substances obtained from root exudate. *Science* **145**, 1066–1067.

Kende, H. (1965). Kinetin-like factors in the root exudate of sunflowers. *Proc. Nat. Acad. Sci. U.S.* **53**, 1302–1307.

Keren, A. (1970). Autecology of *Nitraria retusa*. (Forssk.) Asch. M.Sc. Thesis, Tel-Aviv Univ., Tel-Aviv.

Kershaw, K. A. (1966). "Quantitative and Dynamic Ecology," 183 pp. Edward Arnold, London.

Kessler, B., Engelber, N., Chen, D., and Greenspan, H. (1964). Studies on physiological and biochemical problems of stress in higher plants. Volcani Inst. (Israel). *Agr. Res. Spec. Bull.* 64.

Kipp-Goller, A. (1940). Über Bau und Entwicklung der viviparen Mangrovekeimlinge. *Z. Bot.* 35, 1–40.

Kirkham, M. B., Gardner, W. R., and Gerloff, G. C. (1969). Leaf water potential of differentially salinized plants. *Plant Physiol.* 44, 1378–1382.

Klepper, B., and Barrs, H. D. (1968). Effects of salt secretion on psychrometric determination of water potential of cotton leaves. *Plant Physiol.* 43, 1138–1140.

Klinge, M., and Fischer, K. (1967). Relation between CO_2 exchange and transpiration in *Bryophyllum daigremontiana* Beig. *Planta* 77, 212.

Klotz, I. M. (1958). Protein hydration and behaviour. *Science* 128, 815–822.

Klugh, A. B. (1909). Excretion of sodium chloride by *Spartina glabra* and *S. alterniflora*. *Rhodora* 11, 237–238.

Knapp, R. (1958). "Arbeitsmethoden der Pflanzensoziologie und die Eigenschaften der Pflanzengesellschaften." Eugen Ulmer Verlag, Stuttgart.

Knodel, H. (1938). Über die Abhängigkeit des osmotischen Wertes von der Saugkraft des Bodens. *Jahrb. Wiss. Bot.* 87, 557–564.

König, D. (1948). *Spartina townsendii* an der Westkuste von Schleswig-Holstein. *Planta* 36, 34–70.

Koller, D. (1955). Regulation of germination in seeds. *Bull. Res. Counc. Israel* 5D, 85–108.

Koller, D. (1957). Germination-regulating mechanism in some desert seeds. 4. *Atriplex dimorphostegia* Kar. et Kir. *Ecology* 38, 1–13.

Kovda, V. A. (1935). Die Typen der Alkaliböden (Solontzi). *Proc. 3rd Intern. Congr. Soil Sci.* 3, 99–102.

Kovda, V. A. (1954). "Geochemistry of USSR Deserts." USSR Academy of Sciences, Moscow.

Kovda, V. A. (1961). Principles of the theory and practice of reclamation and utilization of saline soils in the arid zones. *UNESCO Arid Zone Res. Salinity problems in Arid Zones. Proc. Teheran Symp.* 14, 201–213.

Kozlowski, T. T. (1961). The movement of water in trees. *Forest Sci.* 7, 177–192.

Kozlowski, T. T. (1964). "Water Metabolism in Plants." Harper & Row, New York.

Kramer, P. J. (1959). The role of water in the physiology of plants. *Advan. Agron.* 11, 51–57.

Kramer, P. J. (1969). "Plant and Soil Water Relationships. A Modern Synthesis." McGraw Hill, New York.

Kramer, P. J., and Coile, T. S. (1940). An estimation of the volume of water made available by root extension. *Plant Physiol.* 15, 743–747.

Kramer, P. J., and Kozlowski, T. T. (1960). "Physiology of Trees." McGraw Hill, New York.

Kramer, P. J., and Wiebe, H. H. (1952). Longitudinal gradient of ^{32}P absorption in roots. *Plant Physiol.* 27, 661–674.

Kreeb, K. (1964). Lecture presented before the *10th Intern. Congr. Edinburgh, 1964.*

Kunitake, G., Stitt, C., and Saltman, P. (1959). Dark fixation of CO_2 by tobacco leaves. *Plant Physiol.* 34, 123–127.

Kunkel, G. (1966). Über die Struktur und Sukzession der Mangrove, Liberias und deren Randformationen. *Ber. Schweitz. Bot. Ges.* 75, 20–40.

Kylin, A. (1953). The uptake and metabolism of sulphate by deseeded wheat plants. *Physiol. Plant.* 6, 775–795.

Kylin, A. (1960a). The influence of the external osmotic conditions upon the accumulation of sulphate in leaves. *Physiol. Plant.* 13, 148–154.

Kylin, A. (1960b). The apparent free space of green tissues. *Physiol. Plant.* **13**, 385–397.

Kylin, A. (1964). An outpump balancing phosphate-dependent sodium uptake in *Scenedesmus. Biochem. Biophys. Res. Commun.* **16**, 497–500.

Kylin, A. (1966). Uptake and loss of Na$^+$, Rb$^+$ and Cs$^+$ in relation to an active mechanism for extrusion of Na$^+$ in *Scenedesmus. Plant Physiol.* **41**, 579–584.

Kylin, A., and Gee, R. (1970). Adenosine triphosphatase activities in leaves of the mangrove *Avicennia nitida* Jacq. *Plant Physiol.* **45**, 169–172.

Laetsch, W. M. (1968). Chloroplast specialization in dicotyledons possessing the C_4-dicarboxylic acid pathway of photosynthetic CO_2 fixation. *Amer. J. Bot.* **55**, 875–883.

Läuchli, A., and Lüttge, U. (1968). Untersuchungen der Kinetik der Ionenaufnahme in das Cytoplasma von *Mnium*–Blattzellen mit Hilfe der Mikroautoradiographie und der Röntgen Mikrosonde. *Planta* **83**, 80–98.

Lagerwerff, J. V. (1969). Osmotic growth inhibition and electrometric salt tolerance evaluation of plants. *Plant Soil* **31**, 77–96.

Lagerwerff, J. V., and Eagle, H. E. (1961). Osmotic and specific effects of excess salts on beans. *Plant Physiol.* **36**, 472–477.

Lagerwerff, J. V., and Holland, J. P. (1960). Growth and mineral content of carrots and beans as related to varying osmotic and ionic composition effects in saline-sodic sand cultures. *Agron. J.* **52**, 603–608.

LaHaye, P. A., and Epstein, E. (1969). Salt toleration by plants: enhancement with calcium. *Science* **166**, 395–396.

Lamar, J. E., and Shrode, R. S. (1953). Water soluble salts in limestones and dolomites. *Econ. Geol.* **48**, 97–112.

Lange, O. L. (1959). Untersuchungen über Wärmehaushalt und Hitzresistenz mauretanisches Wüsten und Savannenpflanzen. *Flora (Jena)* **147**, 595–651.

Langlois, J. (1961). Aspects morphologiques et ecophysiologiques de la germination de trois varietés de *Salicornia herbacea* L. *Bull. Soc. Linn. Normandie* **2**, 160–174.

Langlois, J. (1971). Influence de l'immersion sur le metabolisme glucidique de *Salicornia stricta* Dumort. *Oecologia Plant.* **6**, 15–24.

Laties, G. G. (1954). The osmotic inactivation *in situ* of plant mitochondria enzymes. *J. Exp. Bot.* **5**, 49–70.

Laties, G. G. (1969). Dual mechanisms of salt uptake in relation to compartmentation and long distance transport. *Ann. Rev. Plant Physiol.* **20**, 89–116.

Lawrence, D. B. (1949). Self-erecting habit of seedling of red Mangroves. *Amer. J. Bot.* **36**, 426–427.

Lawrence, G. H. M. (1947). The genus *Armeria* in North America. *Amer. Midl. Natur.* **37**, 757–779.

Lazaroff, N., and Pitman, M. G. (1966). Calcium and magnesium uptake by barley seedlings. *Aust. J. Biol. Sci.* **19**, 991–1005.

LeClerc, J. A., and Breazeale, J. F. (1908). Plant food removed from growing plants by rain or dew. *U.S. Agr. Dept. Yearb.* 389–402.

Leggett, J. E., and Epstein, E. (1956). Kinetics of sulphate absorption by barley roots. *Plant Physiol.* **31**, 222–226.

Leggett, J. E., and Olsen, R. A. (1964). Anion absorption by Baker's yeast. *Plant Physiol.* **39**, 387–390.

Leggett, J. E., and Stolzy, L. H. (1961). Anaerobiosis and sodium accumulation. *Nature (London)* **192**, 991–992.

Lehr, J. J. (1942). The importance of sodium for plant nutrition. *Soil Sci.* **53**, 399–411.

Lehr, J. J., and Wybenga, J. M. (1958). Exploratory pot experiments on sensitiveness of different crops to sodium. *Plant Soil* 9, 237–253.

Lerner, H. R., Mayer, A. M., and Evenari, M. (1959). The nature of the germination inhibitors present in dispersal units of *Zygophyllum dumosum* and *Trigonella arabica*. *Physiol. Plant.* 12, 245–250.

Levitt, J. (1956). Significance of hydration to the state of protoplasm. *Encycl. Plant Physiol.* 3, 650–651.

Levitt, J. (1962). A sulfhydryl disulfide hypothesis of frost injury and resistance in plants. *J. Theor. Biol.* 3, 355–391.

Lewis, G. G., and Juve, R. L. (1956). Some aspects of irrigation water quality and soil characteristics. *Soil Sci.* 2, 125–137.

Lewis, T., and Taylor, L. R. (1967). "Introduction to Experimental Ecology," 401 pp. Academic Press, New York.

Lineweaver, H., and Burk, D. (1934). The determination of enzyme dissociation constants. *J. Amer. Chem. Soc.* 56, 658–666.

Linser, H. and Herwig, K. (1963). Zur abhängigkeit der Nahrstoffaufnahme von Osmotischen Druk der Aussenlösung. *Protoplasma* 57, 588–600.

Liphschitz, N., and Waisel, Y. (1970a). Effects of environment on relations between extension and cambial growth of *Populus euphratica* Oliv. *New Phytol.* 69, 1059–1064.

Liphschitz, N., and Waisel, Y. (1970b). The effect of water stresses on the radial growth of *Populus euphraitca* Oliv. *La-Yaaran* 20, 80–84.

Litwak, M. (1957). The influence of *Tamarix aphylla* on soil composition in the northern Negev of Israel. *Bull. Res. Counc. Israel* 6D, 39–45.

Livne, A., and Levin, N. (1967). Tissue respiration and mitochondrial oxidative phosphorylation of NaCl treated pea seedlings. *Plant Physiol.* 42, 407–414.

Lötschert, W., and Liemann, F. (1967). Die Salzspeicherung im Keimling von *Rhizophora mangle* L. während der Entwicklung auf der Mütterpflanze. *Planta* 77, 142–156.

Loewengart, J. (1958). Geochemistry of waters in northern and central Israel and the origin of their salts. *Bull. Res. Counc. Israel* 7G, 176–205.

Loneragan, J. F., and Snowball, K. (1969). Calcium requirements of plants. *Aust. J. Agr. Res.* 20, 465–478.

Long, W. G., and Levitt, J. (1952). Distribution of ^{45}Ca and ^{32}P in protein fractions as a result of absorption by potato slices. *Physiol. Plant.* 5, 610–619.

Love, D. (1963). Dispersal and survival of plants. *In* "North Atlantic Bida and Their History" (A. Love and D. Love, eds.), pp. 189–205. Pergamon Press, Oxford.

Love, A., and Love, D. (1954). Vegetation of prairie marsh. *Bull. Torrey Bot. Club* 81, 16–34.

Lüttge, U. (1962). Über die Zusammensetzung des Nektars und den Mechanismus seiner Sekretion. III. Die Rolle der Rückresorption und der spezifischen Zuckersekretion. *Planta* 59, 175–194.

Lüttge, U. (1966). Funktion und Struktur pflanzlicher Drüsen. *Naturwissenschaften* 53, 96–103.

Lüttge, U. (1969). Aktiver Transport (Kurzstreckentransport bei Pflanzen). *Protoplasmatologia* VIIIb, 146 pp.

Lüttge, U., and Laties, G. G. (1966). Dual mechanisms of ion absorption in relation to long distance transport in plants. *Plant Physiol.* 41, 1531–1539.

Lüttge, U., and Laties, G. G. (1967). Selective inhibition of absorption and long distance transport in relation to the dual mechanisms of ion absorption in maize seedlings. *Plant Physiol.* 42, 181–185.

Lüttge, U., and Osmond, C. B. (1970). Ion absorption in *Atriplex* leaf tissue. III. Site of metabolic control of light dependent chloride secretion to epidermal bladders. *Aust. J. Biol. Sci.* **23**, 17–25.

Lüttge, U., and Weigl, J. (1962). Mikroautoradiographische Untersuchungen der Aufnahme und des Transportes von $^{35}SO_4$- und $^{45}Ca^{++}$ in Keimwurzeln vom *Zea mays* L. und *Pisum sativum* L. *Planta* **58**, 113–126.

Lüttge, U., Ball, E., and Willert, von K. (1971). A comparative study of the coupling of ion uptake to light reactions in leaves of higher plant species having the C-3 and C-4 pathway of photosynthesis. *Z. Pflanzenphysiol.* **65**, 336–350.

Lundegårdh, H. (1945). Absorption, transport and exudation of inorganic ions by the roots. *Ark. Bot.* **32**, 1–139.

Lundegårdh, H., and Burström, H. (1933). Untersuchungen über die Salzaufnahme der Pflanzen. III. Quantitative Beziehungen zwischen Atmung und Anionenaufnahme. *Biochem. Z.* **261**, 235–251.

Lundegårdh, H., and Burström, H. (1935). Untersuchungen über die Atmungsvorgänge in Pflanzenwurzeln. *Biochem. Z.* **277**, 223–249.

Luther, H. (1950). Beobachtungen über die Fruktifikative Vernehrung von *Phragmites communis* Trin. *Acta Bot. Fenn.* **46**, 1–18.

MacArthur, R. H., and Conell. J. H. (1967). "The Biology of Populations," 200 pp. Wiley, New York.

MacKay, J. B., and Chapman, V. J. (1954). Some notes on *Suaeda australis* Moq. var. *nova Zelandica* var. nov. and *Mesembryanthemum australe* sol. ex. Forst. *Trans. Roy. Soc. N.Z.* **82**, 41–47.

Macke, A. J., and Ungar, I. A. (1971). The effect of salinity on germination and early growth of *Puccinellia nuttalliana. Can. J. Bot.* **49**, 515–520.

McMillan, C. (1959). Salt tolerance within a *Typha* population. *Amer. J. Bot.* **46**, 521–526.

McMillan, C., and Moseley, F. N. (1967). Salinity tolerances of five marine spermatophytes of Redfish Bay, Texas. *Ecology* **48**, 503–506.

Macnae, W. (1963). Mangrove swamps in South Africa. *J. Ecol.* **51**, 1–25.

Macnae, W. (1966). Mangroves in eastern and southern Australia. *Aust. J. Bot.* **14**, 67–104.

Macnae, W. (1968). A general account of the fauna and flora of mangrove swamps and forests in the Indo-West-Pacific region. *Advan. Mar. Biol.* **6**, 73–270.

Macnae, W., and Kalk, M. (1962). The mangrove swamps of Inhaca Island. *J. Ecol.* **50**, 19–34.

McNaughton, S. J. (1966). Ecotype function in the *Typha* community type. *Ecol. Monogr.* **36**, 297–325.

MacRobbie, E. A. C. (1964). Factors affecting the fluxes of potassium and chloride ions in *Nitella translucens. J. Gen. Physiol.* **47**, 859–877.

MacRobbie, E. A. C. (1965). The nature of the coupling between light energy and active ion transport in *Nitella translucens. Biochim. Biophys. Acta* **94**, 64–73.

MacRobbie, E. A. C. (1966). Metabolic effects on ion fluxes of potassium and chloride ions in *Nitella translucens. Aust. J. Biol. Sci.* **19**, 363–370.

McRoy, C. P., and Barsdate, R. J. (1970). Phosphate absorption in eelgrass. *Limnol. Oceanogr.* **15**, 6–13.

Magistad, O. C. (1945). Plant growth relations on saline and alkali soils. *Bot. Rev.* **2**, 181–230.

Mangum-Shields, D. (1950). Leaf xeromorphy as related to physiological and structural influences. *Bot. Rev.* **16**, 399–447.

Marchant, C. J., and Goodman, P. J. (1969). *Spartina maritima* (Curtis) Fernald. In biological flora of the British Isles. *J. Ecol.* **57**, 287–291.

Marloth, R. (1887). Zur Bedeutung der Salz abscheidenen Drüsen der Tamariscineen. *Ber. Deut. Bot. Ges.* **5**, 319–324.

Marshall, J. K. (1965). *Corynephorus canescens* (L) P. Beauv. as a model for the *Ammophila* problem. *J. Ecol.* **53**, 447–463.

Mason, E. (1928). Note on the presence of mycorrhiza in the roots of salt marsh plants. *New Phytol.* **27**, 193–195.

Matile, P., Moo, H., and Mühlethaler, K. (1967). Isolation and properties of the plasmalemma in yeast. *Arch. Mikrobiol.* **58**, 201–211.

Maximov, N. A. (1929). "The Plant in Relation to Water." Allen and Unwin, London.

Mayer, A. M., and Poljakoff-Mayber, A. (1963). "The Germination of Seeds," 236 pp. Pergamon Press, Oxford.

Means, T. H. (1929). Fog precipitation by trees. *Science* **66**, 402–403.

Meiri, A., and Poljakoff-Mayber, A. (1969). Effect of variations in substrate salinity on the water balance and ionic composition of bean leaves. *Israel J. Bot.* **18**, 99–112.

Menchikovsky, F. (1925). Composition of rain falling at Tel-Aviv. *Bull. Agr. Exp. Sta. Tel-Aviv, No. 2.*

Merkel, D. L., and Hopkins, H. H. (1957). Life history of salt cedar (*Tamarix gallica* L.) *Trans. Kansas Acad. Sci.* **60**, 360–369.

Metcalfe, C. R., and Chalk, L. (1950). "Anatomy of the Dicotyledons," Vol. I and II. Clarendon Press, Oxford.

Michel, B. E., and El Sharkawi, H. M. (1970). Investigation of plant water relations with divided root systems of soybean. *Plant Physiol.* **46**, 728–731.

Milburn, T. R., Pearson, D. J., and Ndegwe, N. A. (1968). Crassulacean acid metabolism under tropical conditions. *New Phytol.* **67**, 883–897.

Milner, H. W., Hiesey, W. M., and Nobs, M. A. (1959). Physiology of climatic races. *Carnegie Inst. Wash. Ann. Rept. Dept. Plant Biol.*, pp. 346–350.

Milton, W. E. J. (1939). Occurrence of buried viable seeds in soils at different elevations and on a salt marsh. *J. Ecol.* **27**, 149–159.

Miyamoto, T. (1962). Increasing the resistance of wheat to high concentrations of salt by treating the seed with trimethylammonium chloride. *Naturwissenschaften* **49**, 213.

Miyewaki, A. and Ohba, T. (1965). Studien über Strand-Salzwiesengesellschaften auf Ost-Hokkaido (Japan). *Sci. Rept. Yokohama Nat. Univ.* **12**, 1–25.

Monk, R. W., and Wiebe, H. H. (1961). Salt tolerance and protoplasmic salt hardiness of various woody and herbaceous ornamental plants. *Plant Physiol.* **36**, 478–482.

Montfort, C. (1926). Physiologische und Pflanzengeographische Seesalzwirkungen. I. Einfluss ausgeglichender Salzlösungen auf Mesophyll und Schliesszellen; Kritik der Iljinschen Hypothese der Salzbeständigkeit. *Jahrb. Wiss. Bot.* **65**, 502–550.

Montfort, C. (1927). Über Halobiose und ihre Aufstufung: Versuch einer Synthetischen Verknüpfung isolierter analytischer Probleme. *Flora (Jena)* **121**, 434–502.

Montfort, C. (1931). Assimilation und Stoffgewinn der Meersalgen bei Aussüßung und Rückversalzung I. *Ber. Deuts. Bot. Ges.* **49**, 58, (59)–(66).

Montfort, C., and Brandrup, W. (1927). Physiologische und Pflanzengeographische Seesalzwirkungen. I. Ökologische Studien über Keimung und erste Entwicklung bei Halophyten. *Jahrb. Wiss. Bot.* **66**, 902–946.

Montfort, C., and Brandrup, W. (1928). Physiologische und Pflanzengeographische Seesalzwirkungen. III. Vergleichende Untersuchungen der Salzwachstumsreaktion von Würzeln. *Jahrb. Wiss. Bot.* **67**, 105–173.

Mooring, M. T., Cooper, A. W., and Seneca, E. D. (1971). Seed germination response and evidence for height ecophenes in *Spartina alterniflora* from North Carolina. *Amer. J. Bot.* **58**, 48–55.

Moser, H. (1934). Untersuchungen über die Blattstruktur von *Atriplex* Arten und ihre Beziehungen zur Systematik. *Beih. Bot. Centralbl.* **52**, 378–388.

Mozafar, A., and Goodin, J. R. (1970). Vesiculated hairs: a mechanism for salt tolerance in *Atriplex halimus* L. *Plant Physiol.* **45**, 62–65.

Mozafar, A., Goodin, J. R., and Oertli, J. J. (1970a). Na and K interaction in increasing the salt tolerance of *Atriplex halimus* L. I. Yield characteristics and osmotic potential. *Agron. J.* **62**, 478–481.

Mozafar, A., Goodin, J. R., and Oertli, J. J. (1970b). Sodium and potassium interaction in increasing the salt tolerance of *Atriplex halimus* L. II. Na^+ and K^+ uptake characteristics. *Agron. J.* **62**, 481–484.

Muir, J. (1937). The seed drift of South Africa and some influences of ocean currents on the strand vegetation. *Mem. Bot. Survey, S. Africa* **16**, 108 pp.

Mullan, D. P. (1932). Observations on the biology and physiological anatomy of some Indian halophytes. *J. Indian Bot. Soc.* **11**, 103–118, 285–302.

Mullan, D. P. (1933). Observations on the biology and physiological anatomy of some Indian halophytes. *J. Indian Bot. Soc.* **12**, 165–182, 235–253.

Negbi, M., and Evenari, M. (1961). The means of survival of some desert summer annuals. *UNESCO, Arid Zone Res.: Plant Water Relationships in Arid and Semiarid Conditions* **16**, 249–259.

Neuwohner, W. (1938). Der tägliche Verlauf von Assimilation und Atmung bei einigen Halophyten. *Planta* **28**, 644–679.

Nielsen, P. T. (1963). Light promoted uptake of chloride in *Chlorella*. *Plant Physiol. Suppl.*, p. IV.

Nieman, R. H. (1962). Some effects of sodium chloride on growth, photosynthesis and respiration of twelve crop plants. *Bot. Gaz. (Chicago)* **123**, 279–285.

Nishida, K. (1963). Studies on stomatal movement of Crassulacean plants in relation to the acid metabolism. *Physiol. Plant.* **16**, 281–291.

Nitsos, R. E., and Evans, H. J. (1969). Effects of univalent cations on the activity of particulate starch synthetase. *Plant Physiol.* **44**, 1260–1266.

Noggle, J. C., De Wit, C. T., and Fleming, A. L. (1964). Interrelations of calcium and rubidium absorption by excised roots of barley and plantain. *Proc. Amer. Soc. Soil Sci.* **28**, 97–100.

Norkrans, B., and Kylin, A. (1969). Regulation of the potassium to sodium ratio and of the osmotic potential in relation to salt tolerance in yeasts. *J. Bacteriol.* **100**, 836–845.

Önal, M. (1964). Zusammensetzung des Zellsaftes einiger Salzmarshen und Dünenpflanzen in der Umgebung von Neapel. *Beit. Phytol.* (Walter-Festschrift) 89–100.

Önal, M. (1966). Vergleichende ökologische Untersuchungen bei Halophyten und Glycophyten in der Nähe von Neapel. *Rev. Fac. Sci. Univ. Istanbul. Ser.* **B31**, 209–248.

Önal, M. (1971). Der Einfluss steigernder Natriumchlorid-Konzentrationen auf den Transpirationskoeffizient einigen Halophyten. *Rev. Fac. Sci. Univ. Istanbul Ser.* **B36**, 1–8.

Oertli, J. J. (1966). Effects of external salt concentrations on water relations in plants. III. Concentration dependence of the osmotic differential between xylem and external medium. *Soil Sci.* **104**, 56–62.

O'Leary, J. W. (1969). The effect of salinity on permeability of roots to water. *Israel J. Bot.* **18**, 1–9.

Oliver, F. W. (1925). *Spartina townsendii;* its mode of establishment, economic uses and taxonomic status. *J. Ecol.* **13**, 74–91.

Oosting, H. J. (1945). Tolerance of salt spray of plants of coastal dunes. *Ecology* **26**, 85–89.

Oosting, H. J. (1956). "The Study of Plant Communities. An Introduction to Plant

Ecology," 2nd Ed. Freeman, San Francisco, California.

Oosting, H. J., and Billings, W. D. (1942). Factors affecting vegetational zonation on coastal dunes. *Ecology* **23**, 131–142.

Orshan, G. (1953). Note on the application of Raunkaier's system of life forms in arid regions. *Palestine J. Bot. Jerusalem Ser.* **6**, 120–122.

Orshan, G., and Zohary, D. (1955). Vegetation of the littoral salt marshes in Israel. *Bull. Res. Counc. Israel* **4**, 363–369.

Ortenburger, A. I., and Bird, R. D. (1931). The ecology of the western Oklahoma salt plains. *Okla. Biol. Surv. Bull.* **3**, 49–64.

Osaga, T. (1939). Effect of alkaline salts on the germination of seeds. I. Effect of NaCl solutions on the germination of soybeans especially on the influence at high and low temperatures. *Rept. Inst. Sci. Res. Manchoukuo* **3**, 303–315 (Chem Abstr. **36**, 2588, 1942).

Osipova, O. P. (1947). Ob izvlekaemosti Khlorefilla iz zelenykh rastenii (Extraction of chlorophyll from green plants). *Dokl. Akad. Nauk SSSR* **57**(8), 799–801.

Osmond, C. B. (1963). Oxalates and ionic equilibria in Australian saltbushes (*Atriplex*). *Nature (London)* **198**, 503–504.

Osmond, C. B. (1965). Divalent cations absorption and interaction in *Atriplex*. *Aust. J. Biol. Sci.* **19**, 37–48.

Osmond, C. B. (1967). Acid metabolism in *Atriplex*. I. Regulation of oxalate synthesis by the apparent excess cation absorption in leaf tissue. *Aust. J. Biol. Sci.* **20**, 575–587.

Osmond, C. B. (1968). Ion absorption in *Atriplex* leaf tissue. I. Absorption by mesophyll cells. *Aust. J. Biol. Sci.* **21**, 1119–1130.

Osmond, C. B. (1970). Carbon metabolism in *Atriplex* leaves. *In* "The Biology of *Atriplex*" (R. Jones, ed.), pp. 17–21. Div. Plant Industry, C.S.I.R.O. Canberra, Australia.

Osmond, C. B., and Laties, G. G. (1968). Interpretation of the dual isotherm for ion absorption in beet tissues. *Plant Physiol.* **43**, 745–755.

Osmond, C. B., Troughton, J. H., and Goodchild, D. J. (1969). Physiological, biochemical and structural studies of photosynthesis and photorespiration in two species of *Atriplex*. *Z. Pflanzenphysiol.* **61**, 218–237.

Ovadia, S. (1969). Autecology of *Suaeda monoica* Forssk. ex J. F. Gmel. M.Sc. Thesis, Tel-Aviv Univ., Tel-Aviv, Israel.

Pannier, F. P. (1959). El efecto dez distintas concentraciones salinas sobre el desarrollo de *Rhizophora mangle* L. *Acta Cient. Venez.* **10**, 68–78.

Pannier, F. P. (1962). Estudio fisiologico sobre la viviparia de *Rhizophora mangle* L. *Acta Cient. Venez.* **13**, 184–197.

Pardee, A. B., and Prestidge, L. S. (1966). Cell-free activity of sulphate binding sites involved in active transport. *Proc. Nat. Acad. Sci. U.S.* **54**, 189–191.

Penfound, W. T. (1952). Southern swamps and marshes. *Bot. Rev.* **18**, 413–446.

Penman, H. L. (1948). Natural evaporation from open water, bare soil and grass. *Proc. Roy. Soc. London* **A193**, 120–145.

Peterson, H. B. (1961). Some effects on plants of salt and sodium from saline and sodic soils. *UNESCO Arid Zone Res. Salinity Problems in Arid Zones. Proc. Teheran Symp.* **14**, 163–167.

Pettersson, S. (1960). Ion absorption in young sunflower plants. I. Uptake and transport mechanisms for sulphate. *Physiol. Plant.* **13**, 133–147.

Pettersson, S. (1961). Ion absorption in young sunflower plants. II. The sulphate uptake in the apparent free space. *Physiol. Plant.* **14**, 124–132.

Phillips, R. C. (1960). Observations on the ecology and distribution of the Florida seagrasses. *Fla. State Bd. Conserv. Marine Lab. Prof. Papers* **2**, 1–72.

Pielou, E. C. (1960). A single mechanism to account for regular, random and aggregated population. *J. Ecol.* **48**, 575–587.

Pitman, M. G. (1970). Active H⁺ efflux from cells of low salt barley roots during salt accumulation. *Plant Physiol.* **45**, 787–790.

Place, G. A., Siddique, M. A., and Wells, B. R. (1971). Effects of temperature and flooding on rice growing in saline and alkaline soil. *Agron. J.* **63**, 62–66.

Pokrovskaya, E. I. (1954). Obmen veshchestv rastenii na zasolennykh pochvakh (Metabolism of plants on saline soils). Avtoreferat Kandidatskoi dissartatsii. Moskva, 1954.

Pokrovskaya, E. I. (1957). Nekotorye dannye ob okislitel'no–vosstanovitel'nykh protsessakh u khalofitov (Some data on the oxido–reductive processes in halophytes). *In* "Pamyati akademika N.A. Maksimova," Izdatel'stvo Akademii Nauk. SSSR. pp. 268–274.

Poljakoff-Mayber, A., and Meiri, A. (1969). The response of plants to changing salinity. *Final Tech. Rept. U.S. Dept. Agr. Proj.* A10-SWC-7.

Pollak, G. (1967). Autecology of *Aeluropus litoralis* (Willd) Parl. M.Sc. Thesis, Tel-Aviv Univ., Tel-Aviv.

Pollak, G., and Waisel, Y. (1968). Salt secretion in *Aeluropus litoralis* (Willd) Parl. *Israel J. Bot.* **17**, 126–127.

Pollak, G., and Waisel, Y. (1970). Salt secretion in *Aeluropus litoralis* (Willd) Parl. *Ann. Bot. (London)* **34**, 879–888.

Pollak, G., and Waisel, Y. (1972). Germination and vegetative reproduction of *Aeluropus litoralis* (Willd) Parl. *Ann. Arid Zone Res.,* In Press.

Polunin, N. (1960). "Introduction to Plant Geography," 640 pp. McGraw Hill, New York.

Poma, G. (1922). L'Influence de la salinité de l'eau sur la germination et la croissance des plantes halophytes. *Acad. Roy. Belg. Bull. Cl. Sci.* **8**, 81–97.

Porath, E., and Poljakoff-Mayber, A. (1968). The effect of salinity in the growth medium on carbohydrate metabolism in pea root tips. *Plant Cell Physiol.* **9**, 195–203.

Praeger, R. L. (1913). On the buoyancy of seeds of some Britannic plants. *Proc. Roy. Dubl. Soc. Sci.* **14**, 13–62.

Purer, E. A. (1942). Plant ecology of the coastal salt marsh lands of San Diego County, California. *Ecol. Monogr.* **12**, 81–111.

Rains, D. W., and Epstein, E. (1967). Preferent absorption of potassium by leaf tissue of the mangrove *Avicennia marina:* an aspect of halophytic competence in coping with salt. *Aust. J. Biol. Sci.* **20**, 847–857.

Raleigh, G. E. (1948). Effects of the sodium and chloride ions in the nutrition of table beet in culture solutions. *Proc. Amer. Soc. Hort. Sci.* **51**, 433–436.

Randall, R. E. (1970a). Salt measurements on the coast of Barbados, West Indies. *Oikos* **21**, 65–70.

Randall, R. E. (1970b). Vegetation and environment on the Barbados coast. *J. Ecol.* **58**, 155–172.

Ranson, S. L., and Thomas, M. (1960). Acid metabolism in Crassulaceae. *Ann. Rev. Plant Physiol.* **11**, 81–110.

Ranwell, D. S. (1961). *Spartina* salt marshes in southern England. I. The effects of sheep grazing at the upper limits of *Spartina* marsh in Bridgewater Bay. *J. Ecol.* **49**, 325–340.

Ranwell, D. S. (1964). *Spartina* salt marshes in southern England. III. Rates of establishment succession and nutrient supply at Bridgewater Bay, Somerset. *J. Ecol.* **52**, 95–105.

Rasmussen, H. P. (1970). Electron microprobe X-ray analysis in horticulture. *Proc. 8th Intern. Hort. Congr. Tel-Aviv* pp. 168–169.

Raunkiaer, C. (1934). "The Life Forms of Plants and Statistical Plant Geography." Oxford

Univ. Press, Oxford.

Rauser, W. E., and Hanson, J. B. (1966). The metabolic status of RNA in soybean roots exposed to saline media. *Can. J. Bot.* **44**, 759–776.

Rebhun, M. (1965). Salinity increments in municipal sewage, their significance and sources and methods of reduction. *Intern. J. Air Water Pollut.* **9**, 253–261.

Rechav, Y. (1967). Ecotypic differentiation in *Phragmites communis* Trin. M.Sc. Thesis, Tel-Aviv Univ., Tel-Aviv.

Reed, J. F. (1947). The relation of the Spartinetum glabrae near Beaufort, North Carolina, to certain edaphic factors. *Amer. Midl. Natur.* **38**, 605–614.

Reese, G. (1958). Cytosystematische Notizen zur Gattung *Nitraria*. *Flora (Jena)* **146**, 478–488.

Repp, G. (1939). Ökologische Untersuchungen im Halophytengebiet am Neusidler See. *Jahrb. Wiss. Bot.* **88**, 554–632.

Repp, G. (1961). The salt tolerance of plants: basic research and tests. *UNESCO Arid Zone Res. Salinity problems in arid zones. Proc. Teheran Symp.* **14**, 153–161.

Repp, G. (1963). Die Kupferresistenz des Protoplasmas höherer Pflanzen auf Kupfer-erzböden. *Protoplasma* **57**, 643–659.

Repp, G., McAllister, D. R., and Wiebe, H. H. (1959). Salt resistance of protoplasm as a test for the salt tolerance of agricultural plants. *Agron. J.* **51**, 311–314.

Rho, R. J. (1959). Some aspects of the metabolism of the marine diatom *Nitzchia closterium* (Ehrenberg) Wn. Smith. *Intern. Ocean Con. AAAS, Wash. Preprints*, pp. 199–200.

Richards, L. A. (Ed.). (1954). Diagnosis and improvement of saline and alkali soils *U.S. Dept. Agr. Handb. No. 60.*

Richards, L. A., and Wadleigh, C. H. (1952). Soil water and plant growth. *In* "Soil Physical Conditions and Plant Growth" (B. T. Shaw, ed.), pp. 73–251. Academic Press, New York.

Ridley, N. H. (1930). "The Dispersal of Plants Throughout the World." Reeve & Co., Ashford.

Robertson, J. H. (1955). Penetration of roots of tall wheatgrass in wet saline-alkali soil. *Ecology* **36**, 755–756.

Robertson, R. N. (1958). The uptake of minerals. *Encycl. Plant Physiol.* **4**, 243–275.

Robinson, E. R., and Brown, R. (1952). The development of the enzyme complement in growing root cells. *J. Exp. Bot.* **3**, 356–374.

Rodin, L. E. (1954). The effect of plants on the build-up of the Takyr salines and their complexes. (In Russian). *Pustyni SSSR i ich oswojenije* **2**, 392–413.

Rosenblum, I., and Waisel, Y. (1969). Autecology of *Atriplex halimus*. Research Report. Tel Aviv University.

Rovira, A. D., and Bowen, G. D. (1968). Anion uptake by the apical region of seminal wheat roots. *Nature (London)* **218**, 685–686.

Rovira, A. D., and Bowen, G. D. (1970). Translocation and loss of phosphate along roots of wheat seedlings. *Planta* **93**, 15–25.

Rudinsky, J. A., and Vité, J. P. (1959). Certain ecological and physiological aspects of the pattern of water conduction in conifers. *Forest Sci.* **5**, 159–266.

Rufelt, H. (1963). Rapid changes in transpiration of plants. *Nature (London)* **197**, 985.

Ruhland, W. (1915). Untersuchungen über die Hautdrüsen der Plumbaginaceen. Ein Beitrag zur Biologie der Halophyten. *Jahrb. Wiss. Bot.* **55**, 409–498.

Rumyantseva, A. S. (1959). The effect of habitat on stem structures of bushy halophytes in Turkestan. *Tr. Inst. Bot.* imeni Komorova. *Akad. Nauk. SSSR* **5**. (Cited after Strogonov, 1964).

Sachs, J. (1875). "Textbook of Botany" (Engl. ed. trans. by A. W. Bennett and W. T. Thiselton Dyer). Oxford Univ. Press, Oxford.

Saltman, P., Lynch, V. H., Kunitake, G. M., Stitt, C., and Spolter, H. (1957). The dark fixation of CO_2 by succulent leaves: metabolic changes subsequent to initial fixation. *Plant Physiol.* 32, 197–200.

Sandström, B. (1950). The ion absorption in roots lacking epidermis. *Physiol. Plant.* 3, 496–505.

Schaedle, M., and Jacobson, L. (1966). Ion absorption and retention by *Chlorella pyrenoidosa*. II. Permeability of the cell to sodium and rubidium. *Plant Physiol.* 41, 248–254.

Schaedle, M., and Jacobson, L. (1967). Ion absorption and retention by *Chlorella pyrenoidosa*. III. Selective accumulation of rubidium, potassium and sodium. *Plant Physiol.* 42, 953–958.

Schaffner, J. H. (1898). Notes on the salt marsh plants of northern Kansas. *Bot. Gaz. (Chicago)* 25, 255–260.

Schardakov, V. S. (1928). The physiological significance of guttation. *Bull. Inst. Recherches Biol. Perm.* 6, 193–208.

Schimper, A. F. W. (1891). "Die Indo-Malaiische Strandflora." Fischer, Jena.

Schimper, A. F. W. (1903). "Plant Geography upon a Physiological Basis." Clarendon, Oxford.

Schmalfuss, K. (1945). Zur Kenntnis der Stickstoffernährung der Runkelrüben. *Bodenk. Pflanzenernärh.* 36, 10–26.

Scholander, P. F. (1967). Osmotic mechanism and negative pressure. *Science* 156, 67–69.

Scholander, P. F. (1968). How mangroves desalinate seawater. *Physiol. Plant.* 21, 251–261.

Scholander, P. F., van Dam, L., and Scholander S. (1955). Gas exchange in the roots of mangroves. *Amer. J. Bot.* 42, 92–98.

Scholander, P. F., Hammel, H. T., Hemmingsen, E., and Garey, W. (1962). Salt balance in mangroves. *Plant Physiol.* 37, 722–729.

Scholander, P. F., Hammel, H. T., Bradstreet, E. D., and Hemmingsen, E. A. (1965). Sap pressure in vascular plants. *Science* 148, 339–346.

Scholander, P. F., Bradstreet, E. D., Hammel, H. T., and Hemmingsen, E. A. (1966). Sap concentrations in halophytes and some other plants. *Plant Physiol.* 41, 529–532.

Schrader, H. A. (1809). Über Palla's Halophyta mit besonderer Rücksicht auf die Gattungen *Salsola* und *Suaeda*. *Schrad. Neues J. Bot.* 3, 58–92.

Schratz, E. (1934). Beiträge zur Biologie der Halophyten. I. Zur Keimungsphysiologie. *Jahrb. Wiss. Bot.* 80, 112–142.

Schratz, E. (1936). Beiträge zur Biologie der Halophyten. III. Über Verteilung, Ausbildung und NaCl–Gehalt der Strandpflanzen und Ihrer Abhängigkeit vom Salzgehalt des Standortes. *Jahrb. Wiss. Bot.* 83, 133–149.

Schratz, E. (1937). Beiträge zur Biologie der Halophyten. IV. Die Transpiration der Strand und Dünenpflanzen. *Jahrb. Wiss. Bot.* 84, 593–638.

Schtscherback, J. (1910). Über die Salzausscheidung durch die Blätter von *Statice gmelini*. (Vorläufige Mitteilung). *Ber. Deut. Bot. Ges.* 28, 30–34.

Scott, B. I. H., and Martin, D. W. (1962). Bioelectric fields of bean roots and their relation to salt accumulation. *Aust. J. Biol. Sci.* 15, 83–100.

Scott, G. T., and Hayward, H. R. (1954). Evidence for the presence of separate mechanisms regulating Na and K distribution. *J. Gen. Physiol.* 37, 601–620.

Scott, G. T., and Hayward, H. R. (1955). Sodium and potassium regulation in *Ulva lactuca* and *Valonia marrophysa*. In "Electrolytes in Biological Systems" (A. M. Shames, ed.), pp. 35–64. Amer. Physiol. Soc. Wash., Washington, D.C.

Sculthorpe, C. D. (1967). "The Biology of Aquatic Vascular Plants." Edward Arnold, London.

Setchell, W. A. (1929). Morphological and phenological notes on *Zostera marina* L. *Univ.*

Calif. Publ. Bot. **14**, 389–452.

Seybold, A. (1930). Untersuchungen über die Transpirations, widerstände und über die Temperatur Ägyptisch-Arabischer Wüstenpflanzen. *Planta* **9**, 270–314.

Shachar-Hill, B., and Hill, A. E. (1970). Ion and water transport in *Limonium*. VI. The induction of chloride pumping. *Biochim. Biophys. Acta* **211**, 313–317.

Shah, C. B., and Loomis, R. S. (1965). Ribonucleic acid and protein metabolism in sugar beet during drought. *Physiol. Plant.* **18**, 240–254.

Shakhov, A. A. (1952). Salt resistance of plants. "An Oecologo-Biological and Oecologo-Physiological Study," 35 pp. Akad. Nauk. SSSR Moscow.

Shalhevet, J., and Bernstein, L. (1968). Effects of vertically heterogenous soil salinity on plant growth and water uptake. *Soil Sci.* **106**, 85–93.

Shastri, G. G. K. (1962). Origin of the desert gypsum in Rajasthan. *Geol. Surv. India Rec.* **87**, 781–786.

Sherill, C. G. (1966). Boron deficiency in white clover (*Trifolium repens*) seedlings grown in an organic soil. *N.Z. J. Agr. Res.* **9**, 1025–1031.

Sherstnev, E. A., and Ragomova, M. V. (1965). The effect of boron deficiency on the ribonuclease activity in young leaves of sunflower plants. *Agrochimica* **9**, 348–350.

Shimony, C., and Fahn, A. (1968). Light and electron microscopical studies on the structure of salt glands of *Tamarix aphylla* L. *J. Linn. Soc. London Bot.* **60**, 283–288.

Shimshi, D. (1963). The effect of soil moisture and phenylmercuric acetate upon stomatal aperture, transpiration and photosynthesis. *Plant Physiol.* **38**, 713–721.

Shinozaki, K., and Kira, T. (1956). Intraspecific competition among higher plants. VII. Logistic theory of the C-D effect. *J. Inst. Polytech. Osaka City Univ. Ser. D,* **7**, 35–72.

Shive, J. W. (1916). The effect of salt concentration on the germination of seeds. *N.J. Agr. Exp. Sta. Ann. Rept.,* pp. 455–457.

Shmueli, E. (1948). The water balance of some plants of the Dead Sea salines. *Palestine J. Bot.* Jerusalem Ser. **4**, 117–142.

Siew, D., and Klein, S. (1968). The effect of sodium chloride on some metabolic and fine structural changes during the greening of etiolated leaves. *J. Cell Biol.* **37**, 590–596.

Simmoneau, P. (1952). "Le Végétation halophile de la plaine de Perregaux." Gov. Gén. de l'Algerie Servis des études Scien., Alger.

Sinclair, C. (1969). The level and distribution of amines in barley as affected by potassium nutrition, arginine level and temperature fluctuation and mildew infection. *Plant Soil* **30**, 423–433.

Skelding, A. D., and Winterbotham, J. (1939). The structure and development of the hydathodes of *Spartina townsendii* Groves. *New Phytol.* **38**, 69–79.

Skou, J. C. (1964). Enzymatic aspects of active-linked transport of Na^+ and K^+ through the cell membrane. *Progr. Biophys. Mol. Biol.* **14**, 131–166.

Skou, J. C. (1965). Enzymatic basis for active transport of Na^+ and K^+ across cell membrane. *Physiol. Rev.* **45**, 596–617.

Slatyer, R. O. (1958). The significance of the permanent wilting percentage in studies of plant and soil water relations. *Bot. Rev.* **23**, 586–636.

Slatyer, R. O. (1961). Effects of several osmotic substrates on the water relations of tomato. *Aust. J. Biol. Sci.* **14**, 519–540.

Slatyer, R. O. (1967). "Plant-Water Relationships." Academic Press, New York.

Slatyer, R. O. (1970). Comparative photosynthesis, growth and transpiration of two species of *Atriplex. Planta* **93**, 175–189.

Sloane, H. (1695). Catalogue of plants of Jamaica.

Smith, F. A. (1970). The mechanism of chloride transport in Characean cells. *New Phytol.* **69**, 903–917.

Spanswick, R. M., and Williams, E. J. (1964). Electrical potentials and Na, K and Cl concentrations in the vacuole and cytoplasm of *Nitella translucens. J. Exp. Bot.* **15**, 193–200.

Springfield, H. W. (1970). Germination and establishment of fourwing salt bush in the Southwest. *US. Forest Serv. Res. Paper* RM-55.

Stadelmann, E. (1957). Über Resistenz und Zell-Nekrobiose nach Salzplasmolyse bei Pflanzen der westalgerischen Stein-wüste. *Protoplasma* **48**, 452–498.

Stalter, R., and Batson, W. T. (1969). Transplantation of salt marsh vegetation, Georgetown, South Carolina. *Ecology* **50**, 1087–1089.

Stebbins, G. L. (1950). "Variation and Evolution in Plants," 641 pp. Columbia Univ. Press, New York.

Stebbins, G. L. (1951). Natural selection and the differentiation of angiosperm families. *Evolution* **5**, 299–324.

Steiner, M. (1935). Zur Oekologie der Salzmarschen der nordöstlichen Vereinigten Staaten von Nordamerika. *Jahrb. Wiss. Bot.* **81**, 94–202.

Steiner, M. (1939). Die Zusammensetzung des Zellsaftes bei höheren Pflanzen in ihrer ökologischen Bedeutung. *Ergeb. Biol.* **17**, 151–254.

Stenlid, G. (1956). Salt losses and redistribution of salts in higher plants. *Encycl. Plant Physiol.* **4**, 615–637.

Stephens, C. G. (1960). The Australian soil landscape. *Trans. 7th Intern. Congr. Soil Sci.* **3**, 20–26.

Stern, W. L., and Voigt, G. K. (1959). Effect of salt concentration on growth of red mangrove in culture. *Bot. Gaz. (Chicago)* **121**, 36–39.

Steward, F. C., and Martin, J. C. (1937). The distribution and physiology of *Valonia* at the dry Tortugas, with special reference to the problem of salt accumulation in plants. Papers from Tortugas Laboratory. *Carnegie Inst. Wash. Publ. No. 475*, 87–170.

Steward, F. C., and Sutcliffe, J. F. (1959). Plants in relation to inorganic salts. *In* "Plant Physiology, A Treatise" (F. C. Steward, ed.), Vol. 2, pp. 253–478. Academic Press, New York.

Steward, F. C., Prevot, P., and Harrison, J. A. (1942). Absorption and accumulation of rubidium bromide by barley plants. Localization in the root of cation accumulation and of transfer to the shoot. *Plant Physiol.* **17**, 411–421.

Stewart, J. (1898). Effect of alkali on seed germination. *9th Ann. Rept. Utah Agr. Exp. Sta.* pp. 26–35.

Stocker, O. (1925). Beiträge zum Halophytenproblem. II. Standort und Transpiration der Nordseehalophyten. *Z. Bot.* **17**, 1–24.

Stocker, O. (1928). Das Halophytenproblem. *Ergeb. Biol.* **3**, 265–353.

Stocker, O. (1930). Über die Messung von Bodensaugkräften und ihrem Verhältnis zu den Wurzelsaugkräften. *Z. Bot.* **23**, 27–56.

Stocker, O. (1933). Salzpflanzen. *Handb. Naturwiss.* **8**, 699–712.

Stocker, O. (1954). Der Wasser und Assimilationshaushalt südalgerischer Wüstenpflanzen. *Ber. Deut. Bot. Ges.* **67**, 288–298.

Stocker, O. (1960). Physiological and morphological changes in plants due to water deficiency. *UNESCO Arid Zone Res. Plant-Water Relationships in Arid and Semi-Arid Conditions* **15**, 63–104.

Strogonov, B. P. (1964). "Physiological Basis of Salt Tolerance of Plants (as affected by various types of salinity)". Akad. Nauk. SSSR. Translated from Russian, Israel Progr. Sci. Transl., Jerusalem.

Stuart, L. S. (1938). Isolation of halophilic bacteria from soil, water and dung. *Food Res.* **3**, 417–420.

Sutcliffe, J. F. (1962). "Mineral Salt Absorption in Plants." Pergamon Press, London.

Sutcliffe, J. F., and Sexton, R. (1968). Cell differentiation in the root in relation to physiological functions. *In* "Root Growth." Univ. of Nottingham (W. J. Whittington, ed.), pp. 80–102. Proc. 5th Easter School in Agr. Sci. Butterworth, London.

Sutherland, G. K., and Eastwood, A. (1916). The physiological anatomy of *Spartina townsendii. Ann. Bot. (London)* **30**, 333–351.

Tadmor, N. H., Koller, D., and Rawitz, E. (1958). Experiments in the propagation of *Juncus maritimus* Lam. II. Germination and field trials in the Arava. *Ktavim* **9**, 177–205.

Takaoki, T. (1957). Relationships between plant hydrature and respiration. 2. Respiration in relation to the concentration and the nature of external solutions. *J. Sci. Hiroshima Univ. Ser. B. Div. 2* **8**, 73–80.

Talling, J. F. (1961). Photosynthesis under natural conditions. *Ann. Rev. Plant Physiol.* **12**, 133–154.

Taylor, N. (1939). Salt tolerance of Long Island salt marsh plants. *Circ. N.Y. State Mus.* **23**, 1–42.

Teakle, L. J. H. (1937). The salt (sodium chloride) content of rain water. *J. Dept. Agr. West Aust.* **14**, 115–133.

Teal, J., and Teal, M. (1969). "Life and Death of the Salt Marsh," 278 pp. Little, Brown, New York.

Thom, B. G. (1967). Mangrove ecology and deltaic geomorphology: Tabasco, Mexico. *J. Ecol.* **55**, 301–343.

Thomas, D. A. (1970). The regulation of stomatal aperture in tobacco leaf epidermal strips. I. The effect of ions. *Aust. J. Biol. Sci.* **23**, 961–979.

Thomson, W. W., and Liu, L. L. (1967). Ultrastructural features of the salt gland of *Tamarix aphylla* L. *Planta* **73**, 201–220.

Thomson, W. W., Berry, W. L., and Liu, L. L. (1969). Localization and secretion of salt by the salt glands of *Tamarix aphylla. Proc. Nat. Acad. Sci. U.S.* **63**, 310–317.

Thuet, M., Thuet, P., and Philippot, J. (1969). Activite de l'ATPase ($Na^+ - K^+$) sur la fonction de la morphologie et de la structure histologique des plépodes ainsi que de la concentration du sodium de l'héolymphe chez *Sphaeroma serratum. C. R. Acad. Sci. (Paris)* **269**, 233–236.

Tinker, P. B. H. (1965). The effects of nitrogen, potassium, and sodium fertilizers on sugar beet. *J. Agr. Sci.* **65**, 207–212.

Toole, E. H., Hendricks, S. B., Borthwick, H. A., and Toole, V. K. (1956). Physiology of seed germination. *Annu. Rev. Plant Physiol.* **7**, 299–324.

Torii, K., and Laties, G. G. (1966). Dual mechanisms of ion uptake in relation to vacuolation in corn roots. *Plant Physiol.* **41**, 863–870.

Troll, W., and Dragendorff, O. (1931). Über die Luftwurzeln von *Sonneratia* L. und ihre biologische Bedeutung. *Planta* **13**, 311–473.

Troughton, A. (1960). Growth correlations between the roots and shoots of grass plants. *Proc. 8th Intern. Grassl. Congr.,* pp. 280–283.

Troughton, A. (1967). Effect of sodium chloride on growth of *Lolium perenne. Plant Soil* **27**, 148–150.

Tsopa, E. (1939). La végétation des halophytes du nord de la Roumanie en connexion avec celle du reste du pays. S.I.G.M.A. **70**, 1–22.

Tukey, H. B., and Morgan, J. V. (1962). The occurrence of leaching from above ground plant parts and the nature of the material leached. *16th Intern. Hort. Congr. Brussels,* pp. 153–160.

Tukey, H. B., Jr., Tukey, H. B., and Wittwer, S. H. (1958). Loss of nutrients by foliar leaching as determined by radioisotopes. *Proc. Amer. Soc. Hort. Sci.* **71**, 496–506.

Tullin, V. (1954). Response of sugar beet to common salt. *Physiol. Plant.* **7**, 810–834.

Turesson, G. (1922a). The species and variety as ecological units. *Hereditas* **3**, 100–113.

Turesson, G. (1922b). The genotypical responses of the plant. *Hereditas* **3**, 211–350.

Turesson, G. (1926). De Bedeutung der Rassenokologie für die system und Geographie der Pflanzen. *Feddes Rept. Beih.* **41**, 15–37.

Turesson, G. (1931). The selective effect of climate upon plant species. *Hereditas* **15**, 99–152.

Turner, W. J. (1956). *Suaeda nova-zelandiae* (Allan sp. nov.). An autecological study. M.Sc. Thesis, Auckland Univ. New Zealand.

Tutin, T. G. (1938). The autecology of *Zostera marina* in relation to its wasting disease. *New Phytol.* **37**, 50–71.

Tutin, T. G. (1942). Biological flora of the British Isles *Zostera* L. *J. Ecol.* **30**, 217–226.

Uhvits, R. (1946). Effect of osmotic pressure on water absorption and germination of alfalfa seeds. *Amer. J. Bot.* **33**, 278–285.

Ulrich, A., and Ohki, K. (1956). Chloride, bromine and sodium as nutrients for sugar beet plants. *Plant Physiol.* **31**, 171–181.

Ungar, I. A. (1962). Influence of salinity on seed germination in succulent halophytes. *Ecology* **43**, 763–764.

Ungar, I. A. (1965). An ecological study of the vegetation of the big salt marsh Stafford County, Kansas. *Univ. Kansas Sci. Bull.* **46**, 1–99.

Ungar, I. A. (1967a). Vegetation-soil relationships on saline soils in northern Kansas. *Amer. Midl. Natur.* **78**, 98–120.

Ungar, I. A. (1967b). Influence of salinity and temperature on seed germination. *Ohio J. Sci.* **67**, 120–123.

Ungar, I. A., and Capilupo, F. (1968). An ecological life history study of *Suaeda degressa* (Pursh.) Wats. *Advan. Front. Plant. Sci.* **23**, 137–158.

Uphof, J. C. T. (1937). Die nördliche Verbreitung der Mangroven im atlantischen Gebiet. *Arch. Hydrobiol.* **31**, 141–144.

Uphof, J. C. T. (1941). Halophytes. *Bot. Rev.* **7**, 1–58.

Vaadia, Y., and Waisel, Y. (1967). Physiological processes as affected by water balance. *In* "Irrigation of Agricultural Lands" (R. M. Hagen, H. R. Haise, T. W. Edminster, eds.), pp. 354–372. Amer. Soc. Agron. Madison, Wisconsin.

van Andel, O. M. (1953). The influence of salts on the exudation of tomato plants. *Acta Bot. Neerl.* **2**, 445–521.

van den Berg, C. (1952). The influence of absorbed salts on growth and yield of agricultural crops of salty soils. *Versl. Landbouwk. Ned.* **58**, 1–118.

van den Honert, T. H. (1937). Over eigenschappen van plantenwortels, welke een rol spelen bij de opname van voedingszouten. *Natur. Tijdschr. Ned.* **47**, 150–162.

van den Tooren, J. (1964). Rietonderzoek in oostelijk Flevoland. Rijksdienst voor de Ijselmeer polders zwool.

van den Tooren, J. (1966). Advies over de in oostelijk Flevoland to oogsten hoeveelheid reitzaad bestemd voor uitzaii in Zuidelijk Flevoland. Rijksdienst voor de Ijselmeer polders, Interne note no. 20.

van der Pijl, L. (1957). The dispersal of plants by bats. *Acta Bot. Neerl.* **6**, 291–315.

van der Pijl, L. (1969). "Principles of Dispersal in Higher Plants." Springer Verlag, Berlin.

van Eijk, M. (1934). Versuche über den Einfluss des Kochsalzgehalt in der Nährlösung auf die Entwicklung vor *Salicornia herbacea* und auf die Zusammensetzung der salze im Zellinnern diese Pflanze. *Proc. Kon. Ned. Akad. Wetensch.* **37**, 556–561.

van Eijk, M. (1938). Einfluss der Salzaufnahme auf die Wurzelatmung bei *Aster tripolium*. *Proc. Kon. Nederl. Akad. Wetensch.* **41**, 1115–1121.

van Eijk, M. (1939). Analyse der Wirkung des NaCl auf die Entwicklung Sukkulenze und Transpiration bei *Salicornia herbacea,* sowie Untersuchungen über den Einfluss der

Salzaufnahme, auf die Wurzelatmung bei *Aster tripolium. Rec. Trav. Bot. Neerl.* **36**, 559–657.

van Fleet, D. S. (1961). Histochemistry and function of the endodermis. *Bot. Rev.* **27**, 165–220.

van Raalte, M. H. (1940). On the oxygen supply of rice roots. *Ann. Jardin Bot. Buitenzorg* **50**, 99–114.

Viets, F. G. (1944). Calcium and other polyvalent cations as accelerators of ion accumulation by excised barley roots. *Plant Physiol.* **19**, 466–479.

Vité, J. P. (1959). Observations on the movement of injected dyes in *Pinus ponderosa* and *Abies concolor. Contrib. Boyce Thompson Inst.* **20**, 17–26.

Vité, J. P., and Rudinsky, J. A. (1959). The water conducting systems in conifers and their importance to the distribution of trunk-injected chemicals. *Contrib. Boyce Thompson Inst.* **10**, 27–38.

Volkens, G. (1884). Die Kalkdrüsen der Plumbaginean. *Ber. Deut. Bot. Ges.* **2**, 334–342.

von Faber, F. C. (1923). Zur Physiologie der Mangroven. *Ber. Deut. Bot. Ges.* **41**, 227–234.

von Faber, F. C. (1925). Untersuchungen über die Physiologie der javanischen Solfataren Pflanzen. *Flora (Jena)* **18/19**, 88–110.

von Lötsch, B., and Kinzel, H. (1971). Zum Calciumbedarf von Oxalatpflanzen. *Biochem. Physiol. Pflanzen* **162**, 209–219.

von Willert, D. J. (1969). Tagesschwankungen des Ionengehalts in *Salicornia europaea* in Abhängigkeit vom Standort und von der Überflutung. *Ber. Deut. Bot. Ges.* **81**, 442–449.

von Willert, D. J. (1970). Der Einfluss höher NaCl Konzentrationen auf die Atmung intakter Keimpflanzen einiger Halophyten und Glykophyten. *Flora (Jena)* **159**, 512–523.

Wadleigh, C. H., and Ayers, A. D. (1945). Growth and biochemical composition of bean plants as conditioned by soil moisture tension and salt concentration. *Plant Physiol.* **20**, 106–132.

Wadleigh, C. H., and Gauch, H. G. (1948). Rate of leaf elongation as affected by the intensity of the total soil moisture stress. *Plant Physiol.* **23**, 485–495.

Wadleigh, C. H., Gauch, H. G., and Magistad, O. C. (1946). Growth and rubber accumulation in guayule as conditioned by soil salinity and irrigation regime. *U.S. Dept. Agr. Tech. Bull. No. 925.*

Wahhab, A. (1961). Effect of saline irrigation waters on some soil properties. *UNESCO Arid Zone Res. Salinity Problems in the Arid Zones. Proc. Teheran Symp.* **14**, 233–237.

Waisel, Y. (1958). Germination behaviour of some halophytes. *Bull. Res. Counc. Israel* **6D**, 187–189.

Waisel, Y. (1959a). Endurance of a drought period beyond the wilting point. *Bull. Res. Counc. Israel* **7D**, 44–46.

Waisel, Y. (1959b). Ecotypic variation in *Nigella arvensis* L. *Evolution* **13**, 469–475.

Waisel, Y. (1960a). Ecological studies on *Tamarix aphylla* (L). Karst. II. The water economy. *Phyton* **15**, 17–27.

Waisel, Y. (1960b). Fog precipitation by trees. *La-Yaaran* **9**, 29.

Waisel, Y. (1960c). Ecological studies on *Tamarix aphylla* (L.). Karst. I. Distribution and reproduction. *Phyton* **15**, 7–17.

Waisel, Y. (1960d). Morpho-ecological differentiation of intraspecific vicariads in the flora of Israel. Ph.D. Thesis, The Hebrew Univ., Jerusalem.

Waisel, Y. (1961). Ecological studies on *Tamarix aphylla* (L.). Karst. III. The salt economy. *Plant Soil* **13**, 356–364.

Waisel, Y. (1962a). The effect of calcium on the uptake of monovalent ions by excised barley roots. *Physiol. Plant.* **15**, 709–724.

Waisel, Y. (1962b). Presowing treatments and their relation to growth and to drought, frost and heat resistance. *Physiol. Plant.* **15**, 43–46.

Waisel, Y. (1962c). Ecotypic differentiation in the flora of Israel. IV. Seedling behaviour of some ecotype pairs. *Phyton* **18**, 151–156.

Waisel, Y. (1971). Patterns of distribution of a few xerophytic species in the Negev of Israel. *Israel J. Bot.* **20**, 101–110.

Waisel, Y., and Bernstein, R. (1959). The effect of irrigation with saline water on the yield and sugar content of forage and sugar beet. *Bull. Res. Counc. Israel* **7D**, 90–92.

Waisel, Y., and Eshel, A. (1971). Localization of ions in the mesophyll cells of the succulent halophyte *Suaeda monoica* Forssk. by X-ray microanalysis. *Experientia* **27**, 230–232.

Waisel, Y., and Friedman, J. (1965). Selection of *Tamarix* trees for planting on exposed coasts. *La-Yaaran* **15**, 1–4.

Waisel, Y., and Ovadia, S. (1972). Biological flora of Israel. *Suaeda monoica* Forssk. ex J. F. Gmel. *Israel J. Bot.* **21**, 42–52.

Waisel, Y., and Pollak, G. (1969a). Estimation of water stresses in the root zone by the double root system technique. *Israel J. Bot.* **18**, 123–128.

Waisel, Y., and Pollak, G. (1969b). Estimation of water stresses in the active root zone of some native halophytes in Israel. *J. Ecol.* **57**, 789–794.

Waisel, Y., and Rechav, Y. (1972). Ecotypic differentiation in *Phragmites communis* Trin. *Hydrobiologia,* in press.

Waisel, Y., Neumann, R., and Eshel, Y. (1966a). The nature of the pH effect on the uptake of rubidium by excised barley roots. *Physiol. Plant.* **19**, 115–121.

Waisel, Y., Neumann, R., and Eshel, Y. (1966b). Mineral uptake of plants. (In Hebrew). *Mada* **10**, 273–279.

Waisel, Y., Hoffen, A., and Eshel, A. (1970a). The localization of aluminium in the cortex cells of bean and barley roots by X-ray microanalysis. *Physiol. Plant.* **23**, 75–79.

Waisel, Y., Neumann, R., and Kuller, Z. (1970b). Selectivity and ion transport in excised bean hypocotyls. *Physiol. Plant.* **23**, 955–963.

Waisel, Y., Liphschitz, N., and Kuller, Z. (1972). Patterns of water movement in trees and shrubs. *Ecology,* in press.

Walker, J. M., and Barber, S. A. (1962). Absorption of potassium and rubidium from the soil by corn roots. *Plant Soil* **17**, 243–259.

Wallace, A., and Hemaiden, N. (1963). Studies on the absorption and distribution of [22] Na in bush bean plants. *In* "Solute Uptake by Intact Plants" (A. Wallace, ed.), pp. 62–74. Edwards Brothers Inc., Ann Arbor, Michigan.

Walter, H. (1936a). Über den Wasserhaushalt der Mangrovenpflanzen. *Ber. Schweiz. Bot. Ges.* **46**, 217–228.

Walter, H. (1936b). Der Wasser und Salzgehalt der Ost-Afrikanischen Mangroven. *Ber. Deut. Bot. Ges.* **54**, 76–79.

Walter, H. (1937). Die Ökologishen Verhältnisse in der Namib Nebelwüste (Südwestafrika). *Jahrb. Wiss. Bot.* **84**, 58–219.

Walter, H. (1955). The water economy and the hydrature of plants. *Annu. Rev. Plant Physiol.* **6**, 239–252.

Walter, H. (1961). The adaptation of plants to saline soils. *UNESCO Arid Zone Res. Salinity Problems in the Arid Zones. Proc. Teheran Symp.* **14**, 129–134.

Walter, H. (1962). "Die Vegetation der Erde in Ökologischer Betrachtung," Band I. Fischer Verlag, Jena.

Walter, H. (1968). "Die Vegetation der Erde in Öko-physiologischen Betrachtung," Band II. Fischer Verlag, Stuttgart.

Walter, H., and Steiner, M. (1936). Die Ökologie der ostafrikanischen Mangroven. *Z. Bot.*

30, 65–193.

Warburg, O. (1898). Einige Bemerkungen über die Litoral-Pantropisten. *Ann. Jardin Bot. Buiten. Suppl.* II.

Warburg, O. (1948). "Schwermetalle als Wirkungsgruppe von Fermenten." Saenger, Berlin.

Warington, W. J. (1923). The effect of boric acid and borax on the broad bean and certain other plants. *Ann. Bot.* **37**, 629–672.

Watson, R. W. (1942). The mechanism of elongation in palisade cells. *New Phytol.* **41**, 206–221.

Weaver, J. E. (1918). The quadrat method in teaching ecology. *Plant World* **21**, 267–283.

Weaver, J. E., and Clements, F. E. (1938). "Plant Ecology," 2nd ed. McGraw-Hill, New York.

Weavind, T. E. F., and Hodgson, J. F. (1971). Iron absorption by wheat roots as a function of a distance from the root tip. *Plant Soil* **34**, 697–705.

Webb, K. L. (1966). NaCl effect on growth and transpiration in *Salicornia bigolovii,* a salt marsh halophyte. *Plant Soil* **24**, 261–265.

Webb, K. L., and Burley, J. W. A. (1965). Dark fixation of CO_2 by obligate and facultative salt marsh halophytes. *Can. J. Bot.* **43**, 281–285.

Wedding, R. T., and Erickson, L. C. (1957). The role of pH in the permeability of Chlorella to 2,4,D. *Plant Physiol.* **32**, 503–512.

Weigl, J., and Lüttge, U. (1962). Microautoradiographische Untersuchungen über die Aufnahme von $^{35}SO_4^-$ durch Wurzeln von *Zea mays* L. Die Funktion der primären Endodermis. *Planta* **59**, 15–28.

Weimberg, R. (1967). Effect of sodium chloride on the activity of a soluble malate dehydrogenase from pea seeds. *J. Biol. Chem.* **242**, 3000–3006.

Weimberg, R. (1968). An electrophoretic analysis of the isozymes of malate dehydrogenase in several different plants. *Plant Physiol.* **43**, 622–628.

Weimberg, R. (1970). Enzyme levels in pea seedlings grown on highly salinized media. *Plant Physiol.* **46**, 466–470.

Weiss, C., and Vaadia, Y. (1965). Kinetin-like activity in root apices of sunflower plants. *Life Sci.* **4**, 1323–1326.

Weissenbock, G. (1969). Einfluss des Bodensalzgehaltes auf Morphologie und Ionen-speicherung von Halophyten. *Flora (Jena)* **158**, 369–389.

Welch, R. M., and Epstein, E. (1968). The dual mechanisms of alkali cation absorption by plant cells: their parallel operation across the plasmalemma. *Proc. Nat. Acad. Sci. U.S.* **61**, 447–452.

Wells, B. W., and Shunk, V. I. (1938). Salt spray: an important factor in coastal ecology. *Bull. Torrey Bot. Club* **65**, 485–492.

West, R. C. (1956). Mangrove swamps of the Pacific coast of Colombia. *Ann. Amer. Geog. Ass.* **46**, 98–121.

West, K. R. (1970). The anatomy of *Atriplex* leaves. *In* "The Biology of *Atriplex*" (R. Jones, ed.), pp. 11–15. Div. Plant Industry, C.S.I.R.O., Canberra, Australia.

Westermaier, M. (1900). "Zur Kenntniss der Pneumatophoren. Botanische Untersuchungen im Anschluss an eine Tropenreise." Heft I. Freiburg.

Whiteman, P. C., and Koller, D. (1964). Saturation deficit of the mesophyll evaporating surfaces in a desert halophyte. *Science* **146**, 1320–1321.

Whitney, R. S., and Peech, M. (1952). Ion activities of sodium clay suspensions. *Soil Sci. Soc. Amer. Proc.* **16**, 117–122.

Whittig, L. D., and Janitzky, P. (1963). Mechanisms of formation of sodium carbonate in soils. I. Manifestations of biological conversions. *J. Soil Sci.* **14**, 322–333.

Wiehe, P. O. (1935). A quantitative study of the influence of tide upon populations of

Salicornia europaea. J. Ecol. **23**, 323–333.

Wilcox, L. V., Blair, C. Y., and Bower, C. A. (1954). Effect of bicarbonate on suitability of water for irrigation. *Soil Sci.* **77**, 259–266.

Wilgus, F., and Hamilton, K. C. (1962). Germination of salt cedar seeds. *Weeds* **10**, 332–333.

Wilkinson, H. F., Lomeragin, J. F., and Quirck, J. P. (1968). Calcium supply to plant roots. *Science* **161**, 1245–1246.

Williams, M. C. (1960). Effect of sodium and potassium salts on growth and oxalate content of *Halogeton. Plant Physiol.* **35**, 500–505.

Williams, W. T., and Barber, D. A. (1961). The functional significance of aerenchyma in plants. *Symp. Soc. Exp. Biol.* **15**, 132–144.

Williamson, F. A., and Wyn Jones, R. G. (1971). On the role of plasma membrane and tonoplast adenosine triphosphatases in ion uptake. IAEA, Symposium on the use of isotopes and radiation in research on soil-plant relationships. IAEA/SM-51/32. Vienna, December 1971.

Willis, A. J., Folkes, B. F., Hope-Simpson, J. F., and Yemma, E. W. (1959). Braunton Burrows: The dune system and its vegetation. *J. Ecol.* **47**, 249–288.

Willis, J. C. (1966). "A Dictionary of the Flowering Plants and Ferns," 7th ed. (Revised by H. K. Airy Shaw.) Cambridge Univ. Press, Cambridge.

Wilson, C. C. (1948). Diurnal fluctuations in growth in length of tomato stem. *Plant Physiol.* **23**, 156–157.

Winter, H. (1961). The uptake of cations by *Vallisneria* leaves. *Acta Bot. Neer.* **10**, 341–393.

Wood, J. G. (1925). The selective absorption of chlorine ions and the absorption of water in the genus *Atriplex. Aust. J. Exp. Biol. Med. Sci.* **2**, 45–56.

Wood, J. G. (1937). Vegetation of South Australia. *In* "Handbook of Flora and Fauna of South Australia." Gov't. Print., Adelaide.

Woodell, S. R. J., Mooney, H. A., and Hill, A. J. (1969). The behaviour of *Larrea divaricata* (Creosote bush) in response to rainfall in California. *J. Ecol.* **57**, 37–44.

Woolley, J. T. (1957). Sodium and silicon as nutrients for the tomato plants. *Plant Physiol.* **32**, 317–321.

Wuenscher, J. E. (1970). The effect of leaf-hairs of *Verbascum thapsus* on leaf energy exchange. *New Phytol.* **69**, 65–73.

Wybenga, J. M. (1957). A contribution to the knowledge of the importance of sodium for plant life. Ph.D. Thesis, Agricultural University, Wageningen. Kantoordruk De Goede-Wageningen.

Wyn Jones, R. G. (1971). A possible new method for improving plant growth in saline and calcareous environments. Symposium on the use of isotopes and radiation in research on soil-plant relationships. IAEA/SM-151/31. Vienna, December 1971.

Wyn Jones, R. G., and Lunt, O. R. (1967). The function of calcium in plants. *Bot. Rev.* **33**, 407–426.

Yaalon, D. H. (1955). Studies of the effect of saline irrigation water on calcareous soils. I. Permeability characteristics and adsorption of sodium. *Bull. Res. Counc. Israel* **5B**, 83–97.

Yaalon, D. H. (1963). On the origin and accumulation of salts in ground water and in soils of Israel. *Bull. Res. Counc. Israel* **11G**, 105–131.

Yaalon, D. H. (1964a). The concentration of ammonia and nitrate in rain water over Israel in relation to environmental factors. *Tellus* **16**, 200–204.

Yaalon, D. H. (1964b). Downward movement and distribution of anions in soil profiles with limited wetting. *In* "Experimental Pedology" (E. G. Hallsworth and D. V. Crawford, eds.), pp. 157–164. Proc. 11th Easter School Univ. of Nottingham. Butterworth, London.

Yaalon, D. H., and Ganor, E. (1968). Chemical composition of dew and dry fall-out in Jerusalem, Israel. *Nature (London)* **217**, 1139–1140.

Yaalon, D. H., and Katz, A. (1962). The chemical composition of precipitation in Israel. *Proc. 4th Congr. Israel Ass. Advan. Sci.,* pp. 189–190.

Yaalon, D. H., and Lomas, J. (1970). Factors controlling the supply and the chemical composition of aerosols in a near shore and coastal environment. *Agr. Meteorol.* **7**, 443–454.

Yih, R. Y., Hille, F. K., and Clark, H. E. (1966). Requirements of the *Ginkgo* pollen-driven tissue cultures for boron and effects of boron deficiency. *Plant Physiol.* **41**, 815–820.

Yoda, K., Kira, T., and Hozumi, K. (1957). Intraspecific competition among higher plants. IX. Further analysis of the competitive interaction between adjacent individuals. *J. Inst. Polytech. Osaka* **8D**, 161–178.

Zellner, J. (1926). Zur Chemie der Halophyten. *Ber. Akad. Wiss. Wien. Math-Naturwiss. Kl. Abt. IIb* **135**, 585–592.

Zholkevitch, V. N., and Koretskaya, T. F. (1959). Metabolism of pumpkin roots during drought. *Fiziol. Rast.* **6**, 690–700.

Zhukovskaya, N. V., and Matukhin, G. R. (1964). Postuplenie ^{32}P v asteniya yachmenya i aktinost' v nikh adnozintrifosfatazy v zavisimosti of tipa zasoleniya pochvy. *In* "Rol mineral'nykh elementov v obmene veshchestv i produktivnosti rastenii Moscow, Izdatel'stvo 'Nauka'," pp. 332–336.

Ziegler, H. (1965). Die Physiologie pflanzicher Drüsen. *Ber. Deut. Bot. Ges.* **78**, 466–477.

Ziegler, H., and Lüttge, U. (1966). Die Salzdrüsen von *Limonium vulgare.* II. Mitteilung: Die Feinstruktur. *Planta* **70**, 193–206.

Ziegler, H., and Lüttge, U. (1967). Die Salzdrüsen von *Limonium vulgare.* II. Mitteilung: Die Lokalisierung des Chlorids. *Planta* **74**, 1–17.

Zobell, C. E., and Feltham, C. B. (1942). The bacterial flora of a marine mud flat as an ecological factor. *Ecology* **23**, 69–78.

Zobell, C. E., Anderson, D. G., and Smith, W. W. (1937). The bacteriostatic and bactericidal action of Great Salt Lake water. *J. Bacteriol.* **33**, 253–262.

Zohary, M. (1937). Die verbreitungsökologischen Verhältniss der Pflanzen Palastinas. Die antitelochorischen Erscheinungen. *Beih. Bot. Centralbl.* **56**, 1–155.

Zohary, M. (1944). Vegetational transects through the Desert of Sinai. *Palestine J. Bot. Jerusalem Ser.* **3**, 57–78.

Zohary, M. (1956). The genus *Tamarix* in Israel. *Trop. Woods* **104**, 24–60.

Zohary, M. (1962). "Plant Life of Palestine." Ronald Press, New York.

Zohary, M., and Fahn, A. (1952). Ecological studies on east Mediterranean dune plants. *Bull. Res. Counc. Israel* **1**, 38–53.

Zohary, M., and Orshan (Orshansky), G. (1949). Structure and ecology of the vegetation in the Dead Sea region of Palestine. *Palestine J. Bot. Jerusalem Ser.* **4**, 177–206.

Zohary, M., and Waisel, Y. (1956). The ecology of non-irrigated cultivated forest trees in the Negev. *La-Yaaran* **6**, 32–33.

Author Index

Genus and Species Index

Subject Index

A

Accumulating organs, 145
Accumulation, *see also* specific substances
  mechanism, 155
  patterns, 115
Acetate, 223
Acid metabolism, 96
Acid phosphatase, 218, 280
Acidity, 140
Aconitic acid, 224
Active uptake, *see* Uptake, Ions, Salt
Adaptation
  to anaerobic conditions, 279
  anatomical, 312, 316, 317
  to marine environments, 270, 285–289
  morphological, 185, 312–316
  phenological, 276, 312, 316
  physiological, 312, 317–321
  to salinity, 26, 130, 170, 171, 191, 195, 198, 205, 208, 210–221, 244, 279, 280, 292, 302, 313, 316, 319

Adaptive value, 184, 260, 287, 321
ADP (Adenosine diphosphate), 219
Adsorption, 26, 28, 100
Aeration, 34, 35, 57, 108, 109, 192, 256
Aerenchyma, 179, 259, 267, 276, 279, 294, 299
Aerial plant organs, *see* Plant(s)
Aerobiosis, 103, 139
Aerohalophytes, *see* Halophyte(s)
Africa, 265, 266, 295, *see also* specific countries
  east, 61, 274, 277
  north, 71, 294, 310
  south, 126, 262, 269, 270, 297, 310
  tropical, 262
  west, 264, 267, 274, 282
Air spaces, intercellular, 173, 174
Albumin, 161, 244
Alfalfa, 179, 180
Algae, 123, 125, 127, 131, 139, 225, 266, 278, 282
Alaska, 313